D0938015

MECHANISM AND THEORY IN FOOD CHEMISTRY

MECHANISM AND THEORY IN FOOD CHEMISTRY

DOMINIC W.S. WONG, Ph.D.

Cornell University

An AVI Book
Published by Van Nostrand Reinhold
New York

An AVI Book
(AVI is an imprint of Van Nostrand Reinhold)

Copyright © 1989 by Van Nostrand Reinhold

Library of Congress Catalog Card Number 88-27987
ISBN 0-442-20753-0

Printed in the United States of America

Van Nostrand Reinhold
115 Fifth Avenue
New York, New York 10003

Van Nostrand Reinhold International Company Limited
11 New Fetter Lane
London EC4P 4EE, England

Van Nostrand Reinhold
480 La Trobe Street
Melbourne, Victoria 3000, Australia

Macmillan of Canada
Division of Canada Publishing Corporation
164 Commander Boulevard
Agincourt, Ontario M1S 3C7, Canada

16 15 14 13 12 11 10 9 8 7 6 5 4 3 2 1

Library of Congress Cataloging-in-Publication Data

Wong, Dominic W. S.
  Mechanism and theory in food chemistry.

  Includes index.
  1. Food—Analysis. 2. Food—Composition. 3. Chemical reactions. 1. Title.
TX541.W66 1989   664   88-27987
ISBN 0-442-20753-0

To
My Wife, Eleanor,
and Children, Jacob, Joshua, Jessica

FOREWORD

This is a unique book on food chemistry. Textbooks on food chemistry are written generally in a descriptive manner, with tables of comparative data and sometimes from the functional aspects of the food ingredients. They rarely place emphasis on modern mechanisms underlying the chemical reactions that occur in food during processing and storage nor do they treat interactions among the components of foods. The present book, *Mechanism and Theory in Food Chemistry*, is an exception. The author has stressed the principles of the reaction mechanisms, carefully detailing what is known to occur or is expected to occur based on his detailed understanding of organic chemical reactions. This unifies the themes of oxidation, reduction, hydrolysis, structure, polymerization, emulsification, etc., that are key to the conceptual approach used. Students are challenged to understand and explain the chemical reactions in food systems, based on their background in chemistry. The textbook covers the important topics of food chemistry, based on the more traditional organization around the general types of components found in foods—the proteins, lipids, carbohydrates, enzymes, vitamins—but also expands the content to include both naturally occurring as well as added colorants, flavors, sweeteners, and other food additives. Each chapter includes sections on how the constituents covered interact with other compounds in the food, as well as discussions of effect of temperature, pH, metal ions, oxygen, and other constituents on the roles of the reactions and interactions.

The book should be a valuable text for a more advanced course in food chemistry, probably at the graduate level, or as an adjunctive text for an undergraduate junior or senior course, as well as a valuable reference for researchers.

Robert E. Feeney
John R. Whitaker
Department of Food Science and Technology
University of California, Davis

PREFACE

The primary objective of this book is to focus on the reaction mechanism and theory essential to understanding the many chemical processes occurring in food and food systems. Too often, food chemistry courses tend to be descriptive and require mere memorization of unrelated chemical structures. It is the author's firm belief that an adequate comprehension of the principles of food chemistry is based on a thorough knowledge of the reaction mechanisms and theories under a coherent theme. For every phenomenon or change observed in food or food systems, there is a corresponding chemical equation or model. Most of the presentations in this book are found in reaction mechanisms explained in equations or figures.

The book is organized in ten chapters: (1) Lipids, (2) Proteins, (3) Carbohydrates, (4) Colors, (5) Enzymes, (6) Flavors, (7) Sweeteners, (8) Natural Toxicants, (9) Additives, and (10) Vitamins. The first three chapters constitute the foundation of food chemistry and hence require a broad coverage. Other chapters are slightly more selective in the presentation of the materials. The references at the end of each chapter is by no means encyclopedic, but provide a guide to those who would like to obtain more information on a particular area discussed in the text. Selected references are included mainly for certain topics the author finds interesting and deserving further reading.

There are three appendixes: (1) General kinetics of olefin autoxidation, (2) Singlet oxygen, and (3) Where do the radicals come from? These were written to cover certain chemical principles that students are normally expected or assumed to, but rarely actually understand. The latter two are especially important, since they contain information basic to the numerous chemical reactions often discussed in food systems.

I am very grateful to Professors Robert E. Feeney and John R. Whitaker (University of California, Davis) for their comments, criticisms, encouragement and help in the preparation of this book. I would also like to thank Professors Donald R. Babin (Creighton University), Benito O. de Lumen

(University of California, Berkeley), Dr. Ram T. Shet, Mr. Wayne M. Camirand (Western Regional Research Center, USDA), for their valuable comments and suggestions. The librarians, Julia F. Cookey, Carol M. Nybro-Stevens, and Rena Schonbrun, have always been helpful during my two years stay at WRRC. Finally, my deep admiration and sincere appreciation go to all scientists for their contributions to the many progresses in the field of food chemistry, without which this book could not have been written.

ACKNOWLEDGMENTS

I would like to thank the many publishers for their generosity in granting permissions to use their copyrighted materials: Academic Press, Inc., Agricultural Chemical Society of Japan (Japan); American Association of Cereal Chemists; American Chemical Society; American Oil Chemists' Society; American Society of Biological Chemists, Inc.; American Society of Plant Physiologists; Annual Review Inc.; The Biochemical Society (London); Cambridge University Press; Centre for Agricultural Publishing and Documentation (The Netherlands); CRC Press, Inc.; Elsevier Applied Science Publishers Ltd. (Essex); Elsevier Science Publishers Biomedical Division (The Netherlands); Elsevier Scientific Publishers Ireland Ltd. (Ireland); Federation of European Biochemical Societies (Belgium); FMC Corporation; Institute of Food Technologists; International Journal of Food Science and Technology (London); International Union of Pure and Applied Chemistry (Oxford); Macmillan Magazine Ltd. (London); Marcel Dekker, Inc.; McGraw-Hill Company; Miles Laboratories, Inc. Biotech Products Division; Monsanto Company; National Academy of Sciences; National Technical Information Service, U.S. Department of Commerce; Prentice Hall; Society of Chemical Industry (London); VCH Verlagsgesellschaft (Federal Republic of Germany).

Thanks are also due to the authors whose publications have been referenced in this book.

Credits for Equations

Eq. 1-6. Reprinted with permission from Frankel (2), courtesy of American Oil Chemists' Society.

Eq. 1-19. Reprinted with permission from Porter, N. A., Lehman, L. S., Weber, B. A., and Smith, K. J. Unified mechanism for polysaturated fatty acid autoxidation. Competition of peroxy radical hydrogen atom abstraction, β-scission, and cyclization. *J. Am. Chem. Soc.* **103**, 6447–6455. Copyright 1981 American Chemical Society.

Eq. 2-40. Adapted from Payens, T. A. J. Casein micelles: the colloid-chemical approach. *J. Dairy Res.* **46**, 303, 304. Copyright 1979 Cambridge University Press.

Eq. 5-2. Reprinted with permission from Lewis, S. D., Johanson, F. A., and Shafer, J. A. Effect of cysteine-25 on the ionization of histidine-159 in papain as determined by proton nuclear magnetic resonance spectroscopy. *Biochemistry* **20**, 48–51. Copyright 1981 American Chemical Society.

Eq. 5-11. Reprinted with permission from Winkler, M. E., Lerch, K., and Solomon, E. I. Competitive inhibitor binding to the binuclear copper active site in tyrosinase. *J. Am. Chem. Soc.* **103**, 7001–7003. Copyright 1981 American Chemical Society.

Eq. 5-12. Reprinted with permission from Wilcox, D. E., Porras, A. G., Hwang, Y. T., Lerch, K., Winkler, M. E., and Solomon, E. I. Substrate analogue binding to the coupled binuclear copper active site in tyrosinase. *J. Am. Chem. Soc.* **107**, 4015–4027. Copyright 1985 American Chemical Society.

Eq. 5-14. Reprinted with permission from Palmer (26), courtesy of American Society of Plant Physiologists.

Eq. 5-16. Reprinted with permission from Weibel and Bright (30), courtesy of The American Society of Biological Chemists, Inc.

Eq. 5-17. Reprinted with permission from Chan, T. W., and Bruice, T. C. One and two electron transfer reactions of glucose oxidase. *J. Am. Chem. Soc.* **99**, 2387–2389. Copyright 1977 American Chemical Society.

Eq. 5-18. Reprinted with permission from Chan, T. W., and Bruice, T. C. One and two electron transfer reactions of glucose oxidase. *J. Am. Chem. Soc.* **99**, 2387–2389. Copyright 1977 American Chemical Society.

Eq. 5-22. Reprinted with permission from Quinn, D. M. Solvent isotope effects for lipoprotein lipase catalyzed hydrolysis of water-soluble *p*-nitrophenyl esters. *Biochemistry* **24**, 3148. Copyright 1985 American Chemical Society.

Eq. 5-23. Reprinted with permission from Tsujisaka (63), courtesy of Elsevier Science Publishers, Biomedical Division, Amsterdam, The Netherlands.

Eq. 10-6. Reprinted with permission from Ohloff, G. 1970. Proceeding, Third International Congress on Food Science & Technology, p. 376, copyright 1971 by Institute of Food Technologists.

Eq. 10-8. Reprinted with permission from Dwivedi, B. K., and Arnold, R. G. Chemistry of thiamine degradation in food products and model systems: a review. *J. Agric. Food Chem.* **21**, 54–60. Copyright 1973 American Chemical Society.

Eq. 10-9. Reprinted with permission from Dwivedi, B. K., and Arnold, R. G. Chemistry of thiamine degradation in food products and model systems: a review. *J. Agric. Food Chem.* **21**, 54–60. Copyright 1973 American Chemical Society.

Eq. 10-44. Reprinted with permission from Martell (*20*), pp. 157, 163. Copyright 1982 American Chemical Society.

Eq. 10-52. Reprinted with permission from Gruger and Tappel (*25*), courtesy of American Oil Chemists' Society.

Eq. 10-55. Reprinted with permission from Gorman, A. A., Gould, I. R., Hamblett, I., and Standen, M. C. Reversible exciplex formation between singlet oxygen, $^1\Delta g$, and vitamin E. Solvent and temperature effect. *J. Am. Chem. Soc.* **106**, 6958. Copyright 1984 American Chemical Society.

CONTENTS

Foreword   vii

Preface   ix

Acknowledgments   xi

Credits for Equations   xiii

1. LIPIDS   1

Lipid Oxidation   2
Thermal and Oxidative Thermal Reactions   12
Radiolysis of Lipids   15
Hydrogenation   18
Interesterification   21
Polymorphism of Triglycerides   23
Plasticity of Fat   26
Emulsions   27
Emulsifiers   33
Liquid-Crystalline Mesophase
    in Emulsifier-Water System   38
Functions of Emulsifiers in Stabilization   40
Antioxidants   44

2. PROTEINS   48

Protein Structure   49
Protein-Stabilized Emulsification and Foaming   60

Gel Formation 62
Chemical Reaction 63
Meat Proteins 78
Milk Proteins 86
Wheat Proteins 92
Collagen 97

3. CARBOHYDRATES 105

Glycosidic Linkage 105
Action of Alkali on Monosaccharides 108
Action of Acid on Monosaccharides 111
Nonenzymatic Browning (Maillard Reaction) 113
Complexes of Sugars with Metal Ions 123
Starch 124
Alginate 129
Pectin 132
Carrageenan 134
Cellulose 138
Xanthan Gum 141
Hemicellulose 143

4. COLORANTS 147

Light Absorption 147
Conjugation 149
Substituent Effects 151
Carotenoids 153
Annatto 159
Anthocyanins 160
Betanain 166
Caramel 169
Dyes and Lakes 169
Coordination Chemistry 171
Metalloporphyrin-Electronic Structure 174
Myglobin 178
Chlorophyll 187

5. ENZYMES 194

Papain 194
Lipoxygenase 199
Polyphenol Oxidases 206

Glucose Oxidase 211
Amylases 215
Pectic Enzymes 221
Lipolytic Enzymes 225

6. FLAVORS 231

The Sensation of Taste 231
Odor-The Site-Fitting Theory 232
Character-Impact Compounds 234
Origin of Flavor 234
Beverage Flavor 242
Spice Flavor 247
Fruits and Vegetables 253
Meat Flavor 259
Microencapsulation of Flavors 262

7. SWEETENERS 264

Molecular Theory of Sweetness 264
Amino Acids and Dipeptides 267
The Aminosulfonates 270
Dihydrochalcone 271
Glycyrrhizin 274
Stevioside 275
Sugar Alcohol 276
Corn Sweeteners 277
Sweet Proteins 278
Flavor Potentiators 280

8. NATURAL TOXICANTS 283

Cyanogenic Glycosides 283
Glycoalkaloids 285
Glucosinolates 286
Methylxanthines 289
Amino Acids, Peptides, and Proteins 292
Amines 297
Mycotoxins 299
Polycyclic Aromatic Hydrocarbons 302
Heterocyclic Amines 304
Nitrosamines 306

9. ADDITIVES   314

Phosphates   314
Citric Acid   321
Antimicrobial Short-Chain Acid Derivatives   324
Sulfite   326

10. VITAMINS   335

Vitamin A   335
Vitamin B_1 340
Vitamin B_2 344
Vitamin B_6 351
Vitamin B_{12} 353
Biotin   356
Niacin   358
Vitamin C   359
Vitamin D   365
Vitamin E   369

SELECTED READINGS AND REFERENCES   375

APPENDIX 1. GENERAL KINETICS OF OLEFIN
             AUTOXIDATION   401
APPENDIX 2. SINGLET OXYGEN   404
APPENDIX 3. WHERE DO THE RADICALS
             COME FROM?   408

INDEX   413

MECHANISM AND THEORY IN FOOD CHEMISTRY

1
LIPIDS

Dietary lipids supply approximately 35–40% of the total calories taken by an average adult and exhibit the most efficient energy conversion, yielding 9 calories per gram, twice as many calories supplied by carbohydrates or proteins. The large consumption of fats and oils necessitates better understanding of the basic chemistry that underlines the various changes, both under natural conditions and during food processing.

The oxidation of unsaturated lipids has been one of the most extensively studied areas and will remain so, since it is related to the deterioration of foods, the production of both desirable (e.g., flavor, color) and undesirable breakdown products (e.g., toxic dimers), and the numerous reactions associated with other food constituents. The problem is further complicated by the fact that these reactions can be initiated, inhibited, or altered by many factors, including metal, enzymes, antioxidants, temperature, light, and pH.

The physical chemistry of lipids is another important area of great interest to food chemists. The polymorphic property and crystal habits of fat are of great importance in the formulation of various fat and oil products such as margarine, ice cream, and mayonnaise. Knowledge on the formation and breakdown of emulsions is required for the effective application of emulsifiers in many food-processing systems.

The chemistry of lipids and the mechanisms of the various reactions that occur in processing, including lipid degradation, via oxidation, thermal and radiolytic reactions, hydrogenation, interesterification, and polymorphic changes, are presented. A significant portion of the discussion will be devoted to the theory and applied chemistry of emulsions in food systems.

LIPID OXIDATION

Lipid oxidation in foods is associated with the reaction of oxygen with unsaturated lipids in two different pathways, (a) autoxidation and (b) photosensitized oxidation.

Autoxidation

Autoxidation is a free-radical chain reaction involving the following steps (see Eq. 1-1).

$$LH \xrightarrow[\text{initiator}]{k_i} L\cdot + H\cdot \qquad \text{1-1.1}$$

$$L\cdot + O_2 \xrightarrow{k_o} LOO\cdot \qquad \text{1-1.2}$$

$$LOO\cdot + LH \xrightarrow{k_p} LOOH + L\cdot \qquad \text{1-1.3}$$

Eq. 1-1

$$LOO\cdot + LOO\cdot \xrightarrow{k_t} \left.\begin{array}{c} \\ \\ \\ \end{array}\right\} \qquad \text{1-1.4}$$

$$LOO\cdot + L\cdot \xrightarrow{k_t'} \text{Non-radical products} \qquad \text{1-1.5}$$

$$L\cdot + L\cdot \xrightarrow{k_t''} \qquad \text{1-1.6}$$

1. *Initiation:* Homolytic abstraction of hydrogen to form a carbon-centered alkyl radical in the presence of an initiator.
2. *Propagation:* The free radical reacts with O_2 to form peroxy radical which reacts with more unsaturated lipids to form hydroperoxide. The lipid free radical thus formed can react with O_2 to form peroxy radical. Hence, autoxidation is a radical chain process.
3. *Termination:* The chain reaction can be terminated by the formation of nonradical products.

The overall reactions follow rate equation 1-2. (Refer to Appendix 1.) Under normal oxygen pressure, the alkyl radical reacts rapidly with oxygen to form the peroxy radical (Rxn 1-1.2). The rate of reaction 1-1.2 is fast, and most of the free radicals are in the form of the peroxy radical. Consequently, the major termination takes place via reaction 1-1.4 by peroxy radical combination. The reaction follows rate equation 1-3 (*1*).

$$-\frac{\delta(O_2)}{\delta t} = \frac{\delta(LOOH)}{\delta t} = \left(\frac{R_i}{k_t}\right)^{1/2} \frac{k_p(LH) k_o(O_2)}{k_p(LH) + k_o(O_2)} \qquad \text{Eq. 1-2}$$

R_i = rate of initiation

(LH) = lipid concentration

(O_2) = oxygen concentration

$(LOOH)$ = lipid hydroperoxide concentration

k's = respective rate constants for indicated reaction

t = time

$-\delta(O_2)/\delta t$ = rate of oxygen uptake

$\delta(LOOH)/\delta t$ = rate of hydroperoxide formation

$$-\frac{\delta(O_2)}{\delta t} = \frac{\delta(LOOH)}{\delta t} = \left(\frac{R_i}{k_t}\right)^{1/2} k_p(LH) \qquad \text{Eq. 1-3}$$

The overall rate of autoxidation is independent of oxygen pressure but is determined by the stability of the alkyl radical. This accounts for the differences in the oxidative stability among saturated and unsaturated fatty acids.

The rate of reaction increases with the degree of unsaturation. Linoleate is oxidized 10 times faster than oleate; linolenate 20–30 times faster. In autoxidation, the time period during which there is little or no detectable peroxide formation to the point where there is an abrupt increase in oxidation is termed the inductive period. In unsaturated fatty acids, the initial radical abstraction is stabilized by resonance. Addition of oxygen takes place at different positions as outlined for linoleate. The resulting peroxy radical then abstracts a H· radical from another acid to yield the hydroperoxide (Eq. 1-4) (2).

The product is composed predominantly of the 9- and 13-hydroperoxides, since stability of the conjugated diene system favors the oxygen attack at the end position. Also, the 9- and 12-hydroperoxides assume *trans, cis* configurations. Experimentally, hydroperoxides with *trans, trans* configurations are shown to exist in considerable proportion. Based on the same mechanism, autoxidation of oleate yields the 8-, 9-, 10-, and 11-hydroperoxides. Likewise, linolenate autoxidation results in the formation of the 9-, 12-, 13-, and 16-hydroperoxides (Table 1.1).

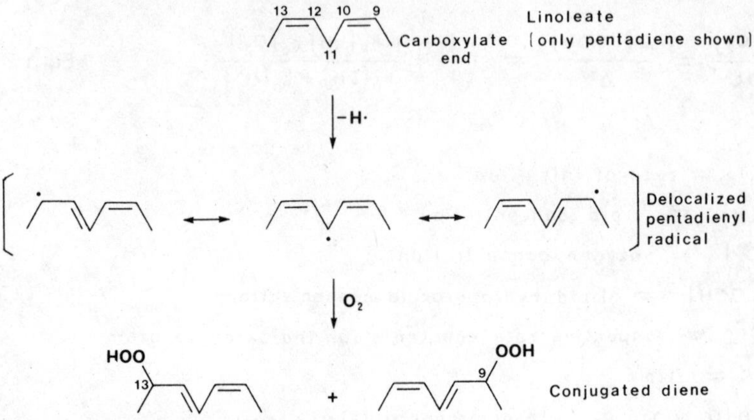

13 12 10 9

11   Carboxylate end

Linoleate (only pentadiene shown)

$-H\cdot$

Delocalized pentadienyl radical   Eq. 1-4

O_2

HOO   13   +   OOH   9   Conjugated diene

Photosensitized Oxidation

In the presence of a suitable sensitizer, such as chlorophyll or flavins, oxidation proceeds via the "ene" reaction (Eq. 1-5) (3). The dioxygen molecule is added to the olefinic (double bond) carbon with a subsequent shift in the position of the double bond. The reaction (1) involves no free radicals, (2) results in changes from *cis* to *trans* configuration of the double bond, (3) is independent of oxygen pressure, (4) shows no measurable inductive period, and (5) is inhibited by singlet oxygen quenchers such as β-carotene and tocopherols but is unaffected by antioxidants. (Refer to Appendix 2.)

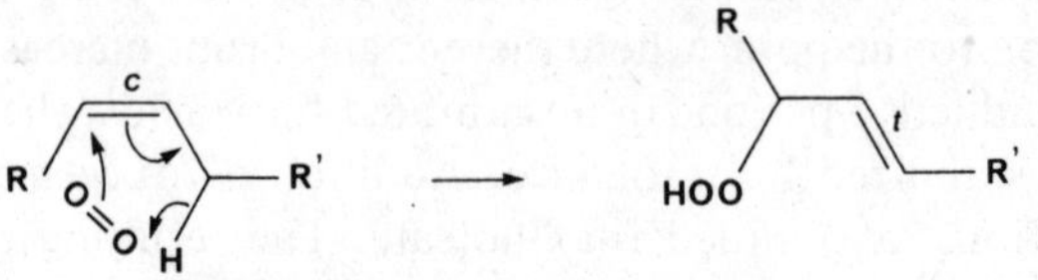

R   c   R'   →   R   t   R'   HOO   Eq. 1-5

Table 1.1. Product Distribution of Autoxidized Oleate, Linoleate, and Linolenate

	DISTRIBUTION OF HYDROPEROXIDES (%)						
	8-OOH	9-OOH	10-OOH	11-OOH	12-OOH	13-OOH	16-OOH
Oleate	27	23	23	27			
Linoleate		50				50	
Linolenate		30			12	12	46

Reprinted with permission from Frankel (2), courtesy of American Oil Chemists' Society.

The reaction mechanism of photosensitized oxidation of linoleate is presented in Eq. 1-6. In photosensitized oxidation, the hydroperoxides are formed at each of the unsaturated carbons. Photosensitized oxidation of linoleate, therefore, yields the 9-, 10-, 12-, and 13-hydroperoxides (Table 1.2).

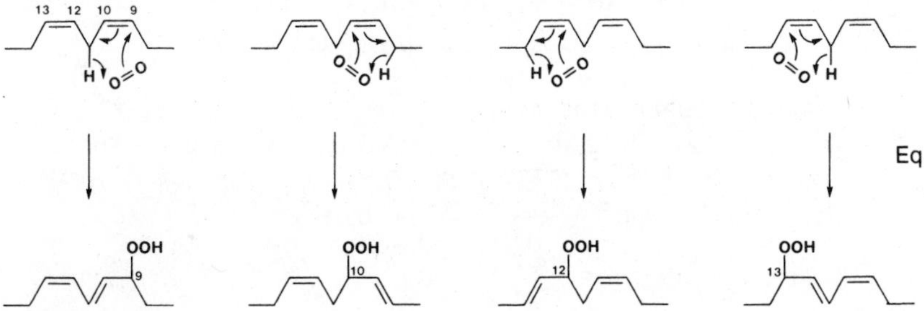

Eq. 1-6

The Role of Metal Ions in Lipid Oxidation

The oxidation of lipids can be catalyzed by metal ions, in the presence of thiols or other common reducing agents in biological systems, such as ascorbate, NADH, and $FADH_2$, in both microsomal and model systems. In these cases, Fe^{3+} induces the autoxidation of the reductant to produce superoxide anion ($O_2^-\cdot$). In the case of thiols, the thiyl radical catalyzes the reduction of O_2 to superoxide anion, which then dismutates to H_2O_2 or undergoes a one-electron reduction of Fe^{3+} to Fe^{2+} (Eq. 1-7). Hydroxy radical ($\cdot OH$) is produced via the Fenton reaction between Fe^{2+} and H_2O_2. Hydroxy radical is a strong oxidant and is suggested to be an active initiator of lipid peroxidation. Metal ions, however, do not exist free in vivo but are bound to ligands such as ATP and DNA as well as metalloproteins. The above reactions therefore, may occur via a chelated iron, ADP-Fe^{2+} (4). The conju-

Table 1.2. Product Distribution of Photosensitized Oxidation of Oleate, Linoleate, and Linolenate

| | DISTRIBUTION OF HYDROPEROXIDES (%) | | | | | |
	9-OOH	10-OOH	12-OOH	13-OOH	15-OOH	16-OOH
Oleate	50	50				
Linoleate	31	18	18	33		
Linolenate	21	13	13	14	13	25

Reprinted with permission from Frankel (2), courtesy of American Oil Chemists' Society.

gate acid of $O_2^-\cdot$, the perhydroxy radical $HO_2\cdot$, has been shown to react with unsaturated fatty acids to form the corresponding hydroperoxides without the mediation of metal ions.

AUTOXIDATION OF THIOL

$$Fe^{3+} + RSH \longrightarrow Fe^{2+} + RS\cdot + H^+$$

$$RSH + RS\cdot + O_2 \longrightarrow RSSR + H^+ + O_2^-$$

RXN OF SUPEROXIDE ANION

Eq. 1-7

$$2H^+ + 2O_2^- \longrightarrow H_2O_2 + O_2$$

$$Fe^{3+} + O_2^- \longrightarrow Fe^{2+} + O_2$$

FENTON REACTION

$$Fe^{2+} + H_2O_2 \longrightarrow Fe^{3+} + H_2O + \cdot OH$$

The mechanism of NADPH-induced microsomal peroxidation has received much attention, and in this case, the NADPH cytochrome P-450 reductase is involved in the formation of a chelated perferryl ion (*4*) (Eq. 1-8). The initiation of peroxidation then proceeds via the reaction between the perferryl ion and the polyunsaturated fatty acid (Eq. 1-9).

$$ADP-Fe^{+++} \xrightarrow[\text{NADPH} \quad \text{NADP}^+]{\substack{\text{Cyt P-450} \\ \text{reductase}}} ADP-Fe^{++} + H^+$$

Eq. 1-8

$$\left(ADP-Fe^{++}-O_2 \longleftrightarrow ADP-Fe^{+++}-O_2^- \right)$$

Perferryl ion

$$LH + ADP-Fe^{+++}-O_2^- \longrightarrow L\cdot + ADP-Fe^{++}-O_2H$$

$$L\cdot + O_2 \longrightarrow LOO\cdot$$

Eq. 1-9

$$LOO\cdot + ADP-Fe^{++}-O_2H \longrightarrow LOOH + ADP-Fe^{+++}-O_2^-$$

The above discussion concerns the role of metal ions in the initiation of lipid oxidation in biological systems. However, metal ions are often involved in the propagation step, by catalyzing homolytic decomposition of lipid hydroperoxides to yield the alkoxy or peroxy radicals which then, in turn, initiate lipid oxidation (Eq. 1-10). In muscle tissue, both heme and nonheme irons are implicated as oxidants. Peroxidation catalyzed by heme compounds consists of homolytic cleavage of the heme-hydroperoxide complex into the peroxy radical and oxy-heme radical. The latter can then abstract a H· radical from another molecule of unsaturated fatty acid to start a new cycle (Eq. 1-11) (5).

$$M^{n+} + LOOH \longrightarrow M^{(n+1)+} + LO\cdot + OH^{\ominus}$$

Eq. 1-10

$$M^{(n+1)+} + LOOH \longrightarrow M^{n+} + LOO\cdot + H^{\oplus}$$

Eq. 1-11

Oxy–heme radical

Secondary Products in Lipid Oxidation

Scission Products. The monohydroperoxides formed in lipid oxidation are very unstable and can undergo carbon-carbon cleavage on either side of the alkoxy radical. The alkoxy radicals also react with other radicals. These reactions account for many of the carbonyls, alcohols, esters, and hydrocarbons responsible for the oxidized flavor of food lipids. The general reaction is presented in Eq. 1-12. A typical example is given for the formation of scission products from oxidized linoleate (Eq. 1-13).

Eq. 1-12

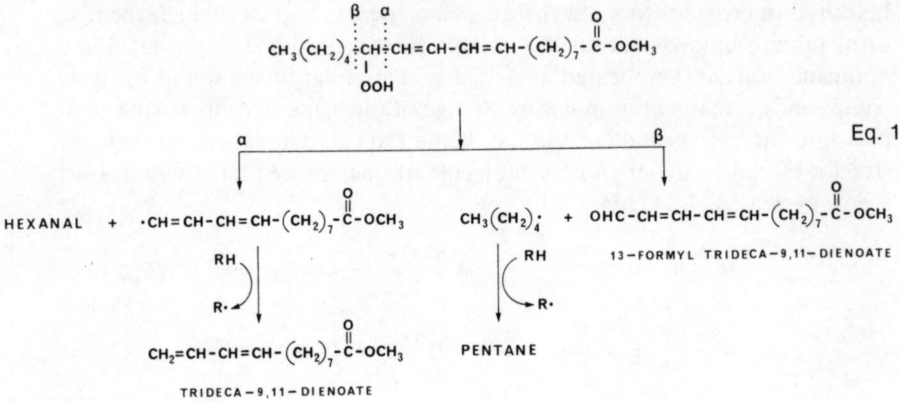

$$\underset{\beta}{CH_3(CH_2)_4} \underset{\alpha}{\overset{\cdot}{\vdots}CH\overset{\cdot}{\vdots}CH}=CH-CH=CH-(CH_2)_7-\overset{O}{\overset{\|}{C}}-OCH_3$$

OOH

Eq. 1-13

HEXANAL   + $\cdot CH=CH-CH=CH-(CH_2)_7-\overset{O}{\overset{\|}{C}}-OCH_3$

$CH_3(CH_2)_4^{\cdot}$ + $OHC-CH=CH-CH=CH-(CH_2)_7-\overset{O}{\overset{\|}{C}}-OCH_3$

13−FORMYL TRIDECA−9,11−DIENOATE

RH

R·

RH

R·

$CH_2=CH-CH=CH-(CH_2)_7-\overset{O}{\overset{\|}{C}}-OCH_3$

PENTANE

TRIDECA−9,11−DIENOATE

Hydroperoxy Cyclic Peroxides.   The linoleate 10- and 12-hydroperoxides from photooxidation are homoallylic and can undergo 1,3- or 1,4-cyclization to form cyclic peroxides (Eq. 1-14) (*6*) Similarily, the 12-, 13-hydroperoxides of linolenate are homoallylic and therefore undergo cyclization to form hydroperoxy peroxides (Eq. 1-15). But the 10- and 15-hydroperoxides from photooxidized linolenate have a neighboring pentadiene, and the hydroperoxy peroxides can cyclize to form the diperoxide (Eq. 1-16) (*7*).

LINOLEATE 10- or 12-HYDROPEROXIDE

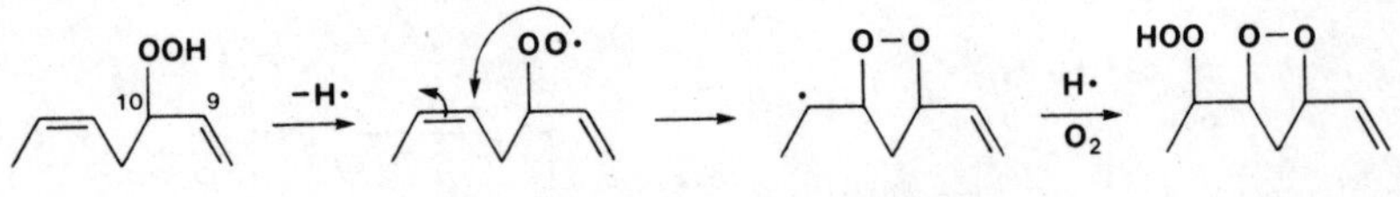

Eq. 1-14

LINOLENATE 12- or 13-HYDROPEROXIDE

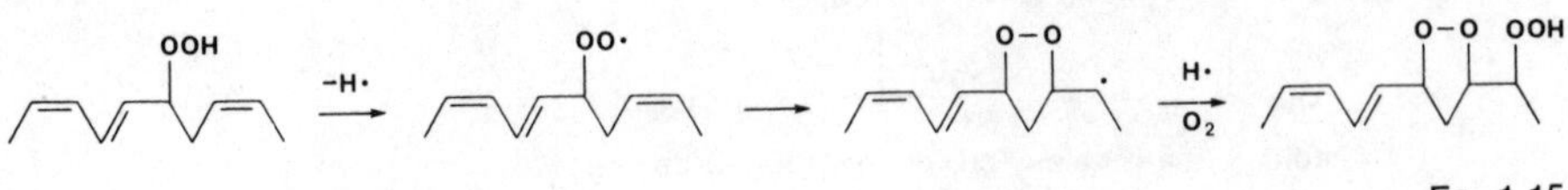

Eq. 1-15

LINOLENATE 10- or 15-HYDROPEROXIDE

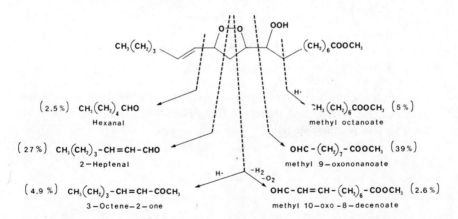

Eq. 1-16

Hydroperoxy cyclic peroxides undergo thermal decomposition mainly through $C-C$ cleavage between the peroxide ring and the carbon with the hydroperoxide group (Fig. 1.1). Types of volatiles produced are similar to those of the corresponding monoperoxides. Cleavage at the peroxide ring results in the formation of unsaturated aldehyde and aldehyde esters, and most uniquely, unsaturated ketones (8).

Formation of Malonaldehyde. In the case of linolenate hydroperoxide, 1,4-cyclization after the initial 1,3-cyclization yields a bicyclic peroxide,

Fig. 1.1. Thermal decomposition of methyl-9-hydroperoxy-10,12-epidioxy-trans-13-octadecenoate. (Reprinted with permission from Frankel et al. (8), courtesy of American Oil Chemists' Society.)

which is stabilized by an allylic radical (Eq. 1-17). Thermal degradation of the bicyclic peroxide gives malonaldehyde as one of the products (*2*). Malonaldehyde is known to cross-link proteins, enzymes, and DNA, and has been shown to react with amino acids to form fluorescent products that absorb at 435 nm.

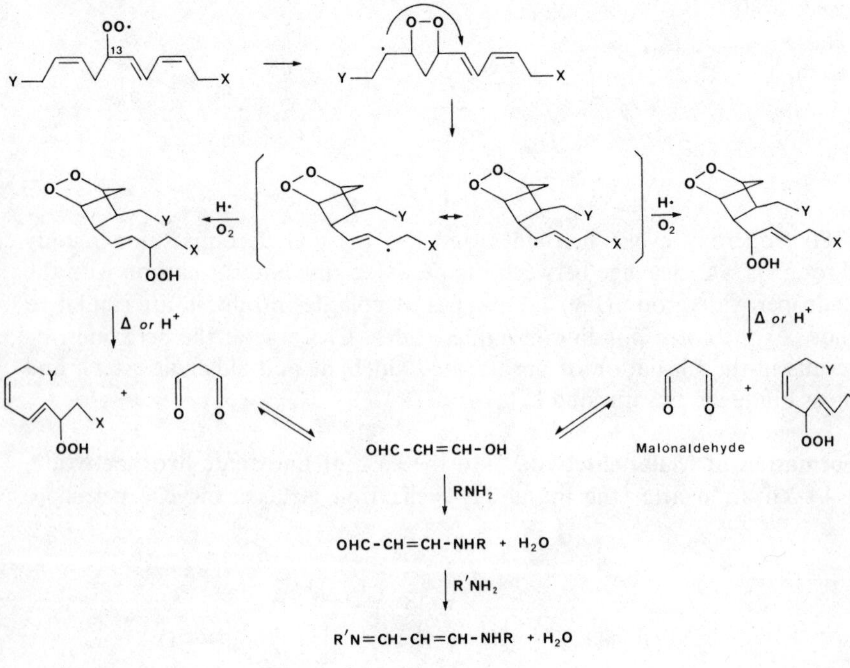

Eq. 1-17

$$OHC-CH=CH-OH$$

Malonaldehyde

$$OHC-CH=CH-NHR + H_2O$$

$$R'N=CH-CH=CH-NHR + H_2O$$

In linoleate, formation of bicyclic peroxide will result in a nonallylic structure (Eq. 1-18), and therefore the yield of bicyclic peroxides in the diene system is much smaller.

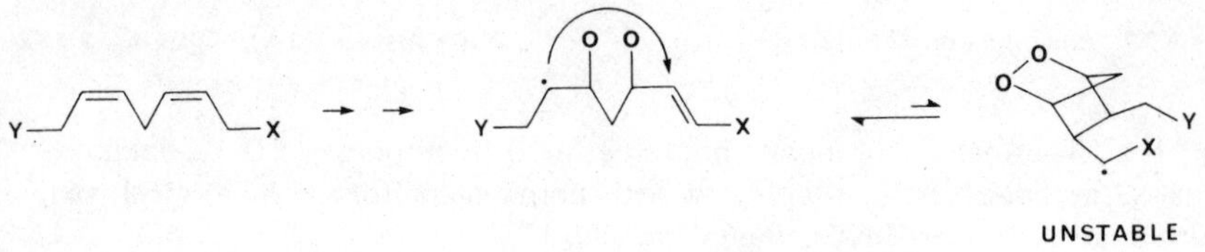

Eq. 1-18

UNSTABLE

The Stereochemistry of Product Hydroperoxides in Autoxidation

The classical mechanism of autoxidation presented in the previous section does not adequately establish the controlling factor for the stereochemical course of the reaction.

Experimentally, in the autoxidation of linoleic acid, four hydroperoxides (13-hydroperoxy-9-*cis*,11-*trans*-octadecadienoic; 13-hydroxyperoxy-9-*trans*, 11-*trans*-octadecadienoic; 9-hydroxyperoxy-10-*trans*, 12-*cis*-octadecadienoic; and 9-hydroperoxy-10-*trans*, 12-*trans*-octadecadienoic) are formed. These four major products comprise ~97% of the product mixture. To account for the stereochemistry of the products, a mechanism is postulated in which the peroxy radical exists in two conformers (A and B) which undergo two competitive pathways (Eq. 1-19) (*9*).

Eq. 1-19

1. Both conformers can abstract hydrogen from the acid to give the *trans, cis* hydroperoxides.
2. Scission of conformer B leads to the formation of a *trans, trans* pentadienyl radical, which can then repeat the autoxidation process to form a *trans, trans* product. However, loss of oxygen from conformer A will give back the *cis* pentadienyl radical.

High temperature favors the scission pathway, and the *trans, cis/trans, trans* ratio is decreased. At high concentration of linoleate, the reaction is directed toward the *trans, cis* hydroperoxides, since the pathway involving transfer of hydrogen becomes more competitive.

Linolenate autoxidation undergoes a similar mechanism, except that there are three independent pathways: hydrogen abstraction, $C-O$ scission, and cyclization.

THERMAL AND OXIDATIVE THERMAL REACTIONS

The course of reactions that occur during heating depends on the composition of the lipid and treatment conditions.

Thermal Reactions

Under anaerobic conditions, a relatively high temperature ($>200°C$) is required to decompose the saturated triglycerides. The products usually consist of a series of normal alkanes and alkenes, C_{2n-1} symmetric ketone, C_n oxopropyl ester, C_n propene and propanediol esters, C_n diglyceride, acrolein, and carbon dioxide (*11*).

Thermal treatment of unsaturated lipids in the absence of oxygen produces predominantly dimers and cyclic compounds (*10*). One major mechanism involves homolytic cleavage of $C-C$ bond α or β to the double bond, with the formation of radical fragments (Eq. 1-20). Direct combination of these radical fragments yields short- and long-chain fatty acids, straight-chain dicarboxylic acid, and hydrocarbons.

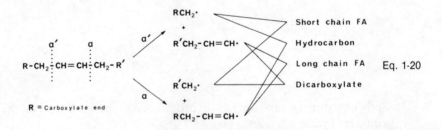

The radical fragments may also abstract hydrogen from the fatty acid (oleate, for example) to give allyl radicals which, in turn, undergo (1) disproportionation to monoenoic and dienoic acids, or (2) intermolecular addition to the $C-C$ double bond of another fatty acid chain to form a dimeric radical. The dimeric radical can disproportionate to monoenoic and dienoic dimers or abstract H· to form a dimer or undergo intramolecular addition to form a cyclic dimer (Eq. 1-21) (*10*).

Eq. 1-21

For linoleate, similar types of reactions are observed, but the products are complex mixtures of dimers—acyclic, bicyclic, and tricyclic of various unsaturation. The conjugated allyl radicals combine to form acyclic dimers or undergo intermolecular addition to form dimeric radicals. Since the dimer radical in this case contains a conjugated double bond, successive intramolecular addition may yield bicyclic and tricyclic dimers (Eq. 1-22).

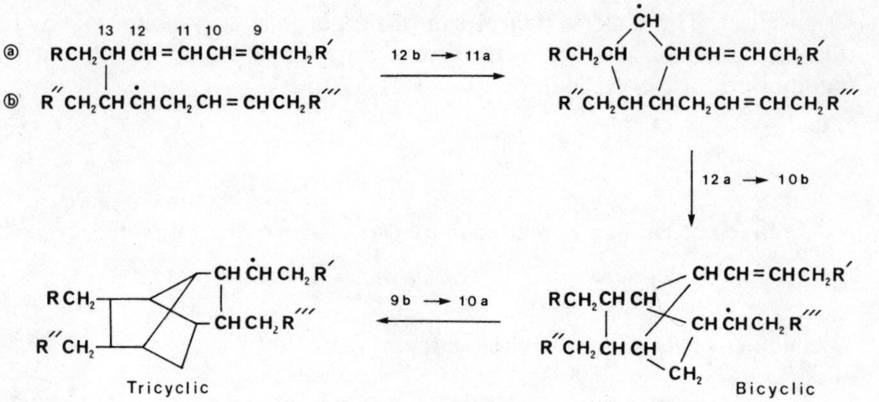

Eq. 1-22

Dimerization to monocyclic structures also occurs by the Diels-Alder reaction, between the conjugated diene in linoleate, for example, and a double-bond dienophile in another molecule of the fatty acid (Eq. 1-23).

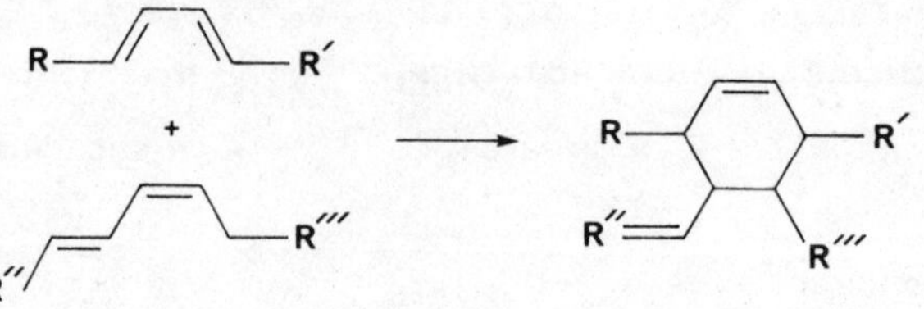

Eq. 1-23

Bicyclic and tricyclic structures can be explained by a free-radical mechanism combined with an intramolecular Diels-Alder cyclization. Some acyclic dimers formed from the combination of free-radical fragments may contain conjugated diene. The acyclic trienoic dimer, for example, can further undergo Diels-Alder reaction to a bicyclic compound (Eq. 1-24).

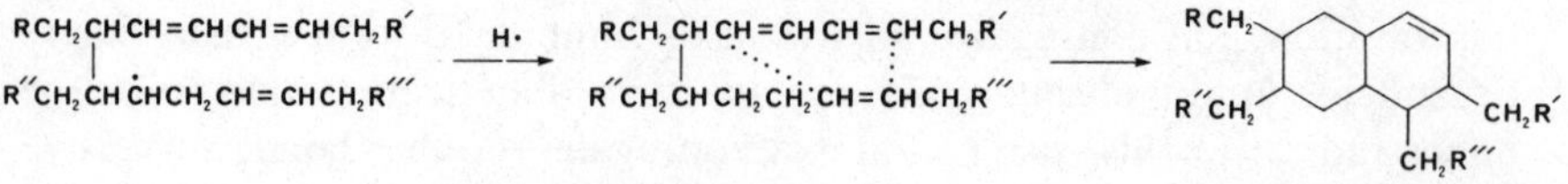

Bicyclic monoenoic dimer

Eq. 1-24

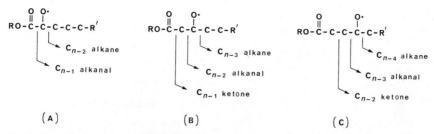

(A) (B) (C)

Fig. 1.2. Thermal cleavage of (A) α-, (B) β-, and (C) γ-oxidation products. (Reprinted with permission from Crnjar, E. D., Witchwoot, A., and Nawar, W. W. Thermal oxidation of a series of saturated triacylglycerols. *J. Agric. Food Chem.* **29**, 41. (*11*). Copyright 1981 American Chemical Society.)

Oxidative Thermal Reactions

Oxidation of saturated fatty acids generally occurs at the α, β, or γ carbon to form the respective alkoxy radicals (*11*). Thermolytic cleavages between the α, β, or γ carbons of the radical produce the various hydrocarbons, carbonyls, and ketones (Fig. 1.2).

Oxidative thermal decomposition of unsaturated fatty acids generally leads to dimers, trimers, and tetramers with polar groups. The hydroperoxide formed from oxidation can decompose into oxy- and peroxy-radicals which can (1) abstract a hydrogen atom from another fatty acid molecule, forming new radicals, or (2) be added to a C−C double bond of a fatty acid molecule to form radical dimers with ether or peroxide bridges (Eq. 1-25). The new radicals from either route may pick up a molecule of oxygen to form peroxy radical, which then undergoes a second addition or combination to form longer polymers.

RADIOLYSIS OF LIPIDS

The general mechanism of radiolysis of lipids involves primary ionization. The simple removal of an electron from the olefin molecule results in an electron deficiency or "hole" localized in the carboxyl group or the double bond. The molecular ion $[R-CH=CH-(CH_2)_7-COOH]^{+\cdot}$ formed can undergo reactions leading to many radiolytic products (Eq. 1-26) (*12*).

Fragmentation proceeds via α-cleavage at the (1) acyl-oxy, or (2) alkyl chain. The former reaction yields the acylium ion, RCO^+, and the latter the fragment ion, $R-O-C\equiv O^+$. These cations, together with the corresponding radical fragments, can abstract hydrogen atoms to form the various stable products. Or the radicals combine to form polymers.

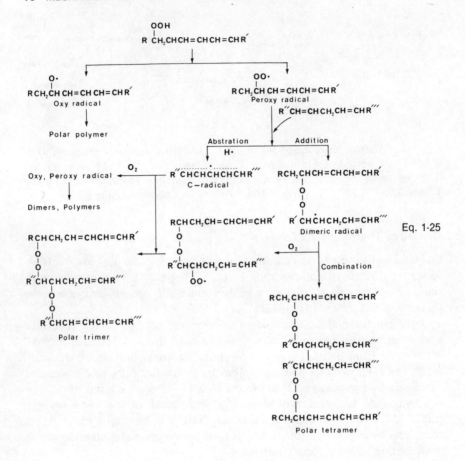

Eq. 1-25

Eq. 1-26

R = GLYCEROL SKELETON

R´ = FATTY ACID CHAIN

Decarboxylation takes place if cleavage occurs at the acyloxy-methylene bond in triglycerides or if intermolecular transfer of H· occurs in fatty acids. The acyloxy radical formed rapidly loses CO_2 to yield a primary alkyl radical that can abstract H· to form the alkane or alkene.

Studies of γ-irradiated fatty acids or triglycerides at low temperatures, usually at 77 K, by electron spin resonance indicate the formation of radical anions from the electron capture by the carboxyl group (13). The radical anion formed from the fatty acid is protonated to form the protonated anion radical or decays by α-cleavage to give the acyl radical. The protonated radical and the acyl radical can then react with the parent fatty acid to generate the stable α-carbon radical (Eq. 1-27).

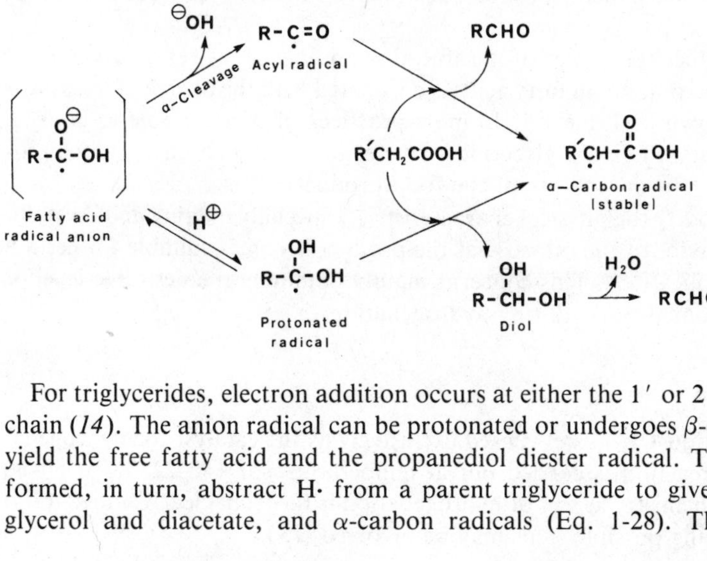

Eq. 1-27

For triglycerides, electron addition occurs at either the 1' or 2' fatty acid chain (14). The anion radical can be protonated or undergoes β-cleavage to yield the free fatty acid and the propanediol diester radical. The radicals formed, in turn, abstract H· from a parent triglyceride to give aldehyde, glycerol and diacetate, and α-carbon radicals (Eq. 1-28). The radicals

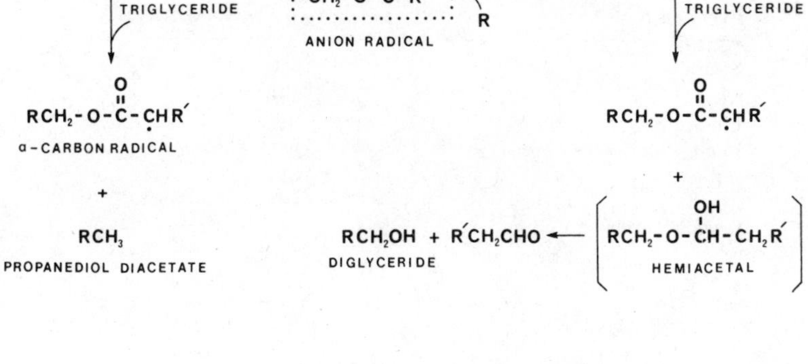

Eq. 1-28

formed in all these reactions may enter into termination reactions by dispro-portionation or by combination to form dimers.

HYDROGENATION

Hydrogenation is used extensively in the oil industry to manipulate the chemistry and composition of the fatty acids to produce various oil or fat stocks with specific functional characteristics. The basic chemistry of oils and fats, unrelated to their source or composition, is always that of triesters of glycerol and fatty acids. These oil or fat stocks are often directly blended in various combinations to formulate a wide range of products from liquid to plastic to sharp melting solid. There are two main objectives in hydrogenation.

1. To reduce the degree of unsaturation and hence the rate of oxidation. The rate of oxidation in fatty acids is correlated with the degree of unsaturation, as shown in Table 1.3. In most practices, it is preferable to convert polyenic acids to give triglyceride mixtures of C18:2 or C18:1, thereby increasing the flavor stability of the food product.

2. To modify the physical characteristics, especially melting and crystallization behavior of the oil, so that the product becomes suitable for certain specific applications. This concerns mainly obtaining the desirable level of cis-trans isomerization during hydrogenation.

Mechanism

In the oil industry, nickel is used exclusively as the catalyst for hydrogenation. The reaction is carried out at temperatures of 140–225°C at 50–60 psig. The chemistry aspect of hydrogenation is far from clearly understood. The following possible steps may be involved (15).

Table 1.3. Relative Oxidation Rates of Various Fatty Acids

FATTY ACID	RELATIVE OXIDATION RATES
Stearic	1
Oleic	10
Linoleate	100
Linolenate	200
α-Eleostearic	800

From Backmann (16).

1. The double bond is absorbed (π-bonding interaction) onto the surface of the metal catalyst (Fig. 1.3A).
2. A hydrogen atom from the metal surface is transferred to one of the carbons of the double bond, and the other carbon binds (σ-bonding) with the metal surface.
3. A second hydrogen transfer liberates the saturated product.

The first reaction step is reversible, with the hydrogen atom retained to the metal surface, and the molecule "desorbed." *Cis-trans* isomerization usually occurs with rotation around the $C-C$ bond. A shift of the double-bond position also occurs, if the reverse reaction proceeds at the neighboring methylene group of the double bond (Fig. 1.3B).

The changes in the degree of saturation and in the isomeric configuration of the fatty acids resulting from hydrogenation have a profound effect not only on the increased stability, but also on the melting properties, and indirectly the crystal habits of the triglycerides.

Cis-Trans Isomerization

Changes in the isomeric configuration of individual fatty acid chains indirectly alter the spatial structure of the triglyceride molecule. This is evident from considering the isomerization of *cis*-C18:1 (Eq. 1-29.1) and *cis*-C18:2 (Eq. 1-29.2) (*16*).

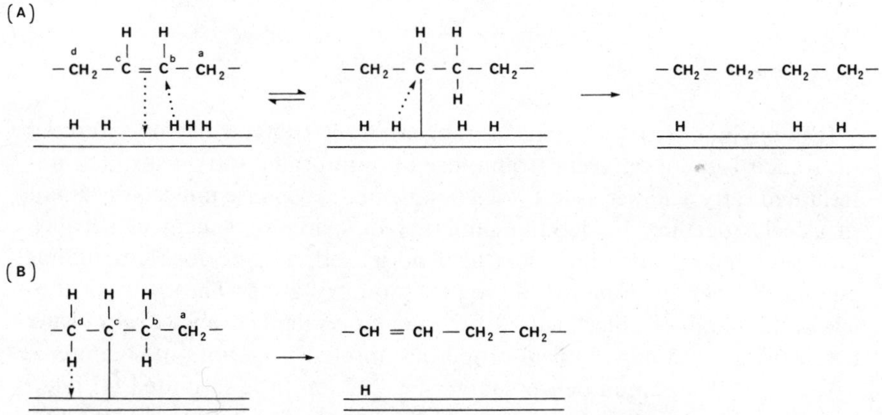

Fig. 1.3. Mechanism of hydrogenation. (Reprinted with permission from Larsson (*15*), courtesy of American Oil Chemists' Society.)

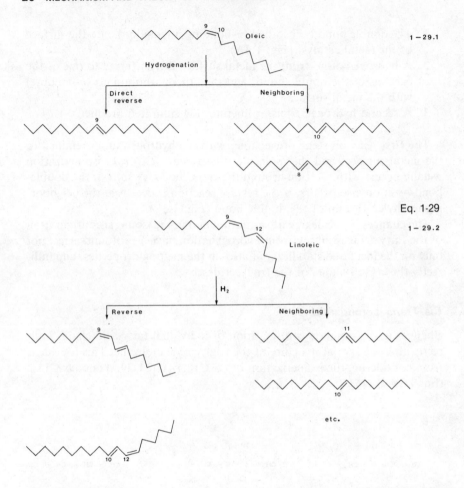

Eq. 1-29

It is obvious that the spatial arrangement of triglycerides of saturated fatty acids is quite different from those of unsaturated fatty acids. The unsaturated fatty acid bends at the *cis* double bond, so these molecules cannot fit closely together. Notice the similarity in spatial arrangement between oleic acid and *cis-trans* linoleic acid. And indeed, a *trans* double bond has physical properties similar to those of a saturated bond: The *cis-trans* linoleic acid is similar to oleic acid rather than *cis-cis* linoleic acid. *Trans* oleate is fairly hard and has physical properties approaching those of stearate.

Contrary to common belief, margarine contains little saturated fat but a high percentage of *trans* fatty acids. The vegetable oil is selectively hydrogenated to produce high *cis-trans* isomerization, retaining a maximum amount of linoleic and oleic acids. The product therefore solidifies at a desirable low temperature. Fully saturated stearate is avoided because its

Table 1.4. Factors Affecting Selectivity in the Hydrogenation Process

INCREASING	SELECTIVITY RATIO	ISOMERIZATION	RATE
Temperature	+	+	+
Pressure	−	−	+
Agitation	−	−	+
Catalyst	+	+	+

From Allen (*17*).

presence produces a sandy or grainy taste, a result of the higher melting characteristics of tristearin.

Selectivity

In order to produce products suitable for various purposes, hydrogenation must be selective and requires rigorous control of reaction conditions (*17*). Table 1.4 summarizes the effects of different parameters, such as pressure, temperature, agitation, and catalyst load, on the rate of hydrogenation and the degree of isomerization.

Selectivity is explained by assuming the following simplified picture of sequences of reactions in hydrogenation in Eq. 1-30. Selectivity ratio is defined as the ratio of the hydrogenation of linoleic acid compared to the hydrogenation of oleic acid, that is, $k_2/k_3 = S$. The higher the ratio, the higher the selectivity. If $S = 0$, all molecules react straight through to stearic acid. When $S = \infty$, no stearic acid is formed.

$$\text{Linolenic} \xrightarrow{k_1} \text{Linoleic} \xrightarrow{k_2} \text{Oleic} \xrightarrow{k_3} \text{Stearic} \qquad \text{Eq. 1-30}$$

INTERESTERIFICATION

Interesterification changes the distribution pattern of the fatty acids in the triglyceride, producing fats and oils with desirable melting and crystallization characteristics. In practice, a catalyst, alkali metals or their alkoxides are used (in the range of 0.2–0.4%). The reaction starts at low temperatures of 50–75°C and is usually processed at ~100°C.

The migration of fatty acids within the same glyceride molecule is referred to as intraesterification. The random migration and replacement of fatty acids among glyceride molecules ultimately attains an equilibrium with all possible combinations of fatty acids in glyceride structures. The random rearrangement of the fatty acids among glyceride molecules is termed interesterification. Take a triglyceride, ABC, as an example (Fig. 1.4). (For lipase-catalyzed interesterification, see Chapter 5.)

Mechanism

The reaction mechanism of interesterification probably involves the nucleophilic substitution at the carbonyl carbon (Eq. 1-31) (*18*). The diglycerinate generated can react with another glyceride, forming new triglyceride and regenerating another diglycerinate until an equilibrium mixture of all possible compositions of the fatty acids is obtained.

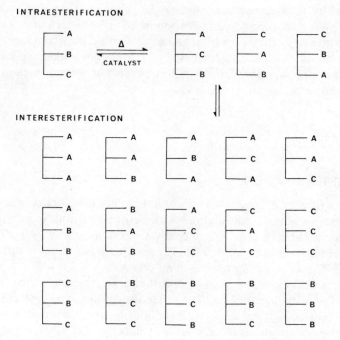

Fig. 1.4. Equilibrium triglyceride mixture of intra- and interesterification.

POLYMORPHISM OF TRIGLYCERIDES

Triglycerides exhibit multiple melting behavior, and this is attributed to a phenomenon known as polymorphism—the existence of alternate crystalline structures for the same substance. For example, tristearin melts at 54.7°, 64.0°, and 73.3°C, representing the transition from the less stable to the stable polymorphic form ($\alpha \rightarrow \beta' \rightarrow \beta$).

Structures of Polymorphic Crystals

It has long been known that long-chain compounds generally exhibit crystalline behavior. Polar end groups tend to associate with each other to form "planes" along the a and b axes as illustrated in a unit cell (a spatial unit in x-ray crystallography). The hydrocarbon chain lies along the c axis (Fig. 1.5).

A cross-sectional view of a crystal with the c axis standing perpendicular to the plane of this paper is shown in Fig. 1.6. Each zigzag period represents the cross section of the hydrocarbon chain. The orientation of these periods determines the stability of the crystalline structure. There are three main types of packing arrangements that concern the oil chemist (Fig. 1.7):

1. Triclinic parallel ($T_{||}$): All zigzag planes (a, b planes) are parallel.
2. Orthorhombic ($O_{\perp}$): The planes are alternate.
3. Hexagonal: The planes are random with rotational freedom.

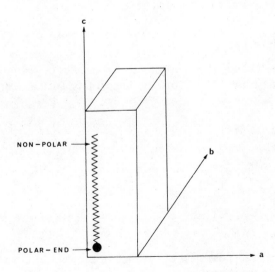

Fig. 1.5. Hydrocarbon in a unit cell crystal.

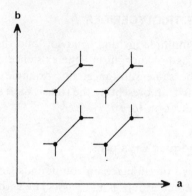

Fig. 1.6. Cross-sectional view of hydrocarbon crystal.

Triglycerides exhibit the same different types of packing arrangement. The triclinic, orthorhombic, and hexagonal packing types are commonly designated as β, β', and α, respectively, and are schematically represented in Fig. 1.8 (19).

Most triglyceride crystals have the chains tilted at a certain angle to accommodate the branching structure. The spatial arrangement of the triglyceride assumes a turning fork configuration. The central and one of the two fatty acid ester chains align along the same axis. The third remaining fatty acid chain parallels the central chain. The arrangement of mixed glycerides in the β form is schematically represented in Fig. 1.9.

Crystal Habit of Fat

Fats with the stable β' forms appear as tiny needles of $\sim 1 \mu$ length microscopically. These are typically of randomized fats, such as rearranged lard,

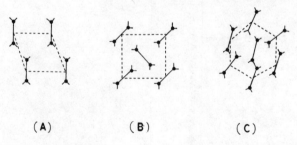

(A)   (B)   (C)

Fig. 1.7. Cross-sectional view of (A) triclinic parallel, (B) orthorhombic, and (C) hexagonal arrangement.

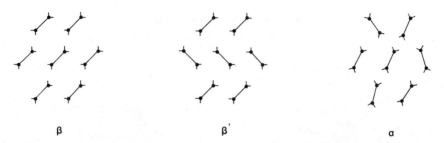

Fig. 1.8.   The β, β', and α triglyceride crystals.

tallow, and the hydrogenated cottonseed oils. Fats that crystallize in the β
forms tend to have large and coarse crystals of 25–50 μ to 100 μ after extended periods of aging. Clumps of these crystals are responsible for grainy
and sandy texture. Alpha crystals are fragile platelets of 5 μ in size. The β'
crystals can incorporate large amounts of tiny bubbles, providing a smooth,
creamy texture to the oil product. For margarine and shortening, the desirable crystal form is β'.

The fatty acid composition of the glyceride molecules determines in part
the crystal habit of the particular fat. Glycerides with uniform fatty acid
composition tend to have β crystals. Fats containing mixed compositions
tend to form β' crystals. For example, soybean oil contains less than 10%
of palmitic acid. Upon increased hydrogenation, it is increasingly composed
of predominantly tristearic acid which can arrange in orderly fashion to
form β crystals. Palm oil and the β'-type oil contain at least twice the
amount of palmitic acid. Palm oil has 50% palmitic acid, and the hydrogenated product, a mixture of stearic and palmitic acid, tends to crystallize in
the β' form. Interesterification usually converts a β'-type to a β-type fat.

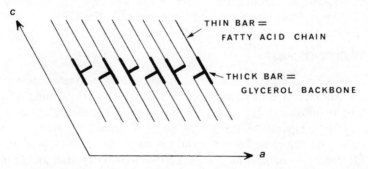

Fig. 1.9.   The β crystal viewed along b axis. (Reprinted with permission from
Lutton (19), courtesy of American Oil Chemists' Society.)

Table 1.5. Hydrogenated Fats
and Crystal Habits

β TYPE	β' TYPE
Soybean	Cottonseed
Safflower	Tallow
Sunflower	Butter oil
Olive	Palm
Lard	Modified lard
Corn	Rapseed
Coconut	
Cocoa butter	
Seasame	

Reprinted with permission from Wiedermann (20), courtesy of American Oil Chemists' Society.

The position of a particular fatty acid in the glyceride also affects the crystal habit. A mixed triglyceride such as 2-stearoyl dipalmitin generally shows a great stability for the β' form, and the stable β form is difficult to obtain. On the other hand, 2-palmitoyl distearin exhibits a stable β form, and the β' form is seldom obtained. Lard is the most extensively studied in this respect. Contrary to common vegetable oils, which tend to have saturated acids in the 1 or 1 and 3 position, natural lard contains 60% palmitic acid in the 2 position. Interesterification changes the palmitic acid porportion in the 2 position, and the randomized lard crystallizes in the β' form instead of the β form for natural lard (19, 20).

Therefore, the process of hydrogenation and interesterification can be controlled to chemically modify the glycerides, producing fats and oils with different melting and crystal habits to fit the demands for a wide variety of food uses. In industrial processing, the polymorphic form of a fat or blend has significant practical meaning. Table 1.5 lists some of the fully hydrogenated common fats in categories of their crystal habits (20).

PLASTICITY OF FAT

The physical characteristics, such as spreadability, softness, and consistency, of commercial solidified edible fats (margarine, shortening, etc.) depend on the following factors: (1) amount of solids (glyceride crystals) present at a given temperature, (2) melting point of component glycerides in the fat, and (3) polymorphic form (crystal habit) of the component glycerides.

A fat can be visualized as a mass of interlocking crystals (solid phase) holding a liquid phase. The plasticity of a fat depends on the proper proportion between the solid and the liquid phase.

It is essential to have balance between the solid and liquid phase, so that

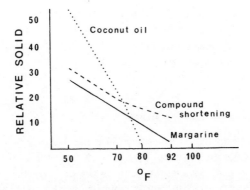

Fig. 1.10. Solid-liquid ratio measurement of margarine, coconut oil, and shortening. (Reprinted with permission from Wiedermann (20), courtesy of American Oil Chemists' Society.)

the solids yield and flow when the external work forces applied exceed the internal bonding forces of the solid mass. The fat regains the original consistency after the external force is lifted. The temperature range in which the fat retains this plastic characteristic (plasticity) is the plastic range. If the fat contains too few solid crystals, the fat will melt, and if the solid content is too high, the fat becomes brittle (21). In practice, the percent solid of edible fats is between 10 and 30%.

Figure 1.10 shows the relative solid content of margarine as compared to coconut oil and shortening at various temperatures (20). Margarine has 30% maximum solid at low temperature, making it less brittle (hence, higher degree of softness) than coconut oil. Compared to shortening, it has a steeper slope at the higher temperature range (70°-90°C); the product therefore possesses a quicker melt in the mouth than shortening.

Fats consisting of largely a single class of glycerides usually have a very narrow plastic range. Coconut oil and butter oil, containing largely a single class of saturated glyceride, melt sharply. Fats made of a mixture of glycerides melt over a wider temperature range, and usually have desirable plastic properties, since each component glyceride will melt at different temperatures. Preparation of margarine oils is by direct blending of selectively hydrogenated oils containing a high amount of *trans* isomers of oleic and linoleic acids.

EMULSIONS

An emulsion is a mixture of two immiscible phases; one is the dispersed or discontinuous phase as droplets or liquid crystals, the other is the nondispersed, continuous phase.

If the oil is dispersed as finite minute droplets in a solution of water (which is the continuous phase), the emulsion is termed the oil-in-water (o/w) type. When water is dispersed in oil, the emulsion type is referred to as water-in-oil (w/o). Most of the discussions here refer to the o/w type of emulsion.

Surface Tension and Surface Area

Surface tension arises from unbalanced intermolecular forces on the surface molecules. Consider an air/water system: A water molecule at the interface, quite different from the water molecule in the bulk of the solution, experiences an uneven effect of intermolecular forces (Fig. 1.11). The molecule tends to move into the bulk of the water solution, since a water molecule in the bulk solution has a lower potential energy than at the interface. This driving force for the water molecule at the interface to move into the bulk is referred to as the surface tension.

Another way of reasoning is to consider that energy is required to move a water molecule from the bulk solution to the surface. The energy is used to increase the surface area of the interface. For this reason, water droplets spontaneously assume a spherical form since a sphere has the smallest surface/volume ratio. The same reasoning can be applied to an o/w system. The oil molecules tend to stay in the oil phase. It requires work to disperse the oil molecules in a water solution.

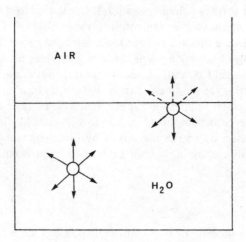

Fig. 1.11. Water molecule at interface.

Formation of an Emulsion

The formation of an emulsion requires energy to form the dispersed droplets in the continuous phase. This energy is represented by Eq. 1-32, where γ = surface tension and σ = surface area.

$$\delta w = \gamma \delta \sigma \qquad \text{Eq. 1-32}$$

Obtaining a fine emulsion means reducing the size of the droplets and hence increasing surface area. Consequently, it takes more energy to maintain a finely dispersed emulsion. This implies that the formation of an emulsion is thermodynamically unfavorable.

However, if the γ term in Eq. 1-32 can be diminished, the energy required to produce a given emulsion is decreased. Lowering the surface tension at the interface enhances emulsification. One of the primary roles of surfactants is precisely the lowering of the surface tension. Surfactants possess both the polar end, which interacts with the water molecules, and the hydrophobic end, which can interact with the oil phase. We will discuss this subject again in a later section.

Breakdown of Emulsions

Three processes contribute to the breakdown of emulsions (*22, 23*).

1. *Sedimentation:* Because of the density difference between the two phases, the droplets tend to rise or sediment through the continuous phase. This process is referred to as "creaming up" or "creaming down."

2. *Disproportionation:* Droplets or bubbles tend to decrease their surface area to minimize the potential energy at the interface. This decrease in surface area is balanced by the rise in internal pressure (Eq. 1-33):

$$\Delta P = 4\gamma/r \qquad \text{Eq. 1-33}$$

where

ΔP = pressure (inside) − pressure (outside)
γ = surface tension
r = radius of the droplet

Equation 1-33, the Laplace equation, suggests that the pressure increases inversely with the size of the droplet. At high dispersed-phase volume, the continuous phase forms the lamella in between the dispersed droplets. The pressure of the droplets causes thinning of the lamella (Fig. 1.12). The process is often viewed as "drainage" of the water past the dispersed phase.

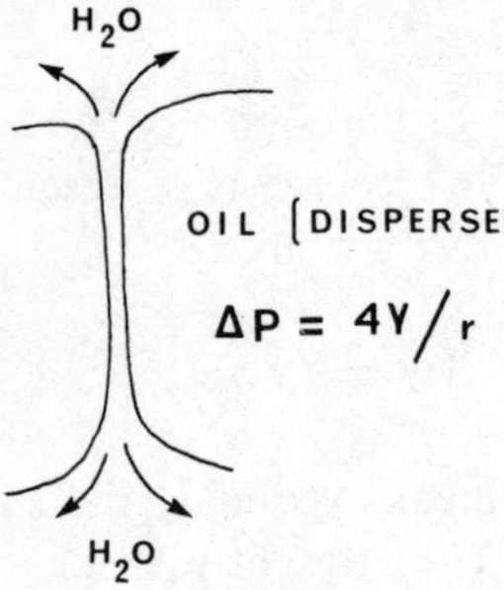

H_2O

OIL [DISPERSED PHASE]

$$\Delta P = 4\gamma/r$$

H_2O

Fig. 1.12. Thinning of the lamella between dispersed droplets.

The critical lamella thickness is reached at 50–150 Å. Diffusion starts from the small droplets to larger droplets or to the bulk phase.

3. *Flocculation and coalescence:* When two droplets in an emulsion (without surfactants present) come close to each other, they tend to flocculate and then coalesce (Fig. 1.13).

The process of flocculation and coalescence is dependent on the balance between two interactions: (1) Van der Waals attraction, and (2) electrostatic repulsion between two droplets.

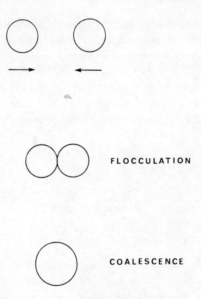

FLOCCULATION

COALESCENCE

Fig. 1.13. Schematic representation of flocculation and coalescence.

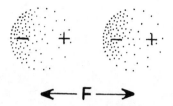

Fig. 1.14. Van der Waals attraction between two adjacent atoms.

Van der Waals Attraction.   The attractive force is derived from the fluctuation in the electron density of a neutral atom. The uneven distribution of electron density momentarily creates a dipole which can then induce the adjacent atom into a similar orientation (Fig. 1.14). This London energy varies inversely with the distance of separation between the atoms (Eq. 1-34).

$$F = -\beta / r^6 \qquad \beta = \text{constant}$$
$$r = \text{distance in between} \qquad \text{Eq. 1-34}$$

The Van der Waals attractions are additive. The attractive potential between particles is equal to the summation of all the attraction between every atom. The attractive potential increases therefore with (1) increasing particle size and (2) decreasing distance between two particles.

Electrostatic Repulsion.   A particle with a uniformly charged surface can cause an uneven distribution of ions in the solution, forming an electric double layer at the surface. Ions with opposite charge (counterions) can be found accumulated near the particle surface, with the concentration gradually decreasing with distance. Such distribution of counterions is described as ion atmosphere (Fig. 1.15A).

Consider two particles approaching each other. Their counterion atmospheres start to interfere and, hence, create a repulsion between the particles. The repulsion force decreases exponentially with increasing distance of separation (Fig. 1.15B).

Stability of an emulsion, therefore, can be considered to depend on the balance between attraction and repulsion. Figure 1.16 is a plot of the repulsive and attractive potential energy as a function of the distance between two particles (22).

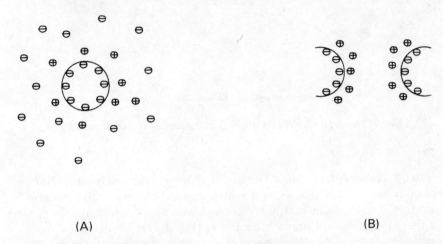

(A)   (B)

Fig. 1.15. Interactions in (A) ion, and (B) counterion atmosphere.

A very important difference between repulsive and attractive forces is their changes with respect to distance. The net potential is always an attraction at long and short ranges. At certain intermediate ranges, repulsion may be greater than attraction and a measure of stability may exist, depending on the magnitude of the net potential.

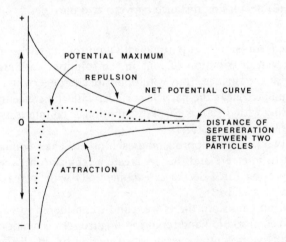

Fig. 1.16. Repulsion and attractive potential energy as a function of the distance between two particles. (Reprinted with permission from Ref. (22), p. 7, by courtesy of Marcel Dekker, Inc.)

EMULSIFIERS

Food grade emulsifiers are partial esters of fatty acids, polyols, and water-soluble organic acids. An emulsifier, therefore, consists of both hydrophilic and hydrophobic segments spatially separated in the same molecule. In food systems, their functions are:

1. To promote stability of the emulsion by control of aggregation of fat globules
2. To improve softness with reduced staling tendencies in baked products
3. To strengthen wheat dough by interaction with gluten
4. To improve the consistency of fat-based products by controlling the crystallization of fats

Monoglycerides

Monoglycerides are the most widely used surfactants in food. The monoglycerides commercially available are combinations of mono- and diglycerides, although monoglycerides are the main active emulsifier. Monoglyceride is prepared by the interesterification of triglycerides with excess glycerol (Eq. 1-35) (24).

$$
\begin{array}{c}
CH_2OH \\
| \\
CH\,OH \\
| \\
CH_2OH
\end{array}
\;+\;
\begin{array}{c}
\overset{O}{\overset{\|}{CH_2-O-C-R}} \\
\overset{O}{\overset{\|}{CH-O-C-R'}} \\
\overset{O}{\overset{\|}{CH_2-O-C-R''}}
\end{array}
\;\xrightarrow[\Delta]{NaOH}\;
\begin{array}{l}
Mono- \\
Di- \quad Glyceride \\
Tr-
\end{array}
\qquad Eq.\ 1\text{-}35
$$

To obtain a high concentration of monoglycerides, the mixture is vacuum distilled to a product of $>90\%$ total monoglyceride. This product is termed distilled monoglyceride. The "soft" varieties, prepared from partially hydrogenated oils, are soft and plastic. The "hard" varieties, obtained from fully hydrogenated oil, are sold in solid powder form. Distilled "hard" monoglycerides are sold as microspersions in water as monoglyceride hydrates. Monoglycerides are nonionic and are used especially as dough softeners.

Monoglyceride Derivatives

The properties of a monoglyceride can be modified by introducing additional functional groups ester-linked to the glycerol backbone (24).

Succinylation of the monoglycerides with succinic anhydride results in a product that is anionic. This alternation in the hydrophilic/lipophilic balance in the molecule changes its degree of solubility in water, and also its functional properties. Succinylated monoglycerides function as both softeners and conditioners in baked products (Eq. 1-36).

$$
\begin{array}{c}
\underset{\substack{| \\ CHOH \\ | \\ CH_2OH}}{\overset{O}{\overset{||}{CH_2O\,C}}\,(CH_2)_{16}CH_3} + O=\underset{}{\diagup}\overset{}{\underset{}{\diagdown}}=O \xrightarrow[80-180\,C]{NaOH} \underset{\substack{Succinylated \\ monoglyceride}}{\overset{O}{\overset{||}{CH_2O\,C}}\,(CH_2)_{16}CH_3} + H_2O
\end{array} \qquad \text{Eq. 1-36}
$$

Similarily, ethoxylated monoglycerides are produced by reacting monoglycerides with ethylene oxide (Eq. 1-37).

$$
\underset{\substack{| \\ CHOH \\ | \\ CH_2OH}}{\overset{O}{\overset{||}{CH_2O\,C}}\,(CH_2)_{16}CH_3} + \underset{CH_2-CH_2}{\overset{O}{\diagup\diagdown}} \xrightarrow[150\,C,\,40-80\,psi]{NaOH} \underset{\substack{| \\ CHOH \\ | \\ CH_2OCH_2CH_2O^{\ominus}\,{}^{\oplus}Na}}{\overset{O}{\overset{||}{CH_2O\,C}}\,(CH_2)_{16}CH_3} \qquad \text{Eq. 1-37}
$$

Diacetyl tartaric acid ester of monoglyceride is prepared by first reacting acetic anhydride and tartaric acid to form acetyl tartaric anhydride, which then reacts with monoglyceride. The product has increased HLB and water solubility.

Ester Derivatives of Alcohols (Nonglycerol)

Sorbitan monostearate is the product of the reaction between sorbitol and fatty acid (Eq. 1-38). It is an antiblooming agent in confectionery products containing cocoa butter substitute and an emulsifier in cakes.

$$
RCOOH + \underset{\substack{CH_2OH \\ CHOH \\ HO\,CH \\ CHOH \\ CHOH \\ CH_2OH}}{} \xrightarrow[225-250\,C]{H^{\oplus}} \qquad \text{Eq. 1-38}
$$

When the product is ethoxylated by reacting with ethylene oxide, the resulting product is polysorbate (Eq. 1-39), which is used as an emulsifying agent in cake and cake mixes, and as a dough strengthener.

$$\text{SORBITAN ESTER} \ + \ \underset{H_2C-CH_2}{\overset{O}{\triangle}} \ \xrightarrow[\substack{140-160\,C \\ 40-80\,psi}]{\ominus OH} \ \text{product} \qquad \text{Eq. 1-39}$$

SUM OF $w, x, y, z = 20$

$$R = (C_{17}H_{33})COO^{\ominus}$$

Stearoyl-2-lactylate, widely used as a dough softener and conditioner, is prepared by reacting stearic acid with lactic acid (Eq. 1-40).

$$\underset{\text{Fatty acid}}{\overset{O}{\underset{\|}{R-C-OH}}} + \underset{\underset{COOH}{|}}{\overset{CH_3}{\underset{|}{CHOH}}} \xrightarrow[180-200\,C]{\ominus OH} M^{\oplus \ominus}O-\overset{O}{\overset{\|}{C}}-\overset{CH_3}{\underset{|}{CH}}\left(O-\overset{O}{\overset{\|}{C}}-\overset{CH_3}{\underset{|}{CH}}\right)_n O-\overset{O}{\overset{\|}{C}}-R \qquad \text{Eq. 1-40}$$

$$M = Na^{\oplus} \ \text{or} \ Ca^{\oplus\oplus}$$

Propylene glycol monostearate is produced by the esterification of propylene glycol with stearic acid (Eq. 1-41).

$$\underset{\substack{\text{Propylene} \\ \text{glycol}}}{\overset{CH_2OH}{\underset{\substack{| \\ CH_3}}{\overset{|}{CHOH}}}} + \underset{\text{Stearic acid}}{RCOOH} \xrightarrow[\Delta]{\ominus OH} \underset{\substack{\text{Propylene glycol} \\ \text{Monostearate}}}{\overset{\overset{O}{\overset{\|}{CH_2-O-C-R}}}{\underset{\substack{| \\ CH_3}}{\overset{|}{CHOH}}}} \qquad \text{Eq. 1-41}$$

Polymerization of glycerol under alkaline conditions at high temperature produces polyglycerols. Condensation of the α-hydroxy forms an ether linkage; the resulting polymer is linear. Direct esterification with fatty acid yields the ester of polyglycerol (Eq. 1-42) (25).

Eq. 1-42

Based on the structural formula of the above described emulsifiers, we can classify them according to their hydrophilic/lipophilic balance (26):

1. Lipophilic (HLB = 3–9)—monoglycerides, propylene glycol stearate, acetylated monoglyceride, ethoxylated, lactylated monoglycerides
2. Very hydrophilic (HLB = 12–20)—polysorbate, stearoyl-2-lactylate, polyglycerol ester
3. Moderately hydrophilic (HLB = 8–12)—diacetylated tartaric acid ester, succinylated monoglyceride

Lecithin

The trade name "lecithin" refers to a mixture of phospholipids including phosphatidylcholine (lecithin specifically), phosphatidylethanolamine (cephalin) phosphatidylinositol, and phosphatidylserine (Fig. 1.17), removed from oil (predominantly from soybean in the United States) in the degumming process. Commercial crude lecithins commonly contain small amounts of triglycerides, fatty acids, pigments, carbohydrates, and sterols. The amount of lecithin used in most food formulations is in the range of 0.1–0.3% (27).

Saturated fatty acids are found primarily in the α-position of phosphatidylcholine (same as in triglycerides) and the β-position of phosphatidylethanolamine. The various phosphatides in lecithin have lipophilic and hydrophilic balance resulting in typical emulsifying properties.

Phosphatidylcholine (PC) stabilizes o/w emulsions, while phosphatidylethanolamine (PE) and phosphatidylinositol (PI) have w/o emulsifying properties. A mixture of these phospholipids in lecithin results in both weak

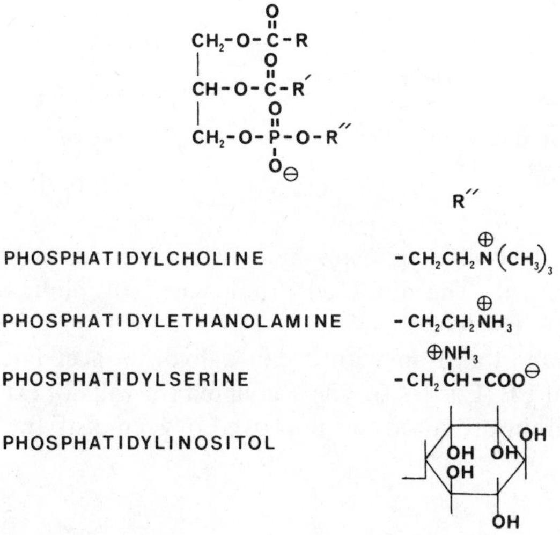

$$CH_2-O-\overset{\overset{O}{\|}}{C}-R$$
$$CH-O-\overset{\overset{O}{\|}}{C}-R'$$
$$CH_2-O-\overset{\overset{O}{\|}}{\underset{\underset{O_{\ominus}}{|}}{P}}-O-R''$$

R''

PHOSPHATIDYLCHOLINE $-CH_2CH_2\overset{\oplus}{N}(CH_3)_3$

PHOSPHATIDYLETHANOLAMINE $-CH_2CH_2\overset{\oplus}{N}H_3$

PHOSPHATIDYLSERINE $-CH_2\overset{\overset{\oplus NH_3}{|}}{C}H-COO^{\ominus}$

PHOSPHATIDYLINOSITOL

R, R' = FATTY ACID CARBON BACKBONE

ONLY α FORMS ARE SHOWN. β FORMS ALSO EXIST.

APPROX. % COMPOSITION OF SOYBEAN LECITHIN [27]

PHOSPHATIDYLCHOLINE	20
PHOSPHATIDYLETHANOLAMINE	15
PHOSPHATIDYLINOSITOL	20
OTHER PHOSPHATIDES	5
TRIGLYCERIDES	35
CARBOHYDRATES, STEROLS, STEROL GLYCOSIDES	5

Fig. 1.17. Composition of "lecithin" mixture. (Reprinted with permission from Van Nieuwenhuyzen (27), courtesy of American Oil Chemists' Society.)

w/o and o/w emulsifying properties. Furthermore, PE flocculates and loses its emulsifying power in the high concentrations of calcium and magnesium salts present in hard water. For more stable emulsions, lecithin is commonly used in combination with other emulsifiers. And in many cases, the lecithin is chemically or enzymatically modified to improve its emulsifying properties and to reduce its reactivity toward metal ions (28).

1. *Alcohol fractionation:* PC dissolves to a greater extent than PE in ethanol. Fractionation of lecithin in 90% ethanol concentrates the PC in the product with the PC to PE ratio increased to 8 to 1. The product has improved o/w emulsifying properties and is commonly used as antispattering agent in margarine.

2. *Hydrolysis:* Phospholipase A2 (used in industrial scale) specifically hydrolyzes the fatty acid at the 2-position of the phospholipids (Eq. 1-43). (Refer to Chapter 5 for details.)

Eq. 1-43

Chemical hydrolysis of phospholipid is less specific and the reaction is difficult to control. The modified product is hydrophilic and has better o/w emulsifying properties.

3. *Acetylation:* Reaction with acetic anhydride acetylates the primary amino group of PE (Eq. 1-44). The acetylated PE will not exist in zwitterion form. The lecithin produced has improved o/w emulsifying properties.

Eq. 1-44

Lecithin has been used extensively as emusifier in varieties of food products. Only two specific applications are discussed to illustrate its wide industrial use.

1. Antispattering in margarine: Margarine is a w/o emulsion, with 18% water (milk) dispersed in a continuous fat phase. Monoglyceride is often used to stabilize the w/o emulsion. In heating, the margarine melts, and the water coalesces into large droplets which evaporate vigorously, causing spattering of hot oil. It is suggested that lecithin functions as nuclei for water droplets where slow evaporation is possible.

2. Dispersion of cocoa powder: Cocoa powder contains a film of cocoa butter on the surface and is not able to disperse in water or milk at low temperature (melting point of cocoa butter $\sim 35°C$). To enhance dispersion and wetting of the cocoa powder in aqueous solution, a thin spray of lecithin is deposited onto the cocoa powder. The lipophilic part of the phospholipids is incorporated in the cocoa butter, and the hydrophilic portion brings the powder into solution.

LIQUID-CRYSTALLINE MESOPHASE IN EMULSIFIER-WATER SYSTEM

When an emulsifier is dispersed in water and heated, the emulsifier crystal starts to melt at a temperature, known as the Kraft point (T_c), before the true melting point is reached (29). The hydrocarbon chains, held by rela-

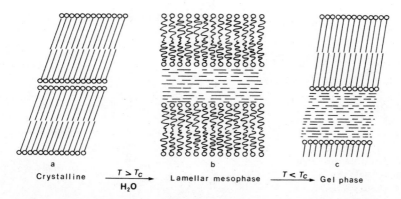

a
Crystalline $\xrightarrow[H_2O]{T > T_c}$ Lamellar mesophase $\xrightarrow{T < T_c}$ Gel phase

Fig. 1.18. Structure models of (A) the crystalline state, (B) the lamellar mesophase, and (C) the gel state. (Reprinted with permission from Ref. (29), p. 84, by courtesy of Marcel Dekker, Inc.)

tively weak Van der Waals forces, melt at a lower temperature than the polar end group, which is hydrogen bonded, to assume a liquid crystalline structure with both liquid (melted chains) and crystal (polar end). A lamella mesophase is formed consisting of bimolecular lipid molecules separated by water. The chains are disordered (Fig. 1.18).

On cooling of the lamellar mesophase, a gel is formed. The structure is still lamellar, but the lipid chains are crystallized in α form and are extended. Upon further heating, the lamellar mesophase breaks down to viscous cubic or reversible hexagonal form (Fig. 1.19). Lamellar mesophase has bimolecular lipid layers alternating with water layers. Two types of hex-

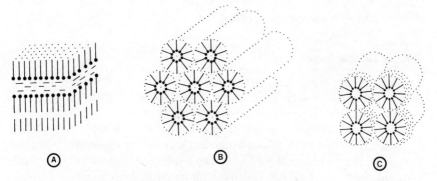

Fig. 1.19. Structure models of (A) lamellar, (B) hexagonal, and (C) cubic mesophase.

Table 1.6. Mesomorphic Behavior of Some Emulsifiers in Water

EMULSIFIER	MESOPHASE FORMATION TEMPERATURE (°C)	TYPES OF MESOPHASE FORMED
Saturated monoglyceride (90% monoester)	55	lamellar $\xrightarrow{68°C}$ cubic
Unsaturated monoglyceride	20	cubic $\xrightarrow{55°C}$ hexagonal II
Sorbitan monostearate	55	lamellar
Polyglycerol monostearate	60	lamellar
Polysorbate	40	hexagonal I or micelles
Sodium stearoyl-2-lactylate	45	lamellar

Reprinted with permission from Lauridsen (30), courtesy of American Oil Chemists' Society.

agonal mesophases have been reported. Hexagonal I has the lipid molecules aggregated in cylinders with the polar groups oriented towards the outer water phase and the hydrocarbon chains filling the core. Hexagonal II exhibits the reverse arrangement, where the polar groups are oriented inward and the hydrocarbon chains radiate outward. The cubic phase consists of spherical aggregates with the polar groups facing the water and the hydrocarbon chains lining the exterior. All the mesophases may be formed in aqueous systems containing food emulsifiers.

Table 1.6 lists some of the mesomorphic behavior of some emulsifiers in water (30). Some of these emulsifiers produce different mesophases at different temperatures.

FUNCTIONS OF EMULSIFIERS IN STABILIZATION

Once an emulsion is formed, the stabilizing action of an emulsifier on an existing emulsion can be ascribed to two functions.

Electric Double Layer

As stated earlier, the process of flocculation and coalescence is governed by the net potential between the Van der Waals attraction and the repulsion between the double layer. In an o/w system without emulsifier, the emulsion is not stable, although an electric double layer is often present. The energy of repulsion is too small to overcome attraction. The oil droplets tend to flocculate and coalesce. When ionic emulsifier is added to the system, it accumulates at the interface, causing an increase in the repulsion potential. Figure 1.20 schematically shows the effect of an anionic emulsifier on the electric double layer.

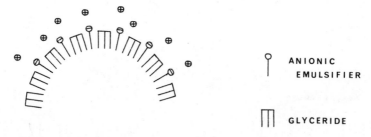

Fig. 1.20. Schematic representation of electric double layer.

Stabilization through an electric double layer, however, is not observed in the case of nonionic emulsifiers or in w/o emulsions that develop no electric double layer.

Adsorption at Interface

In an emulsion, emulsifier molecules orient themselves at the interface, and this adsorption leads to a reduction of the surface tension (Fig. 1.21). The adsorbed molecules form interfacial layers of various types, depending on the lipophilic/hydrophilic properties of the emulsifier (26). In an o/w emulsion, if the emulsifier used is too lipophilic for the system, some molecules tend to partition between the interface and the oil phase, resulting in a loose, discontinuous film.

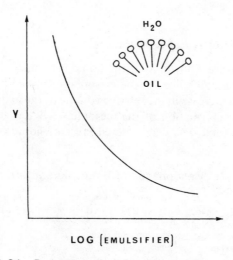

Fig. 1.21. Reduction of surface tension by emulsifier.

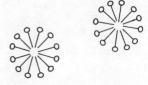

WATER PHASE

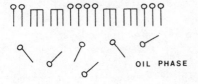

OIL PHASE

EMULSIFIER TOO LIPOPHILIC

TOO   LIPOPHILIC
TOO   CONCENTRATED

Fig. 1.22. Orientation of emulsifier molecules at interface. (Reprinted with permission from Krog (26), courtesy of American Oil Chemists' Society.)

Emulsifiers too hydrophilic for the system tend to form micelles in the water phase. High concentrations also favor the formation of micelles and, in many cases, multilayer films (Fig. 1.22).

Optimum stability is obtained when the emulsifier molecules form a densely packed layer at the interface, with the polar groups facing with the water phase and the hydrocarbon chain interacting with the oil phase. The closely packed interfacial film of emulsifier molecules provides a steric barrier against coalescence.

Formation of Liquid-Crystalline Interface

Many emulsifiers are known to form liquid-crystalline interfaces in an emulsion. The oil is solubilized in the mesophase formed by the surfactant and water, and the increase in the relative volume of the hydrocarbon region often causes a transformation of the mesophase. A few studied examples (31) are listed in Table 1.7. The ordered layer of liquid-crystalline meso-

Table 1.7. Mesophases of Emulsifiers at Interfaces

EMULSIFIER	MESOPHASE FORMED BY EMULSIFIER INTO WATER (1:1)	MESOPHASE FORMED AFTER INTRODUCING SOYBEAN OIL
Glycerol monostearate	lamellar	hexagonal II
Sodium-stearoyl-2-lactylate	lamella	hexagonal II
Polysorbate 60	hexagonal II	lamellar

Reprinted with permission from Krog (31), courtesy of Academic Press, Inc.

phase causes a considerable reduction of attractive forces between droplets. Furthermore, the mesophase is highly viscous (1000 times water alone), forming a steric barrier against coalescence between droplets. Lamellar mesophase is often found in o/w emulsion, while cubic and hexagonal II are present in w/o emulsion.

Complexation with Starch

Staling of bread is related to the crystallization of amylose that is leached out of the starch granules during gelatinization. The function of emulsifier to retain softness and to decrease staling rate is mainly attributed to its complex formation with amylose. The linear chain of amylose coils up to form a helix with the emulsifier, analogous to that with iodine. The C-H groups and the glycosidic oxygen atom orient inward, forming a lipophilic core, with all the hydrophilic OH groups pointing ouward. The inside diameter of the hydrophilic core is 4.5–6.0 Å (Fig. 1.23) (32).

The hydrocarbon chain such as that of saturated monoglycerides fits in the core. Indeed, distilled, saturated monoglycerides are most effective as dough softeners and are used extensively for this purpose.

The complex is insoluble in water and hence is less likely to leach out from the starch granule during heat processing. When the amylose inside the granule is retained, association of the amylose is retained and retrogradation is reduced.

Emulsifier Interaction with Protein

The anionic stearoyl-2-lactylate or organic acid esters of monoglycerides, or nonionic hydrophilic polysorbates, function as dough strengtheners or conditioners through their interactions with the wheat flour protein, gluten.

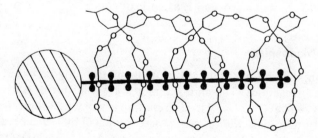

Fig. 1.23. Schematic illustration of monostearin-amylose helical complex with the whole chain inside the helical space. The hydrocarbon chain is extended to give the approximate geometric relations three turns per chain. The hydrogen atoms of the monostearin are indicated. (Reprinted with permission from Carlson et al. (32), copyright 1979, VCH Verlagsgesellschaft, Weinheim, Germany.)

The exact mechanism of the interaction is not clearly understood. It is known that nonpolar lipids are often detrimental to the dough, and polar lipids, especially glycolipids, tend to improve the dough (*33*). (Refer to wheat proteins, Chapter 2.) The final loaf volume depends on the ability of the dough to retain carbon dioxide in the structural network of the gluten in the early stages of baking, and the ordering of these structural networks during dough mixing is facilitated by the lipids.

The emulsifier is bound to gluten by possibly both hydrophobic and hydrophilic interactions during dough mixing. During baking, the bonds are weakened (denaturation of the gluten) as temperature increases. The emulsifier molecules are translocated and bound to the gelatinized starch molecules, forming a protein-lipid-starch complex.

Control of Fat Crystallization

Sorbitan esters have the specific function of stabilizing the β' form of fats and prevent the transformation to β form. During the shelf life of chocolate, or chocolate coating, the β' crystals tend to change to the more stable β form. A gray film known as "bloom" is formed on the surface. Similarly, polymorphic transformation during storage causes "sandiness" in margarine made from sunflower oil. The mechanism of inhibition of this type of transformation by sorbitan esters is not known.

ANTIOXIDANTS

Since autoxidation involves the formation of free radicals, removal of the free radicals should terminate the chain reaction. The most commonly used antioxidants are the substituted phenolic compounds (Fig. 1.24). (Refer to Chapter 10 for vitamins A and E as antioxidants.)

Reaction Mechanism

Antioxidants inhibit autoxidation of lipids by trapping peroxy radicals in two types of reactions. The antioxidant acts by transferring an H atom to the peroxy radical (Eq. 1-45.1). The resulting aryloxy radical of the antioxidant can further react with a second peroxy radical by radical-radical coupling to yield a peroxide product, peroxydienone (*34*). Aryloxy radicals are stable by resonance and are relatively unreactive toward LH and O_2, and therefore are not capable of initiating or propagating oxidative reaction (Eq. 1-46). However, the formation of peroxydienone can limit the effectiveness of the antioxidant at high temperature and exposure to UV light,

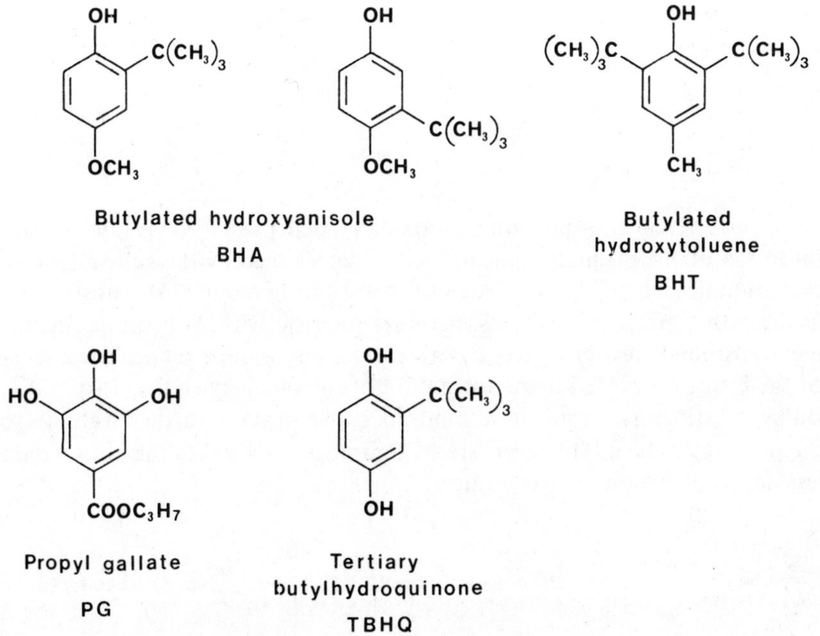

Fig. 1.24. Commonly used phenolic antioxidants.

since under these conditions new free radicals are generated (Eq. 1-45.2) (35).

1 – 45.1

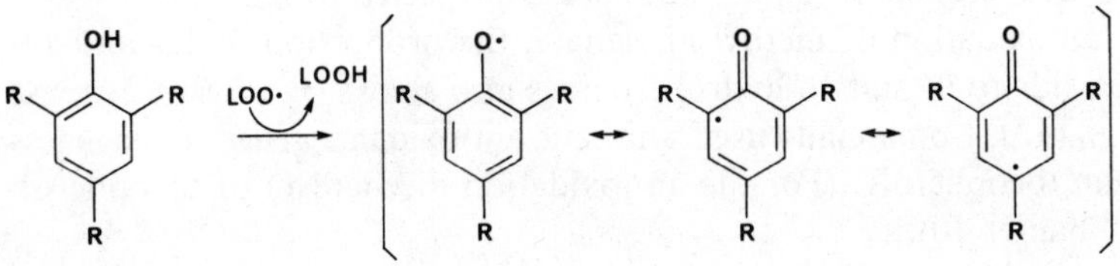

Eq. 1-45

1 – 45.2

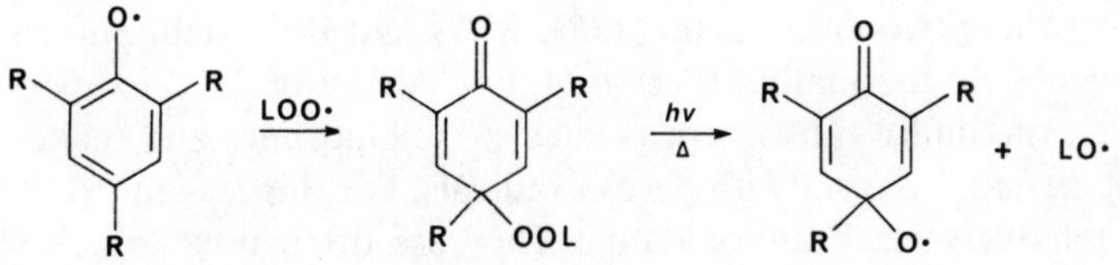

$$\text{(structure)} \quad + \quad RH \quad \xrightarrow{\quad\quad} \quad \text{(structure)} \quad + \quad R\cdot \qquad \text{Eq. 1-46}$$

The effectiveness of phenolic antioxidants depends on the resonance stabilization of the phenoxy radical. This is determined by the substituent on the aromatic ring and by the size of the substituting group (36). Substitution at the ortho and para positions increases the reactivity. Substitution at the meta position is less effective. Electron-releasing groups reduce the energy of the transition state for the formation of the phenoxy radical (Eq. 1-47). Bulky substituents create steric hindrance and provide further stability to the phenoxy radical. However, steric hindrance also makes the antioxidant less accessible to the peroxy radical.

$$\text{(structure)}-OH \xrightarrow{LOO\cdot} \left[\text{(structure)}-O\cdots H\cdots OOL \right] \longrightarrow \text{(structure)}-O\cdot \; + \; LOOH \qquad \text{Eq. 1-47}$$

Since antioxidant acts as a H donor in the initial reaction (Eq. 1-45.1), autoxidation of lipids in the presence of phenolic antioxidant leads not only to the expected slowing of autoxidative consumption of LH, but also an alteration of the distribution of oxidative products. In the two competitive pathways in Eq. 1-19, the conversion of initially formed peroxy radical to the *trans, cis* hydroperoxides is highly favorable. The ratio of *trans, cis/ trans, trans* hydroperoxides is greatly increased, and the proportion of conjugated hydroperoxide to diperoxides is also expected to increase.

In the autoxidation of methyl linolenate, the proportion of 12- and 13-hydroperoxide to 9- and 16-hydroperoxides also shows significant increase (Table 1.8) (37). Comercially used synthetic antioxidants generally have less effect than tocopherols. (For the antioxidation mechanism of tocopherol, refer to Chapter 10).

Most antioxidant addition is applied to finished products, but it is also used during processing. The tertiary alkyl group in BHA and BHT adds stability to the phenoxy radical and is responsible for the carry-through characteristics of these two antioxidants (38). BHA and BHT retain much of their effectiveness in thermally processed foods (baked or fried temperature). However, the bulky substituent creates steric hindrance and makes the antioxidant difficult to react with peroxy radicals. For this reason, BHA and BHT are relatively weak antioxidants. They are often used together

Table 1.8. Effect of Antioxidants on Autoxidation of Methyl Linolenate (100 hr, 40°C)

ANTIOXIDANT	WT. ADDED ($\times 10^2\ \mu g/g$)	CONJUGATED DIENE ($\mu mol/g$ SAMPLE)	MOLAR PROPORTION OF PRODUCT TYPES		
			TOTAL HYDROPEROXIDE	c,t ISOMER	12-, 13- ISOMER
Control	—	434	46	27	6
BHA	2	27	56	36	14
	50	35	73	61	26
	250	69	85	79	41
BHT	2	13	42	27	11
	50	7	51	32	13
	250	8	48	41	14
Control	—	490	49	27	7
α-Tocopherol	2	15	62	34	18
	50	83	88	82	39
	250	354	98	95	50

Reprinted with permission from Peers et al. (37), courtesy of The Society of Chemical Industry and author.

with other antioxidants (such as PG and TBHQ) for synergistic effects and to take advantage of their carry-through effect in baked and fried foods.

PG has relatively high antioxidant potency in oils, due to the three hydroxy groups with no steric hindrance. However, PG forms a highly colored complex with iron. Furthermore, PG is heat sensitive and much of its effectiveness as an antioxidant is lost at high temperatures. For this reason, all formulations of PG include chelating agents. Unlike other antioxidants, PG is water soluble. In water-fat systems, PG tends to leach from the fat into the water phase, causing a decrease in its antioxidant property. The use of high-alkyl gallate (e.g., octyl, dodecyl) serves to change the solubility. TBHQ is the more effective in that it shows no discoloration with metal, and good solubility in fats and oils.

Since the presence of metal accelerates the oxidation process, ethylenediaminetetraacetic acid (EDTA) or citric acid is often included in an antioxidant formulation as a commonly used chelating agent to complex or scavenge trace metals present in the food.

2
PROTEINS

There is a constant and increasing demand for high-quality protein foods for an ever-growing world population. Food scientists are interested not only in the chemical principles underlying protein structures and mechanisms, but the application of this knowledge in exploring the unique characteristics of proteins that are relevant in a mixed food system. The basic chemistry is always related to the functionality of the protein. Thus, the chemistry of muscle contraction is related to rigor mortis and the postmortem tenderness of meat. The chemical structure of collagen and the unfolding and refolding of the helical chains help to explain the formation of gelatin gel. The understanding of the underlying chemistry and mechanisms provides new ideas that challenge food scientists and technologists to improve existing and formulate new food products.

The chemical changes of proteins during food storage and processing are other areas that require unique attention. Chemical or enzymatic modifications occur, intentionally or unintentionally, to a greater extent than is commonly realized and impart desirable but most often unwanted effects. Not only the functional properties of the protein are changed by these reactions, but the nutritional quality is altered. One typical example is the use of alkaline treatment in food processing and its implication in the possible degradation of proteins. On the other hand, modifying food proteins often has the advantage of enhancing functional properties that are necessary for food processing and production.

In this chapter the basic chemistry of proteins is covered, with particular attention given to the reactions and mechanisms that are of importance in food processing. Organized protein systems, such as meat, wheat, and milk proteins, are discussed with emphasis on their functionalities and structures.

48

PROTEIN STRUCTURE

The primary structure of proteins consists of polypeptide chains of a repeating backbone of peptide bonds and various side chains of amino acids. The peptide bond ($-CO-NH-$) is *trans,* planar due to its partial double-bond character resulting from resonance between the oxygen and the nitrogen. There are only two free rotations (about the $C_\alpha - N$ bond axis, and the $C_\alpha - C$ axis) designated by the torsional angles ϕ and ψ. The polypeptide chain can fold into an α-helix or β-pleated sheet to assume a secondary structure.

In the α-helix, the CO group of each amino acid residue is hydrogen bonded to the NH group of the amino acid residue four units apart. The neighboring amino acid residues are 1.5 Å apart with a rotation of 100°, resulting in 3.6 amino acid residues per turn. This stable helix type is denoted as $3.6_{13}r$, where 13 is the number of atoms between the hydrogen bond and r denotes right-handed. Another type commonly found in globular proteins is the $3_{12}r$ helix. The α-helix can be packed by having side chains intercollated between each other. The intercollation occurs only when the axes of the two helices cross at angles of $-82°$, $-60°$, or $+19°$. These are known as class I, II, or III packing.

In the β-pleated sheet, the polypeptide chains are extended and the structure is stabilized by hydrogen bonds formed between adjacent chains. The polypeptide chains can align in the same direction (parallel) or in opposite directions to each other (antiparallel). The backbone connection between β strands can be classified into two categories (*1*) (Figure 2.1A).

1. Hairpin connection—the chain reenters the sheet at the same end it left from.
2. Crossover connection—the chain loops around to reenter the sheet on the opposite end from where it left.

The reentry of the β strand may be the intermediate neighboring strand or separated by one or more strands.

Beta-pleated sheets are commonly right-handed twisted (when viewed along the direction of the polypeptide chains). Apparently, the twist confers a lower free energy than straight or left-handed twist.

Secondary structures are frequently connected by reverse turns, where the polypeptide chain folds back upon itself by forming a hydrogen bond between the CO group to the NH group in the third residue back along the chain. A total of four amino acids is included in the formation of turns. In many globular proteins, about one-third of the amino acid residues are found in turns. Amino acid residues, such as proline, asparagine, and gly-

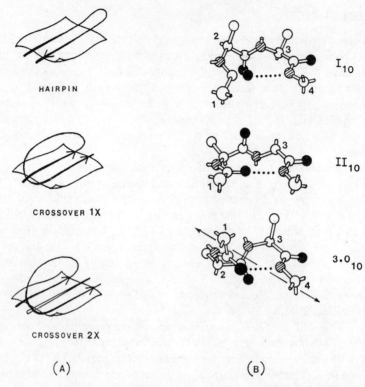

HAIRPIN

I_{10}

CROSSOVER 1X

II_{10}

CROSSOVER 2X

$3{\cdot}0_{10}$

(A)   (B)

Fig. 2.1. Illustration of (A) hairpin and crossover connections in β sheet, and (B) Type I_{10}, II_{10}, $3{\cdot}0_{10}$ helix conformation. (A: Reprinted with permission from Richardson (1), courtesy of National Academy of Science and author. B: Reprinted with permission from Birktoft and Blow (2), courtesy of Academic Press Inc.)

cine, occur in high frequency in these reverse turns. There are three types of turns, characterized by the dihedral angles of the second and third amino acid residues (ϕ_2, ψ_2; ϕ_3, ψ_3) (2) (Fig. 2.1B).

Type I_{10}: $-60°$, $-30°$; $-90°$, $0°$
Type II_{10}: $-60°$, $120°$; $80°$, $0°$
Type $3{\cdot}0_{10}$: $-60°$, $-30°$; $-60°$, $-30°$

Certain folding patterns are commonly found in proteins, involving a number of secondary structure elements but not yet comprising the complete tertiary structure. These assemblies are termed supersecondary structures. The 4-α-helical arrangement is common to a number of proteins.

It is organized as sequentially connected 4-α-helices packed together in an antiparallel pattern at angles of about 18°. The stability of the structure is the result of helix-dipole interaction. Another structural arrangement is coiled-coil α-helix, where two α-helices are wound together, resulting in a left-handed superhelix. The most common arrangement is β-α-β, where two parallel β-sheets are connected by an α-helix. In antiparallel sheets, the structure may wrap around in barrels. The topology assumes the β-meander (+1, +1, +1 as in papain), or Greek-key pattern (−3, +1, +1, −3, or +3, −1, −1, as in chymotrypsin) (3) (Fig. 2.2).

Most proteins can be considered as layered structures of α-helices and β-sheets and classified into four categories: αα (mainly α-helix, e.g., myohaemerythrin, myogen, myoglobin), ββ (mainly β-sheet, e.g., superoxide dismutase, chymotrypsin), α + β (α-helix and β-strand, but not mixed, e.g., papain, insulin, lysozyme), α/β (mixed, mostly alternating α-helix and β-strand, e.g., carboxypeptidase A, alcohol dehydrogenase, triose phosphate isomerase) (4).

Almost all globular proteins consist of locally compact globular regions. These are potentially independent and geometrically separate units with specific functions, such as binding sites. These regions are generally known as domains. The active site of a protein molecule is usually found at the interface between domains.

The polypeptide chain—and, in fact, the protein molecule—is stabilized by many interactions, such as electrostatic interactions, hydrogen bonds, and hydrophobic forces (Fig. 2.3), providing a unique spatial arrangement and orientation of the side chain groups and their special functional properties.

Electrostatic interactions that occur between charged side chains are governed by Coulomb's law (Eq. 2-5) and depend on the dielectric constant of the medium. Interaction between ion pairs may be attractive or repulsive; about one-third of the ion pairs found in proteins may involve repulsive interactions. The most common amino acid residues involved in forming

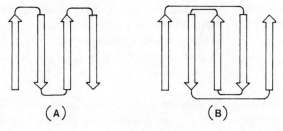

(A)   (B)

Fig. 2.2. (A) β-meander and (B) Greek-key pattern.

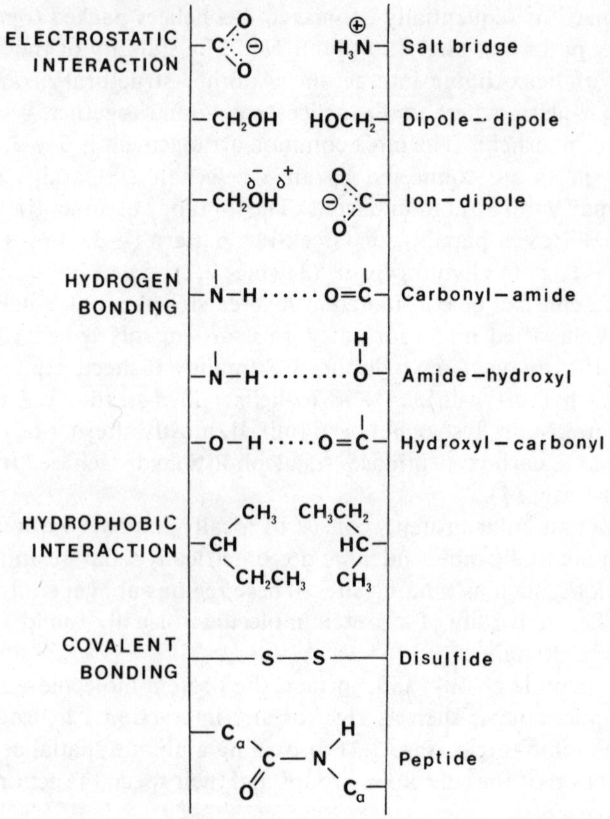

Fig. 2.3. Types of bonding stabilizing protein structure.

ion pairs are Arg, Asp, Glu, Lys, and His. A salt bridge formed between Lys and Asp has a binding energy of -5 kcal/mol. However, ion pair interactions between amino acid residues are not as significant in protein folding as are the ion-dipole interactions between the charge groups and water molecules. Uncharged but polar side chains also participate in electrostatic interactions due to their partial charges expressed as dipole moments. The interaction of dipoles has electrostatic repulsion or attraction depending on the orientation and the extent of polarizability due to induction by neighboring molecules. Dipole-dipole interactions yield energies of ± 0.1–0.2 kcal/mol.

Hydrogen bonding has an intermediate energy between electrostatic interaction and covalent bonding. The energy of the carbonyl-amide hydrogen bond is about -3 kcal/mol. Depending on the type of hydrogen bond, the bond distance between the donor and acceptor atom is about 2.8–3.0 Å. All hydrogen bonds are linear to assume the lowest potential energy.

Hydrophobic interaction constitutes a major stability force in proteins. The interactions are due to unfavorable changes in water entropy when the nonpolar side chains are exposed to the aqueous medium. The details are to be discussed in relation to the role of water in proteins.

Besides the noncovalent forces, disulfide bonds provide further stabilization to many proteins. Disulfide bonds often create intrachain loops. The -Cys-Cys- sequence in a number of proteins links segments of polypeptide chains in close proximity as in the case of wheat germ agglutinin and human serum albumin. The 17 disulfide bonds in human serum albumin align the protein molecule in a series of nine loops of repeated triplets of different sizes. Disulfides may also participate in enzyme catalysis, as in the example of the flavoenzyme glutathione reductase. The active center of the enzyme consists of two redox systems, FAD and the disulfide between Cys-46 and Cys-41, that function in the transfer of reduction equivalent from NADPH to glutathione, GSSG.

The Role of Water

Aside from the covalent and noncovalent interactions, the stability of protein conformation is greatly affected by the solvent. Proteins contain both hydrophilic and hydrophobic side chains. In an aqueous environment, the hydrophobic groups tend to move away from the water phase and aggregate in the interior core of the protein molecule. If the hydrophobic side chains are unfolded and exposed to the aqueous phase, as in the case when a protein is denatured, the process is accompanied by a large negative entropy ($\Delta S^{\ddagger} < 0$) which causes the corresponding free energy to be positive ($T\Delta S^{\ddagger} > \Delta H^{\ddagger}$) (Eq. 2-1). Therefore, this process of hydrophobic hydration is thermodynamically unfavorable. Some of the calculated thermodynamic parameters of transferring hydrophobic amino acid side chains from a hydrophobic medium to water is presented in Table 2.1 (5). The hydrophobicity is linearly correlated with the surface area of the nonpolar side chain, and it follows that hydrophobicity is additive in that the hydrophobicity of a molecule is the summation of that of its constituent residues.

$$\Delta G = \Delta H - T\Delta S \qquad \text{Eq. 2-1}$$

The water molecules located inside the protein molecule are isolated from the surrounding water and also play an important role in stabilizing the conformation of the protein molecule. Three types of water molecules are found within a protein molecule. (1) Water molecules trapped inside the

Table 2.1. Calculated Thermodynamic Parameters of Transfer of Amino Acid Side Chains from a Hydrocarbon Medium to Water at 25°C

AMINO ACID SIDE CHAIN	$\Delta G^{\ddagger}$ (KJ/mol, 25°C)	$\Delta H^{\ddagger}$ (KJ/mol)	$\Delta S^{\ddagger}$ (J/mol-deg)
Alanine	+5.5	− 6.3	−39.5
Valine	+8.0	− 9.2	−57.5
Leucine	+8.0	−10.1	−60.1
Phenylalanine	+7.6	− 4.2	−39.9

Reprinted with permission from Eagland (5), courtesy of Academic Press, Inc.

protein help fill the empty crevices in the side chain packing. (2) Partially ordered water with an altered freezing point is produced by interaction with protein side chain groups. This layer of water molecules covers about one-third of the protein surface. (3) Tightly bound water molecules form the first coordination layer with substantially lower energies than the bulk water. Besides these "structural" waters, the surface of the protein molecule is also solvated by layers of "interfacial" water. The monolayer that covers the immediate surface shows decreased mobility and structural changes. The water molecules are oriented by their own polarity and that of the polar groups on the surface of the protein molecule. The "structural" and "interfacial" waters are included under the category of "bound" water, in contrast to the "bulk" water in the system. Interactions between water and protein molecules involve mainly ion-dipole, dipole-dipole, and hydrogen bonding as illustrated in Fig. 2.4.

Change of Conformation

Alteration of the structural conformation of proteins can be brought about by heat, salts, pH changes, organic solvents, and denaturing agents such as guanidium salt. Two types of changes can occur. (1) Chain-chain interaction (among side chain groups in the polypeptide) resulting in association, aggregation, flocculation, coagulation, and precipitation. (2) Chain-solvent interaction (between solvent molecules and side chain groups) resulting in solubilization, dissociation, swelling, and denaturation.

Heat. The process of folding and unfolding of a protein can be represented by Eq. 2-2, and the equilibrium constant is related to the enthalpy change.

$$\text{Native} \rightleftharpoons \text{Denatured}$$

$$K = \frac{\text{Denatured}}{\text{Native}} = e^{\left(-\Delta H^{\circ} + T\Delta S^{\circ}\right)/RT} \qquad \text{Eq. 2-2}$$

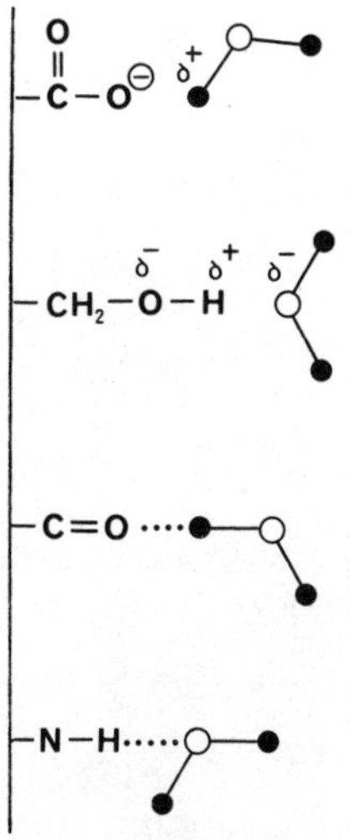

Fig. 2.4. Interaction between water and protein molecules.

For most proteins, the unfolding enthalpy, $\Delta H°$, is in the range of 50–150 kcal/mol and $\Delta G° < 20$ kcal/mol (Table 2.2).

Heating provides energy to break the noncovalent interactions that stabilize the native structure. The process usually occurs in a narrow temperature range. T_m is the melting temperature at which [native] = [denatured] (Fig. 2.5). The sharp transition from the native to the denatured state is indicative of "cooperativity." The many noncovalent interactions "cooperate" by breaking together. It is difficult for a single residue to initiate bond breaking. However, once the initial breaking starts, breaking at adjacent sites occurs more readily, and so does the third site neighboring it, analogous to the operation of a zipper. Thus, the observed $\Delta H°$ and $\Delta S°$ would be the sum of all the contributions from individual bond breaking, so that the exponential term in Eq. 2-2.2 exhibits a large change for a small change in temperature.

Table 2.2. Enthalpy for Unfolding of Various Proteins

PROTEIN	$\Delta H°$ (kcal mole^{-1})
Ribonuclease	96
Lysozyme	88
Cytochrome c	50
Metmyoglobin	68

From Privalov and Khechinashvili. 1974. J. Mol. Biol. 86, 682.

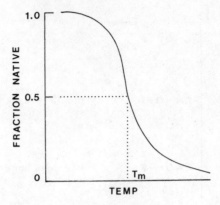

Fig. 2.5. Thermal denaturation of macromolecules of the cooperative type.

pH. Protonation at low pH increases the cationic properties of the protein, while alkaline pH increases the anionic species. For example, in a lysine residue at pH 10.50, half of the ϵ-amino groups are protonated. Increasing the pH will shift the equilibrium to the formation of ϵ-NH$_2$, and decreasing pH will favor protonation of the amino group to ϵ-NH$_3^+$. The pK$_a$ values for the various amino acid side chains are listed in Eq. 2-3.

The change in pH, therefore, changes the distribution of cationic, anionic and nonionic polar sites on the protein molecule, which in turn affects water-protein and protein-protein interactions. At the isoelectric point, the protein molecule has a net charge of zero (Eq. 2-4). The protein exhibits the maximum amount of electrostatic interactions between the charged groups, and the protein molecule becomes shrunken. The charged groups are least available for interaction with water molecules, and the amount of bound water is decreased to a minimum. The protein molecule, therefore, exhibits minimal hydration, swelling, and solubility. Above the isoelectric point, the protein has a net negative charge, and in the acid range, a net positive charge. The like-charge side chains repel each other, and at extreme pH's, the protein tends to unfold.

Eq. 2-4

isoelectric
point

Organic Solvents. The force between two charged sites in a protein molecule is given by Coulomb's law (Eq. 2-5). If Q_1 and Q_2 are of like sign, the force is positive, indicating repulsion. Likewise, if they are of opposite sign, the force is negative, indicating attraction.

$$F = \frac{Q_1 Q_2}{\varepsilon r^2}$$

Q = Charges $\oplus$ or $\ominus$

ε = Dielectric constant of medium

r = Distance between sites

Eq. 2-5

Organic solvents usually have dielectric constants lower than water (Table 2.3). Therefore, adding organic solvents lowers the dielectric constant and thereby increases the force of attraction between the opposite charges in the protein molecule, causing them to come close enough to aggregate and precipitate.

Table 2.3. Dielectric Constants of Selected Solvents

SOLVENT	DIELECTRIC CONSTANT (25°C, 1 atm)
Benzene	22.7
Methanol	32.6
Ethanol	24.3
Water	78.5

Salts.   For most proteins, low salt concentrations (ionic strength < 0.1–0.15) tend to increase the solubility of the protein molecules. This "salting-in" process is the result of the salt effect on the electrostatic interactions. At high concentrations, salts decrease the solubility of the protein molecule. This "salting-out" process is the result of the salt effect on hydrophobic interactions.

The salt effect can be represented by Eq. 2-6 (6). At low salt concentration, $\Delta G_{e.s.}$ is positive, which accounts for the "salting-in" effect. Salt ions interact with the countercharges on the protein (Fig. 2.6), decrease the potential energy for ion-ion interactions, and hence increase the solubility of the protein.

$$\ln \frac{W}{W_o} = \frac{\left[\Delta G_{e.s.} + \Delta G_{cav}\right]}{RT} + constant$$

W = Weight of protein soluble in 1 liter of salt solution.

W_o = Weight of protein soluble in 1 liter of water.

$\Delta G_{e.s.}$ = Free energy change of electrostatic interactions when the protein goes into salt solution.

ΔG_{cav} = Free energy required for formation of a cavity in the solvent to accommodate hydrophobic groups exposed on surface.

Eq. 2-6

However, $\Delta G_{e.s.}$ increases nonlinearly with salt concentration. With increasing salt concentration, eventually all the charged groups are shielded

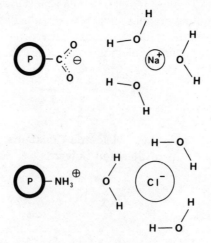

Fig. 2.6. Salt ion interaction with protein.

and the protein is effectively a neutral dipole molecule. The hydrophobic term (ΔG_{cav}) in the equation becomes dominant, and $\Delta G_{cav} = -\Omega\sigma m$, where Ω is related to the hydrophobic area and σm is the surface tension of the salt solution. For most simple salts, σm is positive (increases the surface tension of water). This means that ΔG_{cav} is always increasingly negative and hence W/W_o (the solubility) is decreased.

However, some proteins do not follow the processes predicted by the equation, and exhibit "salting-out" before salting-in. The salting-out effect of ions on hydrophobic interactions can occur at relatively low salt concentrations, especially in proteins that tend to associate in salt solutions. In general, it is the salt effect on the charge profiles of the electrostatic and hydrophobic forces that determines the solubility behavior of a protein molecule.

If we plot the solubility against salt concentration (expressed as ionic strength), the electrostatic and hydrophobic effects counteract each other. Normalization of these two effects results in a curve with intercept β' and slope K_s' (Fig. 2.7). Protein solubility at high salt concentration, therefore, can then be expressed by Eq. 2-7 (6).

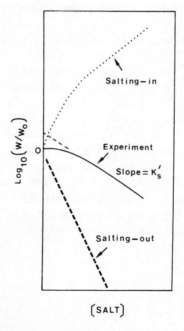

(SALT)

Fig. 2.7. Plot of solubility (log) against salt concentration (molal) for hemoglobin in ammonium sulfate. (Reprinted with permission from Melander and Horváth (6), courtesy of Academic Press, Inc.)

$$\log S = \beta' - \kappa'_s \mu \qquad \qquad \text{Eq. 2-7}$$

where

S = solubility
β' = hypothetical solubility at zero ionic strength
K_s' = salting-out constant
μ = ionic strength

The β' term is dependent on the type of protein, pH, and temperature, while the K_s' term is affected by the nature of the salt.

PROTEIN-STABILIZED EMULSIFICATION AND FOAMING

Protein molecules contain both hydrophilic and hydrophobic groups and, as expected, may act like a surface-active substance. The protein molecule adsorbs at the interface of the emulsion system, denatures, and unfolds with trains of amino acids along the interface and with loops and tails protruding into either phase (Fig. 2.8). The orientation of the loops and tails depends on the hydrophobicity and hydrophilicity of the amino acid chains (7).

Formation of these loops and tails is favored with increasing concentration of the protein. At low concentration, most polypeptides have their amino acid backbone lying along the interface with few loops and tails. As the concentration increases, the polypeptides are more closely packed and hence more loops and tails are formed. Finally, the looped conformation creates enough electrostatic repulsion and steric hindrance that further protein adsorption becomes energetically unfavorable. The protein concentration at the interface is at the saturated level, and more concentrated solutions give rise to multilayer adsorption (Fig. 2.9).

In a colloidal system such as a protein solution, foam formation occurs when air is mechanically incorporated. A monolayer film of surface-denatured protein is adsorbed at the air-water interface of the colloidal solution similar to that discussed in an emulsion system.

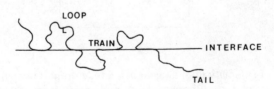

Fig. 2.8. Protein conformation at interface.

INCREASING
CONCENTRATION

Fig. 2.9. Protein conformation at interface with increasing concentration.

When air is mechanically incorporated into the protein solution (e.g., shaking the solution or, in practical application, whipping egg white), cells are formed with air surrounded by the protein film at the air-water interface (Fig. 2.10).

Adsorption of surface-denatured protein continues to occur at the interface to replace the coagulated regions of the film around the air cells. When bubbles come in close contact, drainage occurs from the aqueous lamella formed between the bubbles and finally causes rupture of the films (Fig. 2.11).

The adsorbed protein film at the interface provides stability against coalescence by:

1. Reinforcing the repulsive forces: The charged groups in the protein create, between the air cells, a relatively dense electric double layer that helps prevent the thinning of the lamella.
2. Forming a steric barrier: The rigidity of the protein film helps stabilize against coalescence of the cells and disruption of the lamella.
3. Increasing viscosity: Increasing viscosity in the aqueous solutions in the lamella tends to act against drainage.

It should be expected that the molecular structure of the particular protein, and hence the parameters that affect the unfolded state of the protein (e.g., pH, ionic strength, temperature), all play important roles in determi-

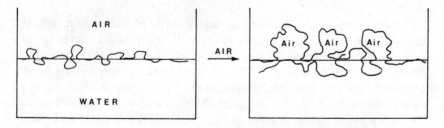

Fig. 2.10. Foam formation at air-water interface.

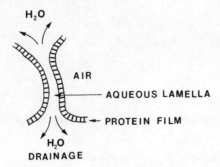

Fig. 2.11. Drainage from aqueous lamella.

nating the stability properties of the protein film. Proteins that have a flexible structure, such as β-casein, are easily surface denatured. Highly structured globular proteins, such as lysozyme, are difficult to surface denature. The degree of unfolding at the interface affects the rheology of the film. Thus, a β-casein film is liquidlike and provides good stability in emulsions. Lysozyme films are rigid and more resistant to deformation. A more denatured protein is a more extended molecule and hence allows the development of better structure and network (8).

GEL FORMATION

Protein gelation is an aggregation of denatured molecules with a certain degree of order, resulting in the formation of a continuous network. The mechanism for the formation of protein gel is represented by Eq. 2-8.

$$x P_n \xrightarrow{\text{Denaturation}} x P_d \xrightarrow{\text{Aggregation}} \left(P_d\right)_x$$

x = Number of protein molecules, P   Eq. 2-8
n = Native state
d = Denatured state

    The first step in gel formation is denaturation of the protein, and the second step is an aggregation process. If k_2 is faster than k_1, a coarse network of protein molecules is formed. The gel is opaque, with large interstices that exhibit syneresis. If k_2 is sufficiently slow, the resulting gel becomes finer, less opaque, and elastic. If the second step is reversible (i.e., the gel melts upon heating), the intermediate state (xP_d) is called a progel. The formation of progel is accompanied by increased viscosity and "sets" to form the gel. This gel type is called the "thermoset" or "reversible" to separate it from the thermoplastic or irreversible gels (9). Gelatin is an example of a thermally reversible gel, whereas milk gel is irreversible.

The kind and nature of intra- or inter-cross-links in the gel network are crucial to the gel characteristic. Both covalent and noncovalent bonds are involved. Noncovalent bondings may include hydrogen bonding and hydrophobic interactions, and covalent bondings include disulfide cross-links. Disulfide cross-links help in bridging and ordering of the gel network, while noncovalent bondings help in stabilizing and strengthening the gel structure.

Conditions affecting gelation include heat, pH, ionic strength, and protein concentration. Heat denatures and unfolds the protein molecules and enhances disulfide exchange, forming new cross-linking disulfide bonds (Eq. 2-9). Contrary to denaturation, where the native structure is most stable in the electric range where the net charge is low, aggregation is suppressed at high and low pH. This is readily explained by the electrostatic repulsion produced by the high net charge. Protein-solvent interaction is favored rather than protein-protein interaction. The addition of salt increases aggregation due to the reduction of the repulsive net charges (10). For every protein, there is a critical concentration below which a gel cannot be formed.

$$RSSR + R'S^{\ominus} \rightleftharpoons RSSR' + RS^{\ominus}$$
$$RSSR' + R''S^{\ominus} \rightleftharpoons RSSR'' + R'S^{\ominus}$$

Eq. 2-9

All the factors described above are interrelated with each other and furthermore with the type and nature of bonding formed during the heat treatment. It is a delicate balance among all these factors that contributes to a proper gel. Under conditions where electrostatic repulsion becomes predominant, the protein will not aggregate. On the other hand, strong attractive forces (various bondings and cross-linkings) tend to precipitate the protein molecule. Too dilute a protein solution fails to provide enough contact for the groups to interact and cross-link.

CHEMICAL REACTION

Alkali Degradation

Alkaline treatment is extensively used in the food industry for (1) peeling of fruits and vegetables; (2) solubilization and texturalization of food proteins; (3) manufacture of sodium caseinate, gelatin, sausage casings, and tortillas. The chemical modification of proteins induced by alkali treatment is extensive, and only a few reactions will be considered (*11*).

Hydrolysis.   The amide groups in asparagine and glutamine, and the guanidino group in arginine, can be hydrolyzed in alkaline solution. The mechanism involves nucleophilic acyl substitution with tetrahedral intermediates (Eqs. 2-10, 2-11).

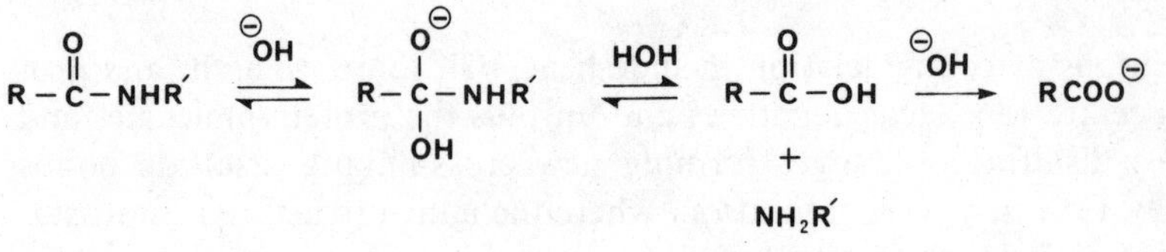

Eq. 2-10

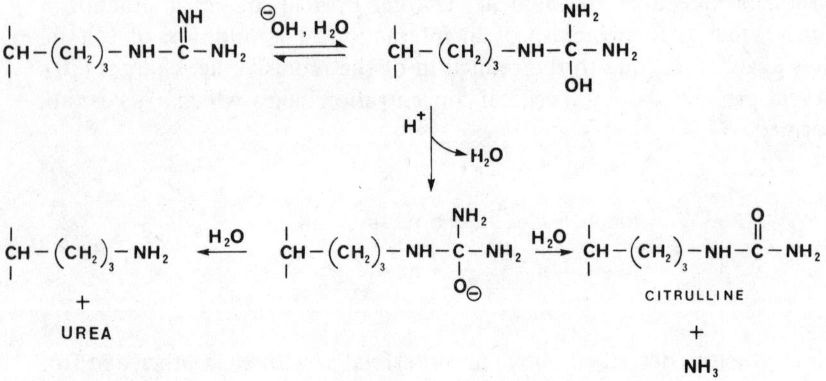

Eq. 2-11

β-Elimination.   The α-hydrogen of an amino acid residue is easily abstracted by the hydroxide ion. In protein-bound cystine, the resulting products are the persulfide and dehydroalanine (Eq. 2-12). Phosphoseryl and phosphothreonyl residues also undergo similar reactions, as does cysteine.

Nucleophilic side groups such as the ϵ-NH_2 group of lysine can then react with the dehydroalanine (which is an α, β-unsaturated compound) to form lysinoalanine (Eq. 2-13). Similarly, dehydroalanine reacts with a cysteinyl residue to form lanthionine, ornithine to form ornithinoalanine, and NH_3 to form β-aminoalanine (Eq. 2-14).

These addition products lead to new cross-linkings in the protein and render the essential amino acids unavailable. Lysinoalanine has been extensively studied for its physiological and toxicological effects.

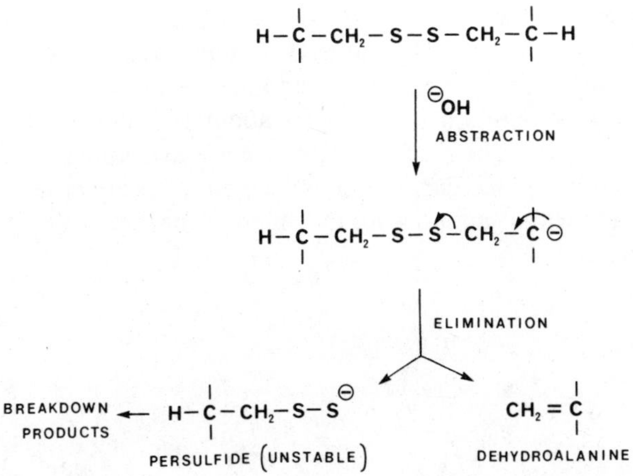

$$H-\overset{|}{\underset{|}{C}}-CH_2-S-S-CH_2-\overset{|}{\underset{|}{C}}-H$$

$\overset{\ominus}{O}H$
ABSTRACTION

$$H-\overset{|}{\underset{|}{C}}-CH_2-S-\overset{..}{S}-CH_2-\overset{|}{C}\ominus \qquad \text{Eq. 2-12}$$

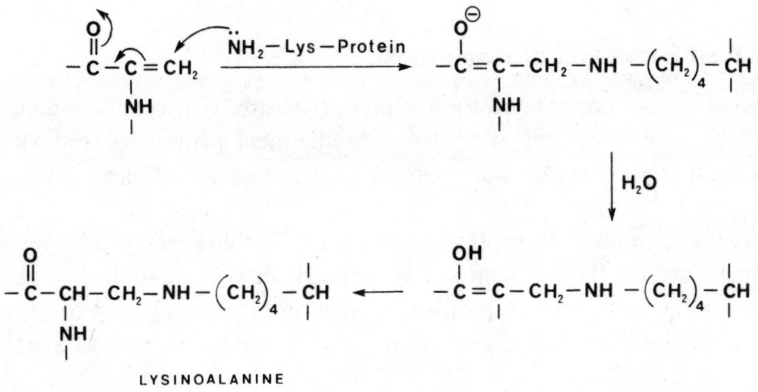

ELIMINATION

BREAKDOWN PRODUCTS ← $H-\overset{|}{\underset{|}{C}}-CH_2-S-S\ominus$ $CH_2=\overset{|}{C}$

PERSULFIDE (UNSTABLE)   DEHYDROALANINE

$$-\overset{O}{\overset{||}{C}}-\overset{|}{C}=CH_2 \quad \xrightarrow{\overset{..}{N}H_2-Lys-Protein} \quad -\overset{O\ominus}{\overset{|}{C}}-C-CH_2-NH-(CH_2)_4-\overset{|}{\underset{|}{CH}}$$
$\underset{NH}{|}$ $\underset{NH}{|}$

H_2O Eq. 2-13

$$-\overset{O}{\overset{||}{C}}-\overset{|}{\underset{NH}{\underset{|}{CH}}}-CH_2-NH-(CH_2)_4-\overset{|}{\underset{|}{CH}} \quad \longleftarrow \quad -\overset{OH}{\overset{|}{C}}=C-CH_2-NH-(CH_2)_4-\overset{|}{\underset{|}{CH}}$$

LYSINOALANINE

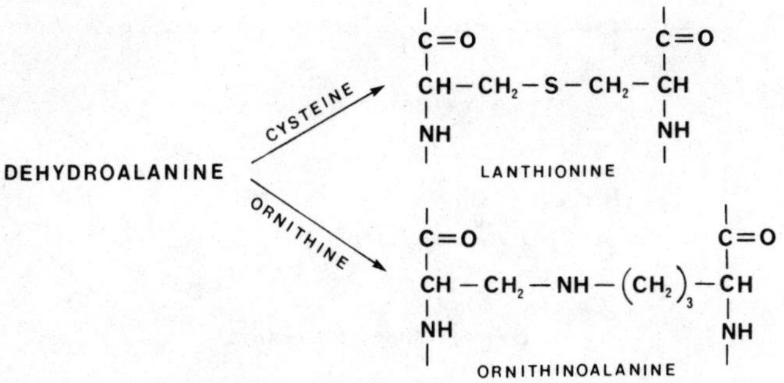

DEHYDROALANINE

$\xrightarrow{CYSTEINE}$

$$\begin{array}{cc} \overset{|}{C}=O & \overset{|}{C}=O \\ \overset{|}{CH}-CH_2-S-CH_2-\overset{|}{CH} \\ \overset{|}{NH} & \overset{|}{NH} \\ | & | \end{array}$$
LANTHIONINE

Eq. 2-14

$\xrightarrow{ORNITHINE}$

$$\begin{array}{cc} \overset{|}{C}=O & \overset{|}{C}=O \\ \overset{|}{CH}-CH_2-NH-(CH_2)_3-\overset{|}{CH} \\ \overset{|}{NH} & \overset{|}{NH} \\ | & | \end{array}$$
ORNITHINOALANINE

Racemization.   Alkali-treated proteins contain D-amino acid residues due to racemization (Eq. 2-15). Amino acid residues in proteins are more susceptible to α-hydrogen abstraction, and hence racemization, than free amino acids. Free amino acids racemize about 10 times more slowly than bound residues (*12*). The D-enantiomers are not available for utilization in the synthesis of proteins. Furthermore, since enzymes are usually stereospecific, substrates with D-amino acids may bind to enzymes to form unreactive intermediates.

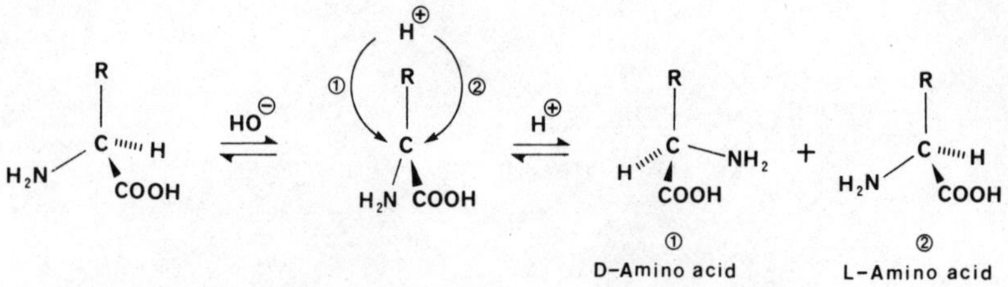

Eq. 2-15

Heat-Induced Formation of Isopeptides

Mild heating causes changes in the tertiary structure of proteins, which in turn influences the physical as well as the chemical properties and alters the functional properties of the proteins that are of significance in food processing.

The nutritional value of proteins when under prolonged (>10hr), and high temperature (>100°C) heating, is reduced, due to cross-linking reactions between the ϵ-amino group of lysine with the carbonyl group of aspartic or glutamic acids (Fig. 2.12) or amide groups of glutamine and asparagine (*13*).

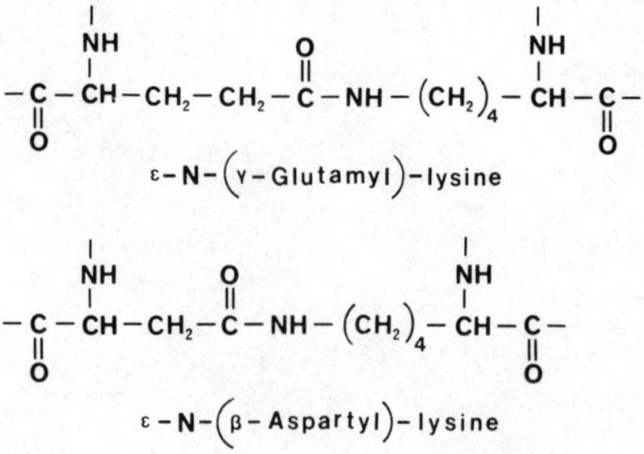

Fig. 2.12.   Heat-induced cross-linked isopeptides.

Radiolysis (14)

Radiation (charged particles, electrons, protons, α-particles, and electromagnetic x-rays, γ-rays) causes ionization of atoms and molecules. In biological and food systems, the damage is usually indirect via the reactive species (e_{aq}^-, ·OH, H·) generated by the radiation of water. (Refer to Appendix 3.)

The *hydrated electron* (e_{aq}^-) reacts with amino acids by (1) nondissociative electron capture resulting in a radical anion, and (2) dissociative electron capture resulting in the elimination of a leaving group. Aliphatic, nonsulfur amino acids have relatively low reactivity to e_{aq}^-; the reaction involves the carbonyl group and yields the radical anion, followed by deamination (Eq. 2-16).

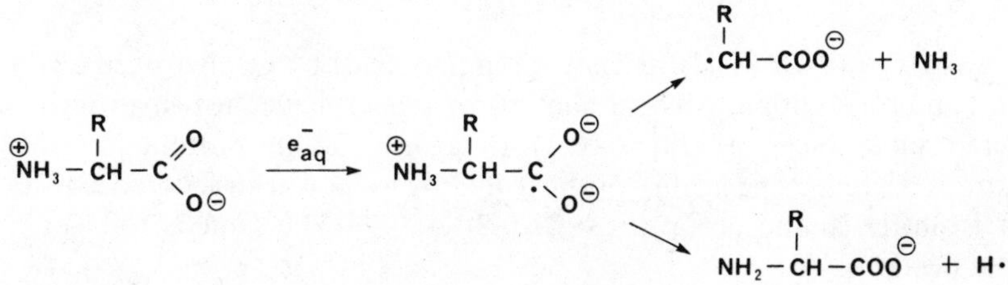

Eq. 2-16

With aromatic amino acids, the e_{aq}^- adds primarily to the imidazole ring in histidine (15) and the benzene ring in tyrosine and phenylalanine, followed by rapid protonation (Eq. 2-17). The rate constant varies depending on the pH. Protonation of the amino group accelerates deamination. Likewise, the reactivity of e_{aq}^- for histidine decreases 100-fold by raising the pH above the pK value of the imidazole (16).

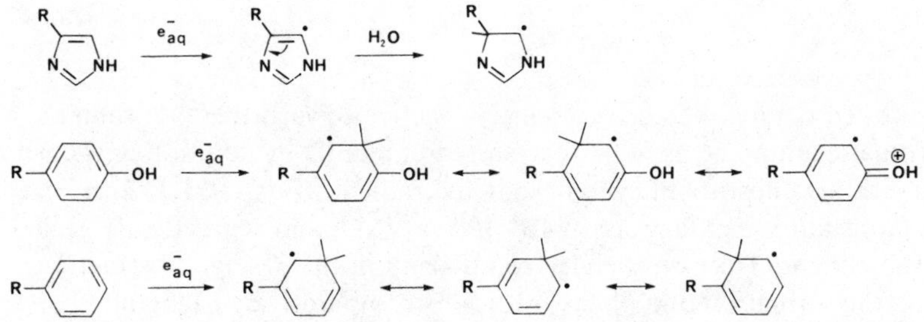

Eq. 2-17

Sulfur-containing amino acids are the most reactive. Cysteine and cystine react with e_{aq}^- near diffusion-controlled rates. Electron addition to the disulfide forms the radical anion intermediate which decomposes unimolecularly to the radical (Eq. 2-18). For cysteine (and most thiols), the reaction is dissociative electron capture and the products are the radicals and anions (Eq. 2-19.1). The amino radicals can undergo "repair" by hydrogen transfer from another molecule (Eqs. 2-19.2, 2-19.3) (17).

$$RSSR + e_{aq}^- \longrightarrow RSSR^{\overline{\cdot}} \rightleftharpoons RS\cdot + RS^- \xrightarrow{H^{\oplus}} RSH \quad \text{Eq. 2-18}$$

$$
\begin{array}{ll}
RSH + e_{aq}^- \longrightarrow R\cdot + SH^- \xrightarrow{H^{\oplus}} H_2S & 2\text{-}19.1 \\[4pt]
R\cdot + RSH \longrightarrow RH + RS\cdot & 2\text{-}19.2 \\[4pt]
RS\cdot + RSH \longrightarrow RS\cdot + RSH & 2\text{-}19.3
\end{array}
\qquad \text{Eq. 2-19}
$$

The *hydroxy radical* ($\cdot$OH) reacts with amino acids by (1) hydrogen abstraction and (2) addition. Most aliphatic amino acids have the α-hydrogen abstracted, under basic pH (Eq. 2-20.1). In acidic solution, protonation of NH_2 groups decreases the acidity of the α-hydrogen, and abstraction occurs farther from the amino group (Eqs. 2-20.2, 2-20.3). The same is true for aliphatic peptides (18).

$$NH_2-\overset{\overset{O}{\|}}{C}H-COO^{\ominus} \xrightarrow{\cdot OH} NH_2-\overset{\overset{R}{|}}{\underset{\cdot}{C}}-COO^{\ominus} + H_2O \qquad 2\text{-}20.1$$

$$NH_2-CHR-\overset{\overset{O}{\|}}{C}-NH-CHR-\overset{\overset{O}{\|}}{C}-NH-CHR-\overset{\overset{O}{\|}}{C}-O^{\ominus} \xrightarrow[\text{pH 9}]{\cdot OH} NH_2-\underset{\cdot}{C}R-\overset{\overset{O}{\|}}{C}-NH-CHR-\overset{\overset{O}{\|}}{C}-NH-CHR-\overset{\overset{O}{\|}}{C}-O^{\ominus} \qquad 2\text{-}20.2$$

$$\overset{\oplus}{N}H_3-CHR-\overset{\overset{O}{\|}}{C}-NH-CHR-\overset{\overset{O}{\|}}{C}-NH-CHR-COO^{\ominus} \xrightarrow[\text{pH 6}]{\cdot OH} \overset{\oplus}{N}H_3-CHR-\overset{\overset{O}{\|}}{C}-NH-CHR-\overset{\overset{O}{\|}}{C}-NH-\underset{\cdot}{C}R-COO^{\ominus} \qquad 2\text{-}20.3$$

$$\text{Eq. 2-20}$$

Unsaturated compounds are extremely reactive toward the $\cdot$OH radical. With aromatic amino acids, $\cdot$OH reacts predominantly by addition to the π system, forming substituted cyclohexadienyl radicals (Eqs. 2-21.1, 2-21.2). The rate constant is of the order of 10^9–$10^{10}\ M^{-1}S^{-1}$, and is greatly affected by the substituent effect on the electron density in the π ring. Abstraction of H$\cdot$ and the formation of a phenyl radical is comparatively negligible. For tryptophan, addition is at the nitrogen ring, and under aerobic conditions the radical formed has been shown to undergo oxygen addition and hydrogen abstraction to form dioxindole-3-alanine (Eq. 2-21.3).

$$\text{Eq. 2-21}$$

All sulfur-containing amino acids, like the aromatics, react with ·OH at diffusion-controlled rates. In cysteine, hydrogen abstraction occurs at the thiol group (Eq. 2-22.1). For cystine, ·OH adds to the $-S-S-$, and the product undergoes cleavage at $C-S$ or $S-S$ (Eqs. 2-22.2, 2-22.3).

$$\text{RSH} \xrightarrow{\cdot OH} \text{H}_2\text{O} + \text{RS}\cdot \qquad 2-22.1$$

$$\text{HO}\cdot + \text{RSSR} \longrightarrow \text{RSOH} + \text{RS}\cdot \qquad 2-22.2 \qquad \text{Eq. 2-22}$$

$$\text{HO}\cdot + \text{RSSR} \longrightarrow \text{RSSOH} + \text{R}\cdot \qquad 2-22.3$$

Cross-linking. The hydroxycyclohexadienyl radicals formed by the addition of ·OH may undergo various reactions. Disproportionation yields the hydroxy phenol and the hydroxycyclohexadiene compounds. The radicals may cross-link by radical-radical binding or substitution-type reactions (*19, 20*) (Eq. 2-23). Thiyl radicals likewise also react with aromatic compounds via substitution (*21*).

Radiation-induced cross-links have been shown to occur between DNA

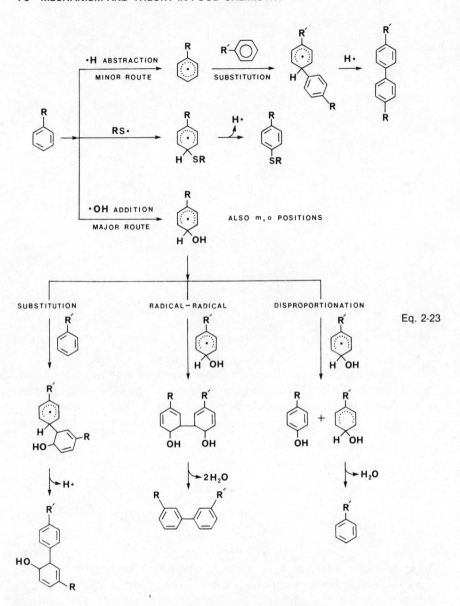

Eq. 2-23

and protein, both in vivo and in vitro. Radiation-generated ·OH radicals in aqueous solutions of thymine and phenylalanine induce cross-linking between the hydroxycyclohexadienyl radical to the thymine at the C-5 position (Fig. 2.13). Crosslinking between phenylalanine and thymine is favored over self-crosslinking among the same radical species (22).

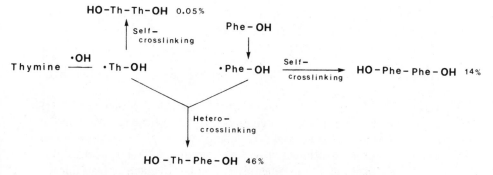

Fig. 2.13. Radiation-induced cross-linking between DNA and protein.

Photolysis (23)

For aliphatic amino acids, absorption of light energy results in excitation, followed by hydrogen abstraction and deamination. Side chain abstraction of $-CH_3$, $-OH$, and $-NH_2$ groups occurs under prolonged UV exposure.

In all aromatic amino acids, the primary photochemical reaction is ionization to hydrated electrons and radicals, involving possibly the triplet excited state, via a biphotonic process (Eq. 2-24.1). For example, the photoionization of phenylalanine can be represented as follows (Eq. 2-24.2) (24).

$$Ar \xrightarrow{hv_1} {}^1Ar \xrightarrow[crossing]{Intersystem} {}^3Ar \xrightarrow{hv_2} Ar\cdot + e_{aq}^- \qquad 2-24.1$$

$$Eq.\ 2\text{-}24$$

$${}^0S\,(Phe) \xrightarrow{hv_1} {}^1S \longrightarrow {}^1T \xrightarrow{hv_2} {}^*T \longrightarrow \begin{cases} e_{aq}^- \\ Dissociation \quad 2-24.2 \end{cases}$$

The photoejected e_{aq}^- can then react with other molecules similar to those discussed in radiation. These secondary reactions account mostly for the role of aromatic amino acids in photosensitized reactions. The major degradative reaction, however, comes from the disruption of the ring structure, forming aliphatic amino acids.

Histidine—aspartic acid, glutamic acid, γ-hydroxyglutamic, and citrulline.

Phenylalanine—tyrosine, serine, alanine, asparagine, ammonia, mono- or dihydroxyphenylalanine (dopa).

Tyrosine—aspartic acid, glycine, alanine, serine, asparagine, acetic acid, ρ-hydroxyphenyl lactic acid, tyramine, dopa.

For the sulfur amino acids, light absorption leads to excitation of the molecule, which then undergoes cleavage at the $-S-S-$ or $-C-S-$ bonds (Eq. 2-25). Photolytic cleavage of the disulfide is complicated by (1) radical-radical recombination to give back the disulfide, (2) deamination, (3) thiol-disulfide exchange, and (4) oxidation to various acids. Subsequent degradation products include pyruvic acid, ammonia, cysteic acid, sulfinic acid, sulfenic acid, alanine, serine, glycine, trisulfide, and tetrasulfide.

$$RSSR \xrightarrow{\;h\nu\;} [RSSR]^* \diagdown \begin{array}{l} RSS \cdot \; + \; R \cdot \\[2ex] 2\,RS \cdot \end{array} \qquad \text{Eq. 2-25}$$

Cross-linking.   A number of proteins have been demonstrated to cross-link DNA by irradiation. Lysine is the most sensitive among the amino acids in this respect, and its reaction with thymine has been well established. The reaction occurs only at alkaline pH (8–12), where the thymine exists as a monoanion. The first step of the photoreaction is a nucleophilic attack of the ϵ- or α-amino group at C-2 of the excited anion of thymine (singlet). Subsequent ring opening and protonation yields the stable conjugated adduct in a *cis, trans* mixture (Eq. 2-26) (*25*).

Eq. 2-26

Intramolecular cyclization followed by β-elimination, when heated in aqueous solution, gives the cyclic conjugate. Other amino acids also react, but at a slower rate and at more alkaline pH (>11).

Photosensitized Oxidation

Photosensitized oxidation is more specific than photolysis. Both type I and type II reactions can occur. Cysteine, methionine, histidine, tryptophan, and tyrosine residues in proteins can be oxidized in the presence of a suitable sensitizer.

Cysteine is oxidized to cysteic acid by the type I reaction, involving the H· abstraction by the triplet sensitizer from the thiol group (Eq. 2-27) (*26*). The thiyl radical is further oxidized to cysteic acid or recombines to give cystine.

$$\begin{aligned}
RSH & \xrightarrow[\text{Sen}]{\text{Sen-H·}} RS· \\
RS· + O_2 & \longrightarrow RSO_2· \\
RSO_2· + O_2 & \longrightarrow RSO_4· \\
RSO_4· + RSH & \longrightarrow RSO_4H + RS· \\
RSO_4H + H_2O & \longrightarrow RSO_3H + H_2O_2
\end{aligned}$$

Eq. 2-27

$$RSH + H_2O \xrightarrow[\text{Sen} \quad \text{Sen-H·}]{2O_2, \, hv} RSO_3H + RS· + H_2O_2$$

Methionine is oxidized to methionine sulfoxide via the type II reaction (Eq. 2-28). In the presence of flavin sensitizer, methionine is converted to methional. The reaction involves electron transfer from the sulfur atom to the photoactivated FMN*, followed by intramolecular migration of the electron, deamination, and then decarboxylation. At alkaline pH (~ 8.5), methional is decomposed to ethylene methyl disulfide and formic acid (Eq. 2-29) (*27*).

$$R-\left(CH_2\right)_2-S-CH_3 \xrightarrow[\text{Sen, }hv]{O_2} R-\left(CH_2\right)-\overset{\overset{\displaystyle O}{\|}}{S}-CH_3$$

Eq. 2-28

$$\text{Eq. 2-29}$$

With aromatic amino acids, photoxidation proceeds via the type II reaction, with subsequent degradation and disruption of the ring structure. The pathway of these photosensitized oxygenations involves complicated reactions of intermediates of hydroperoxides (Eqs. 2-30, 2-31) (28).

N–FORMYLKYNURENINE   KYNURENINE

3–HYDROPEROXYINDOLINE

$$\text{Eq. 2-30}$$

$$\text{Eq. 2-31}$$

Chemical Oxidation

Oxidizing agents such as H_2O_2 can oxidize (1) methionine to methionine sulfoxide and methionine sulfone (Eq. 2-32.1), and (2) cysteine to cysteic acid (Eq. 2-32.2). The rate of oxidation of cysteine decreases at low pH, whereas the oxidation of methionine increases at low pH.

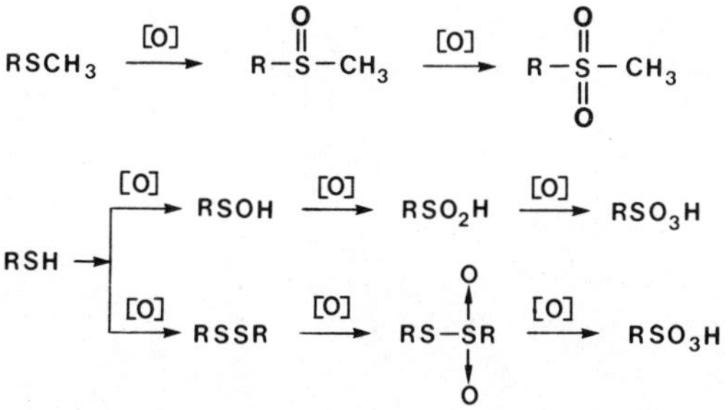

$$RSCH_3 \xrightarrow{[O]} R-\overset{\overset{\displaystyle O}{\|}}{S}-CH_3 \xrightarrow{[O]} R-\overset{\overset{\displaystyle O}{\|}}{\underset{\underset{\displaystyle O}{\|}}{S}}-CH_3 \qquad 2\text{-}32.1$$

$$RSH \rightarrow \begin{cases} \xrightarrow{[O]} RSOH \xrightarrow{[O]} RSO_2H \xrightarrow{[O]} RSO_3H \\ \\ \xrightarrow{[O]} RSSR \xrightarrow{[O]} RS\overset{\displaystyle O}{\underset{\displaystyle O}{\overset{\uparrow}{\underset{\downarrow}{-}}}}SR \xrightarrow{[O]} RSO_3H \end{cases} \qquad 2\text{-}32.2$$

Eq. 2-32

Hydrogen peroxide can further react with organic acids to form acylperoxides (e.g., performic acid from formic) which are extremely potent oxidants and less selective than H_2O_2 itself.

Besides the oxidation of cysteine and methionine, acylperoxides can oxidize tryptophan to formylkynurenine and other products. Tyrosine, serine, and threonine are also destroyed.

Reaction with Carbonyl Compounds

Reducing sugars, carbonyl compounds from the Maillard reaction, and short- and long-chain aldehydes and ketones from lipid oxidation form Schiff-base adducts with the lysyl ϵ-amino groups of proteins (Chapter 3). Bifunctional aldehydes, such as malonaldehyde may cause intra- or intermolecular cross-linking (Chapter 1).

Reaction with Products from Lipid Oxidation

The hydroperoxides formed during lipid oxidation can interact and cause changes in the structural and functional properties of proteins/amino acids. Susceptibility of proteins to lipid oxidation damage depends on the following factors (29):

1. Accessibility of reactive amino acids on the surface of the protein molecule.
2. Hydrophobic interaction or hydrogen bonding between lipid molecules and the protein surface which results in: (a) bringing the reactants to proximity and (b) exposing the buried amino acid side chains for reaction.
3. Presence of radical initiators in the system.

The following scheme depicts, in general, the mechanisms of oxidizing lipid on proteins, which include (1) formation of protein radicals (Eq.

2-33.1), (2) cross-linking of the radicals with lipids (Eq. 2-33.2), and (3) polymerization of protein-lipid (Eq. 2-33.3) (29).

FORMATION OF PROTEIN RADICAL   2−33·1

$$LH \longrightarrow L\cdot + H\cdot$$

$$L\cdot + O_2 \longrightarrow LOO\cdot$$

$$LOO\cdot + LH \longrightarrow LOOH + L\cdot$$

$$LOOH \xrightarrow{M^n \quad M^{n+1}} LO\cdot + OH^{\ominus}$$

$$LOOH \xrightarrow{M^{n+1} \quad M} LOO\cdot + H^{\oplus}$$

$$L\cdot + P \longrightarrow LH + P\cdot$$

$$LO\cdot + P \longrightarrow LOH + P\cdot$$ Eq. 2-33

$$LOO\cdot + P \longrightarrow LOOH + P\cdot$$

LIPID CROSSLINKED TO PROTEIN   2−33·2

$$L\cdot + P \longrightarrow L-P$$

$$LO\cdot + P \longrightarrow LO-P$$

$$LOO\cdot + P \longrightarrow LOO-P$$

POLYMERIZATION   2−33·3

$$L-P + P\cdot \longrightarrow L-PP$$

$$LO-P + P\cdot \longrightarrow LO-PP$$

$$LOO-P + P\cdot \longrightarrow LOO-PP$$

In aqueous solutions, the protein radicals formed can self-cross-link through termination type reactions or cross-link with other protein molecules by displacement reactions and eventually polymerize into insoluble aggregates (Eq. 2-34.1).

$$P\cdot + P\cdot \longrightarrow P-P$$ 2−34·1

$$P\cdot + P \longrightarrow P-P\cdot$$

$$PP\cdot + P \longrightarrow PP-P\cdot \text{ etc.}$$ Eq. 2-34

$$P\cdot + O_2 \longrightarrow P-OO\cdot$$ 2−34·2

In the dry state, the protein radical possibly reacts with O_2 to form peroxide (Eq. 2-34.2). When this occurs at the α-carbon, then scission occurs on either side of the peroxide-bearing carbon.

For the most sensitive amino acids, histidine, cysteine/cystine, methionine, and lysine, exposure to peroxidizing lipids yields a variety of products. For histidine, peroxidizing lipid causes free-radical reaction with deamination and decarboxylation to yield imidazole lactic acid and imidazole acetic acid (Eq. 2-35). The imidazole remains intact, while the carbon-center radical is primarily associated with the α-carbon (30).

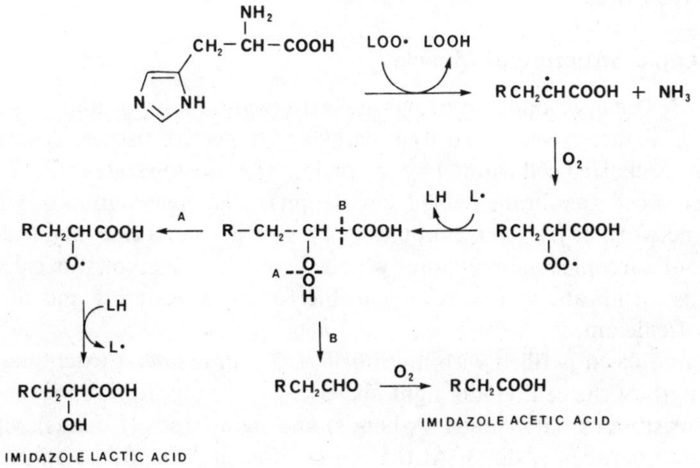

Eq. 2-35

Reaction between peroxidizing methyl linoleate and L-tryptophan results in the formation of formylkynurenine, kynurenine, and dioxinole-3-alanine (Eq. 2-36) (31).

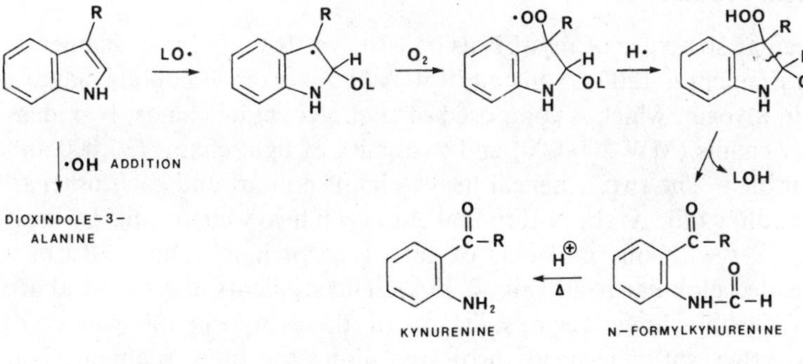

Eq. 2-36

For cysteine, the products include alanine, H_2S, cysteic acid, and cystine (Eq. 2-37).

$$\text{RSH} \xrightarrow{\quad\text{LOO·}\quad\text{LOOH}\quad} \text{RS·} \begin{array}{l} \nearrow \text{RSSR} \\ \searrow \text{RSOH} \rightarrow \text{RSO}_2\text{H} \rightarrow \text{RSO}_3\text{H} \end{array} \qquad \text{Eq. 2-37}$$

ORGANIZED PROTEIN SYSTEMS

MEAT PROTEINS

Macroscopic Structure of Muscle

The muscle organ is made up of parallel arranged bundles of muscle fibers (50–100 μ diameter) separated by sheaths of connective tissues. A muscle fiber is a specialized cell formed by the fusion of many separate cells, with a two-layered cell membrane (called sarcolemma), the invagination of which forms a network of transverse tubules ("T" system) and a fine longitudinal network of sarcoplasmic reticulum which serves as a reservoir for calcium ions. This membrane system is responsible for the seqestering and the releasing of calcium.

The sarcoplasm is filled with myofibrils 1–2 μ in diameter extending the entire length of the cell. Under light microscopy, the myofibril appears striated, consisting of dark bands (A-bands) and light bands (I-bands) with a dark line in between (Z-line). At the center of each A-band is a light zone called the H-zone. In the H-zone there is a dark M-line. The segment between two Z-lines, termed the sarcomere, is the basic contractile unit of the myofibril. In addition, two faintly stained transverse N1- and N2-lines located within the I band are identified by electron microscopy. (Fig. 2.14).

The Muscle Proteins

The molecular structure of myofibrils consists of thick and thin filaments. The thick filament is 1200 Å long and 150 Å in diameter. It contains largely the protein myosin, which is composed of six polypeptide chains, two identical heavy chains (MW 200,000) and two pairs of light chains (20,000 and 16,000 daltons). The two α-helical heavy chains coil around each other to form the rodlike tail. At the N-terminal end, each heavy chain coils by itself and complexes with one molecule of each type of light chain to form a double-headed globular head (Fig. 2.15A). The segments and the head are joined by flexible hinge regions. Many of these myosin molecules are packed together with the heads projected along the thick filament (Fig. 2.15B). The myosin rod is maintained in organized spatial arrangements by myomesin and C-protein. Myomesin, the M-line protein, is a single poly-

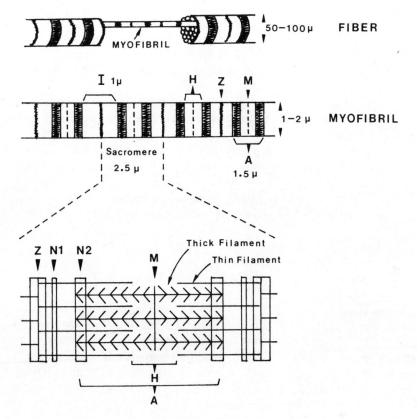

Fig. 2.14. Structural organization of skeletal muscle.

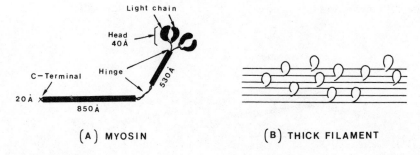

Fig. 2.15. Molecular structure of (A) myosin arranged into (B) thick filament.

peptide of 165,000 daltons, which extends transversely between the myosin filaments. The C-protein with a molecular size of 140,000, wraps around the bundled myosin rod at intervals of 40 Å along the filament.

The thin filament is made up of three major proteins (Fig. 2.16).

1. Actin: The monomer of actin is a globular protein called G-actin. G-actins is arranged in a twisted double-strand, beadlike chain known as F-actin.
2. Tropomyosin: These are long, thin proteins that aggregate end to end along the groove of F-actin, with one protein molecule covering seven actin monomers.
3. Troponin: This protein is associated with tropomyosin in a one-to-one ratio. Troponin is a protein consisting of three separate subunits:

Troponin C (calcium-binding protein)
Troponin I (inhibitory protein)
Troponin T (tropomyosin-binding protein)

Recent studies confirm the existence of another set of longitudinal cytoskeletal filaments in the sarcomere, in addition to the myosin-actin filaments. Titin, a high molecular myofibrillar protein of $\sim 10^6$ daltons, constitutes extremely long (400–700 nm) and flexible filaments in the sarcomere. It may play an important role in maintaining structural integrity of the myofibrils. Nebulin, with a molecular size of $\sim 5 \times 10^5$ daltons, is associated with the N2-line. The protein may also be connected with titin, forming longitudinal continuous filaments that connect Z-lines from within the sarcomere. Along their length, these proteins interact transversely with the thick and thin filaments. Titin and nebulin together constitute ~ 10–15% of the total myofibrillar proteins. These proteins may be similar to the heterogenous protein, connectin, of the gap filaments proposed in earlier literature. A third protein, desmin (55,000 daltons), forms 10 nm "intermediate" filaments and comprises ~ 0.35% of the total myofibrillar proteins in

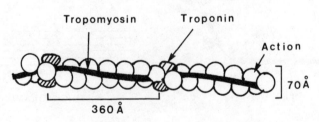

Fig. 2.16. Molecular structure of thin filament.

skeletal muscle. Purified desmin can assemble into filaments *in vitro*. The protein has been shown to locate primarily at or near the Z-line, and may function to link adjacent myofibrils.

Conversion of Chemical Energy to Mechanical Work—The Fate of ATP

The following sequence of steps occurs in muscle contraction during which ATP is utilized to generate work (Fig. 2.17) (*32*).

1. An ATP molecule binds onto the head of the myosin to form myosin-ATP in a 90° conformation, which weakly binds with actin. The free and actin-bound cross-bridges are in rapid equilibrium (M·ATP $\rightleftharpoons$ AM·ATP).
2. Hydrolysis of ATP by Mg^{2+}·ATPase occurs rapidly, with a series of transient conformational changes of the myosin to an activated strong binding state (M·ADP-P_i $\rightleftharpoons$ AM·ADP-P_i).
3. The cross-bridge undergoes the transition from the 90° conformation to a strained 45° conformation. The release of P_i provides the acti-

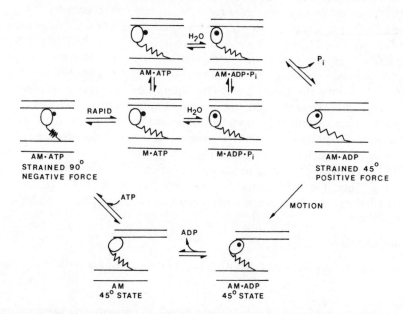

Fig. 2.17. Schematic representation of mechanical and biochemical states in muscle contraction. (Reproduced, with permission, from the Annual Review of Biophysics and Biophysical Chemistry, Volume 15, © 1986 by Annual Reviews, Inc.)

vation energy for the transformation. (AM·ADP-P_i $\rightleftharpoons$ AM·ADP (strained 45°) + P_i).

4. The strained 45° cross-bridge exerts a positive force that pulls the filaments actin and myosin in the shortening direction (~75 Å). The cross-bridge is then relieved of the strain to a stable 45° conformation. The ADP is released sufficiently slowly until the stable 45° conformation is reached.

5. Binding of another molecule of ATP to the cross-bridge causes the formation of a strained 90° conformation, with a negative force countering that of the strained 45° conformation (AM·ADP $\overset{ADP}{\rightleftharpoons}$ AM $\overset{ATP}{\longleftarrow}$ AM·ATP (90° strained).

6. Since the 90° conformation is weakly binding, the cross-bridge rapidly detaches from the actin, releasing the negative strain in the AM·ATP state.

7. The myosin-ATP undergoes hydrolysis to re-form M·ADP·P_i, which may then attach to a new actin site for a new cycle of contraction and detachment.

What are the roles of troponin and tropomyosin? The formation of the activated actin-myosin complex is regulated by low concentrations of Ca^{2+}. The muscle is in a relaxed state in the presence of 10^{-7} M Ca^{2+} or less, and is fully contracted by 10^{-5} M Ca^{2+}. A nerve impulse transmitted to a muscle cell causes the cell membrane to depolarize. This depolarization of the membrane system causes a release of calcium from the sarcoplasmic reticulum tubules into the sarcoplasm. The binding of calcium to troponin C induces a conformational change in the troponin-protein complex. This process, in turn, pulls the tropomyosin toward the groove of the F-actin, allowing the binding of the actin monomers (G-actin) to the activated myosin. Removal of calcium allows the troponin to return to its original conformation and the tropomyosin to a position sterically blocking the actin from binding. Thus, Ca^{2+} functions to control the "on" and "off" state of the actin filament via the conformational changes of the troponin complex.

Rigor Mortis

There are various ways of generating the ATP required for muscle contraction.

1. *Regular pathways:* These include glycolysis, amino acid metabolism, fatty acid oxidation, and ultimately oxidative phosphorylation.

2. *Short-duration supply:* Muscle cells store phosphocreatine for the transfer of a phosphate to ADP to form ATP (Eq. 2-38).

$$
\begin{array}{c}
\overset{\ominus}{COO} \\
| \\
CH_2 \\
| \\
H_3C-N \\
| \\
C=NH \\
| \\
HN-\textcircled{P}
\end{array}
\quad
\xrightarrow[\text{CREATINE KINASE}]{\text{ADP} \quad \text{ATP}}
\quad
\begin{array}{c}
\overset{\ominus}{COO} \\
| \\
CH_2 \\
| \\
H_3C-N \\
| \\
C=NH \\
| \\
NH_2
\end{array}
\qquad \text{Eq. 2-38}
$$

3. *Immediate supply:* Conversion of ADP to ATP can be catalyzed by adenylate kinase (Eq. 2-39).

$$
2\,ADP \xrightarrow[\text{KINASE}]{\text{ADENYLATE}} ATP + AMP \qquad \text{Eq. 2-39}
$$

Immediately after the animal is sacrificed, oxygen is no longer available. The aerobic pathway of producing ATP is stopped. The supply of phosphocreatine is rapidly depleted. The major anaerobic pathway (glycolysis) becomes the remaining source of ATP. However, the conversion of glycogen or glucose to lactic acid with the production of ATP is self-limiting under these conditions, since the buildup of lactic acid causes a drop in pH that is inhibitory to the enzymes involved in glycolysis.

The rapid depletion of ATP is also related to the breakdown in the regulatory system that controls Ca^{2+} levels. The calcium concentration builds up in the sarcoplasm, induces the contraction of the muscle fibers, and consumes the ATP supply.

Since there is no ATP to effect the dissociation of the actin-myosin complex (AM in Fig. 2.17), the muscle loses its natural extensibility. This postmortem change is commonly known as rigor mortis.

The Ultimate pH

The accumulation of lactic acid lowers the pH of muscle, and eventually the muscle reaches an ultimate pH that is critical to postmortem changes. A sharp decrease in pH may cause the denaturation of proteins if the carcass temperature is still high. Denaturation of muscle proteins results in the decrease of the water-holding capacity, which is related to the degree of tenderness in the meat. Minimum water-holding capacity and swelling of meat is observed around pH 5.0. The water-holding capacity increases with either increasing or decreasing pH from the isoelectric point. Addition of salts causes an increase in water-holding capacity and swelling. (Refer to Chapter 9, phosphates.)

Postmortem Tenderness

After the muscle tissue goes into rigor mortis, there is a gradual decrease in the toughness upon postmortem storage. During this period (first 72–96 hr) of increasing tenderness, the following changes are evident:

1. Disintegration of the Z-disk and slow loss of M-line.
2. Change of actin-myosin complex.
3. Gradual degradation of troponin T (33).

It is generally agreed that proteolysis is responsible for postmortem meat tenderness. Muscle proteases can be classified into three groups: (1) the alkaline proteases, (2) the acidic proteases, and (3) the neutral proteases. It is unlikely that the alkaline proteases would have any major role in postmortem tenderization due to the fact that these enzymes have their optimum activity at pH ranges not found in the postmortem condition.

Among the acidic proteases, the lysosomal cathepsins have received the most attention. These proteases are either exopeptidases (the cathepsins A, C, and H) or endopeptidases (B, D, and L). Both cathepsin B and D degrade myosin and actin to fragments of various sizes, with the latter being more active. Cathepsins B and D have their optimum pH at pH 5.2 and 4.0, respectively. Cathepsins L and H have 10 and 5 times more activity than cathepsin B in the degradation of myosin. Cathepsin L also cleaves actin, α-actinin, troponin-T, and troponin-I, but the optimum pH is 4.2, lower than normally found in meat.

The calcium-dependent proteinase (CAF, calcium activated factor) has been extensively studied and linked to the causes of postmortem tenderization. CAF contains two polypeptide chains with molecular weights of 80,000 and 30,000. It has an optimum pH of 6.5–8.0, and requires Ca^{2+} and sulfhydryl groups for activity. The enzyme has an unusual specificity, in that it will not act on actin, α-actinin, troponin, and, under most conditions, myosin, but rapidly disintegrates the Z-disk and M-line and also the protein troponin T. The exact mechanism that causes the disintegration is not clear.

It is likely that other lysosomal enzymes may also be involved. For example, β-glucuronidase, and β-galactosidase have been associated with the breakdown of proteoglycans in connective tissues. The degradation of collagen by collagenase has also been implicated as one of the processes involved in postmortem aging. Proteolysis of titin, nebulin and desmin has also been implicated.

Meat Emulsion

The formation of meat emulsion can be divided into two phases: mechanical comminution and a subsequent heating process. During comminution, the muscle and fatty tissues are reduced to microparticles. The myofibrillar proteins are released, allowing water binding and a higher degree of swelling. The fat is dispersed as fine droplets within the heterogenous aqueous phase. Initial melting is accomplished by the heat generated by comminution ($\sim 18°C$). Adsorption of protein molecules at these newly formed oil-water interfaces thus results in the formation of a protein film surrounding the oil droplets.

Heating causes the protein molecules to denature and aggregate into a gel network. Between fat droplets, the formation of a layer of protein gel matrix sets a steric barrier against coalescence. Simultaneously, the water molecules are held in the interstices by the various protein-water interactions, now including capillary actions between surfaces in the network that physically holds the water molecules in solution or suspension. The result is a three-dimensional network of protein gel that physically and chemically stabilizes dispersed fat droplets and water molecules.

Not all the muscle proteins exhibit the same degree of stabilizing effect on a meat emulsion. Muscle cell proteins can be classified into three fractions: the salt-soluble myofibrillar proteins (50–55%), the water-soluble sarcoplasmic proteins (30–35%), and the connective tissue proteins (10–15%). Only the salt-soluble proteins are chiefly involved in the formation of a gel network in a meat emulsion. Among the myofibrillar proteins, myosin and actomyosin produce the most stabilized emulsions (34) (Table 2.4).

The Role of Myosin in Gelation

Denaturation of muscle proteins involves a series of transition temperatures (T_m), corresponding to heat-induced conformational changes of the many

Table 2.4. Stability of Emulsions of Salt-Soluble Proteins

PROTEIN	pH	IONIC STRENGTH	STABILITY OF EMULSION (DAYS)
Myosin	8.0	0.35	> 4 weeks
Sarcoplasmic	7.0	0.35	12 hours
Actin-myosin	6.7	0.35	> 3 weeks
Actin	7.2	0.35	< 36 hours

Reprinted with permission from Hegarty et al. (34), courtesy of Institute of Food Technologists.

Table 2.5. Heat-Induced Changes of Muscle Proteins

TEMPERATURE (°C)	PROTEIN	ACTION
30–35	tropomyosin	dissociates from actin backbone
38	F-actin	helical strands dissociate into single chains
40–45	Myosin	
	head	some conformational change
	hinge	helix to coil conformational change
45–50	actin-myosin	complex dissociates
50–55	myosin	
	"tail"	helix to coil conformational change
>70	G-actin	major conformational change

Reprinted from Food Technology, 1984. Vol. 38(5):80. Copyright by Institute of Food Technologists.

individual protein molecules in the muscle fiber. Table 2.5 lists many of the events that occur during heat treatment (35).

There are two transition temperatures ($T_m = 43°C$, $55°C$) critical for the formation of gels, implying two types of conformational changes in heat-induced gelation. Studies suggest myosin is the major protein in the myofibrils that is involved in the process of gelation. The myosin heads start to aggregate upon heating at a relatively low temperature, while the rod segments unfold with increasing temperature. The aggregation of the heads involves disulfide exchange and possibly intermolecular association of side chains (36). The head-to-head aggregation provides a junction zone linking the myosin rods to form a gel network, in a way very similar to the polysaccharide gels.

The gel network is further stabilized by noncovalent bonding among the binding sites made available by the unfolding of the myosin protein. The presence of actin enhances the gel formability of myosin. Under the optimum conditions for gel formation (0.6 M salt, pH 6.0, and 65°C), a myosin/actin ratio of 1.5–2.0 has been shown to substantially augment the rigidity of the gel.

MILK PROTEINS

Milk proteins can be classified into two major fractions: the caseins and the whey proteins. We will be concerned mainly with milk caseins. Caseins are the phosphoproteins precipitated from raw skim milk at pH 4.6 and 20°C. The casein fraction comprises ~80% of the total protein content of milk. The principal proteins in this group are the α_{s1}-, β-, and κ-caseins.

α-Caseins

The α-caseins account for 50–55% of all the caseins. They consist of the major α_{s1}-casein and the minor components, α_{s0}-, α_{s2}-, α_{s3}-, α_{s4}-, and α_{s5}-casein, according to their electrophoretic properties. The α_{s1}-casein has a molecular weight of 23,600 and consists of 199 amino acid residues, with 8 phosphoseryl residues distributed between residues 43–79. This same segment also contains 12 carboxyl residues. The phosphate groups bind calcium ion, and the α_{s1}-casein tends to precipitate in the presence of calcium. The rest of the polypeptide chain is hydrophobic and exhibits strong association in the formation of micelles. The α_{s0}-casein is identical to α_{s1}-casein, except for an extra phosphoseryl group at residue 41. The α_{s3}-, α_{s4}-, and α_{s5}-casein differ from α_{s2}-casein only in the degree of phosphorylation.

β-Casein

β-Casein constitutes 30–35% of the total caseins. It is a single polypeptide chain of 209 amino acid residues, with a molecular weight of 24,000. The five phosphoseryl residues are clustered near the N-terminal segment (residues 1–43), while the C-terminal half (136–209) is highly hydrophobic. This concentration in hydrophilic and hydrophobic regains at the terminal ends results in the β-casein being more surfactant-like than the α_{si}-casein. β-Casein associates at a slower rate than the α-caseins, and it is not as readily precipitated by calcium as compared to α_{s1}-casein.

κ-Casein

κ-Casein constitutes about 15% of the total caseins and is the only major casein containing cysteine. The monomers have a molecular weight of 19,000 and form polymers via disulfide bonds in size range from 60,000 to 600,000 daltons. The κ-casein monomer has only one phosphate residue but has 0–5 carbohydrate (trisaccharide) chains. The monomer contains 169 amino acid residues, with a hydrophobic N-terminal segment (1–105) and a hydrophilic C-terminal segment (106–169), known as para-κ-casein and macropeptide, respectively. All the carbohydrates in the monomer, and the one phosphate group, are located in the macropeptide. This uneven distribution, combined with a high content of aspartic acid and glutamic acid, results in an acidic and highly soluble C-terminal segment of the κ-casein monomer. The κ-casein, therefore, is the least sensitive to calcium ion precipitation and functions to stabilize the α-caseins against precipitation with calcium ion in the casein micelle.

Casein Micelle

Caseins exist in large spherical colloidal micelles with calcium phosphate. These micelles comprise 93% (w/w) caseins and range in size from 500 to 3000 Å in diameter and in particle weight from 10^7 to 3×10^{10}. The remaining 7% comprises inorganic calcium (3%), phosphate (3%), and small amounts of magnesium, sodium, potassium, and citrate. The calcium and phosphate play a very important role in maintaining the integrity of the casein micelles, and are commonly referred to as colloid calcium phosphate.

A casein micelle is assembled from submicelles with a particle weight of ~ 3–6×10^5 and a diameter of 10–20 nm, containing 25–30 α-, β-, and κ-caseins. The variations in the size, composition, and particle weight of submicelles is determined by factors such as concentration of individual casein type, pH, and temperature.

Submicelles contain a mixture of α_{s1}-, α_{s2}-, β-, and κ-caseins, with the hydrophobic regions oriented inward and the hydrophilic regions of the caseins located on the surface. Hence the core of the submicelle is hydrophobic and the surface is therefore hydrophilic. Furthermore, the κ-caseins undergo self-association and are restricted to one area of the surface. Therefore, on the surface of the submicelle, there are carbohydrate-rich (macropeptide portion) areas of κ-casein and phosphate-rich (phosphoserine residues) areas of the other caseins. The phosphate groups on the surface interact with calcium to form calcium-phosphate bridges, and therefore, in the presence of calcium, part of the surface of the submicelle is available for hydrophobic interactions.

In building up a micelle, the hydrophobic interactions among the submicelle surfaces tend to align the entering submicelles with the hydrophilic areas oriented outward. Consequently, the resulting micelle must have a heavily hydrophilic surface, rich in κ-casein. Furthermore, the size of the micelle is obviously dictated by the relative concentration of κ-casein in the submicelle (38).

A modification of the above model has been suggested in that the binding of submicelles is facilitated by electrostatic interaction via colloid calcium phosphate rather than hydrophobic bonds. The binding takes place between the negatively charged phosphoseryl residues of the caseins and colloid calcium phosphates in the form of $Ca_9(PO_4)_6$ clusters positively charged with the adsorption of two calcium ions. Since κ-casein is almost phosphate-free, binding occurs only among the α- and β-caseins. Submicelles with a low κ-casein content or with no κ-casein are buried in the interior of the micelles. Micellar growth stops when the micelle surface is covered entirely with κ-casein (39) (Fig. 2.18).

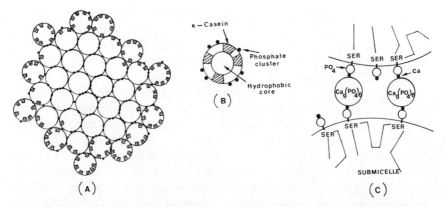

Fig. 2.18. Schematic representation of (A) a casein micelle composed of sub-micelles, (B) a submicelle, and (C) two submicelles bound via Ca$_9$(PO$_4$)$_6$. (Reprinted with permission from Schmidt (*39*), courtesy of Elsevier Applied Science Publishers Ltd., England.)

Milk Coagulation

The coagulation of milk by chymosin includes two separate steps (Eq. 2-40).

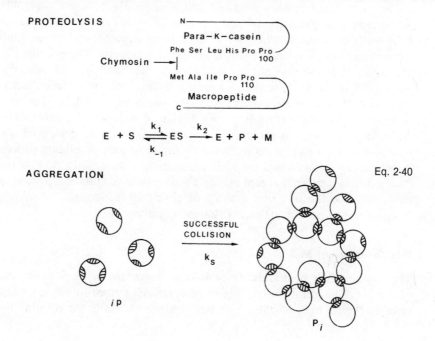

1. Proteolysis of κ-casein: Chymosin hydrolyzes the peptide bond at Phe_{105}-Met_{106} of κ-casein, to give a highly acidic macropeptide, and a basic, hydrophobic para-κ-casein, resulting in a reduction in half of the hydrophilicity of the micelle surface. The reaction follows the Michaelis-Menton mechanism. The release of the hydrophilic macropeptide reduces the surface charges and hence the surface potential that creates the barrier for the micelles to come close. (Refer to Chapter 1, Fig. 1.16.) Consequently, there is an increase in hydrophobic interaction which provides a major force in coagulation. Furthermore, proteolysis destroys the stabilizing action of κ-casein, particularly its resistance to precipitation in the presence of calcium ion. The other caseins, which are calcium sensitive, are now completely susceptible to calcium precipitation. Therefore, in a milk system, the action of chymosin is followed by formation of coagulum in the presence of calcium. Proteolysis of κ-casein changes the functionality of an entire protein system.

2. Aggregation: The reaction is a bimolecular process. There is a lag phase between the addition of chymosin to the milk and observable coagulation. This lag period is commonly about 60% of the clotting time. In the lag period, coagulation will not occur as the concentration of the para-κ-casein produced by the enzymatic hydrolysis is low. It has also been suggested that a critical degree of proteolysis ($\sim 80\%$) must be completed before aggregation can occur (40).

The rate of aggregation does not follow a simple diffusion-controlled mechanism. The biomolecular rate constant (k_s) is lower than the equation $k_s = 4kT/3\eta$ predicts, where k = Boltzmann's constant, T = absolute temperature, and η = viscosity of the dispersing medium. The micelles that are sufficiently close may not interact, because the micelle surface is heterogeneous in reactivity. The para-κ-casein micelles need the surface reactive sites oriented close to each other to register a successful collision. The reactivity of the micelle surface is largely related to the extent of proteolysis. Upon the completion of proteolysis, the micelles are almost completely denuded, and the contribution of the steric factor to the coagulation process is decreased, accompanied by an increase in k_s. The introduction of this "steric factor" offers an explanation for the slow coagulation process in milk, and also the fact that clotting of chymosin-treated micelles almost always forms a porous gel rather than a precipitate (41).

Heat Stability of Milk

The characteristic heat stability of milk is important in dairy processing operations. The colloidal structure of casein micelles constitute a very stable system. Typical milk is stable to a temperature of 140°C for 20 min. The heat coagulation time (usually at 130 or 140°C) of milk is pH dependent. Two types of stability curves are evident. Type A milk shows a stability maximum of pH 6.7 and a minimum at pH 6.9. Type B milk has increasing stability with increasing pH, with no maximum or minimum. The shape of the stability curves is related primarily to the interaction between the whey protein, β-lactoglobulin, and κ-casein.

β-Lactoglobulin is a globular protein with a molecular weight of 18,300 daltons, corresponding to 162 amino acids. The secondary structure of β-lactoglobulin is 15, 51, 17, and 17% α-helix, β sheet, β turn, and aperiodic structure, respectively. The antiparallel β sheet is formed by nine β strands wrapped around to assume a flattened cone shape. There are two disulfide bonds (65–160, 106–119), and a free cysteine-121 buried at the sheet-helix (β strand 115–124 and α helix 129–143) interface (42). The protein is remarkably acid-stable, resisting denaturation at pH 2. In solution, β-lactoglobulin exists as a dimer resulting from the association of the monomer at the respective α-helical segments (129–143) through hydrophobic interactions. Dimerization therefore serves to protect the sulfhydryl group from being reactive. As the pH and temperature increase, the protein undergoes reversible conformational changes, which exposes the polypeptide segment containing the Cys-121. At temperatures above 60°C and pH 6.5, association between the protein and κ-casein occurs via sulfhydryl-disulfide interchange. The stability minimum at pH 6.9 in the heat stability curve of milk has been suggested to be caused by this interaction, producing a complex that is susceptible to heat-induced calcium phosphate precipitation (43). The complexation also renders the casein resistant to the proteolytic action of chymosin, and thereby prevents the use of preheated milk in cheesemaking.

Interestingly, when heated above 70°C, the denaturation temperature, and at pH 7.0, β-lactoglobulin starts to polymerize and aggregate in the absence of other proteins. Two reactions are identified in the process. The primary reaction that occurs at temperatures above 65°C involves the formation of intermolecular disulfide bonds, either by sulfhydryl-disulfide interchange or by sulfhydryl oxidation. The secondary reaction takes place after the initiation of the first reaction and proceeds even at low temperatures, yielding high molecular aggregates. In this reaction, disulfide bonds are not involved and the process is nonspecific. The aggregates show different electrophoretic mobility and higher particle weight than products from the primary reaction. Polymerization also occurs at low temperatures at alkaline pH (>8.5), involving similar disulfide interchange reactions, although the possibility of base-catalyzed hydrolysis of the disulfide bonds may not be excluded.

WHEAT PROTEINS

Wheat proteins are fractionated according to their solubility into albumins (soluble in water), globulins (soluble in 10% NaCl, insoluble in water), gliadins (soluble in 70–90% alcohol), and glutenins (insoluble in water or alcohol, but soluble in acid or alkali).

Commercial gluten is the water-insoluble protein fraction separated from wheat flour. The freshly extracted wet gluten, known as gum gluten, yields a cream-colored powder of high protein content (75–80%) upon drying. The product contains largely the storage proteins, gliadins and glutenins, small amounts of albumins and other nonstorage proteins, 5–10% lipids, and 10–15% carbohydrates. Hydrated gliadin fractions are viscous, but glutenin fractions exhibit high elasticity and cohesiveness, a major factor contributing to dough strength.

Gliadins

The composition of gliadins varies among wheat varieties. Heterogeneity occurs in both the composition and in the sequence of the amino acids. All gliadins consist of exceptionally high contents (38–56%) of glutamic acid (mostly as glutamine), proline (15–30%), and phenylalanine, but are low in basic amino acids, lysine, arginine, and histidine (Table 2.6) (44). Based on the electrophoretic mobility, four groups of gliadins can be distinguished, α-, β-, γ, ω-gliadins (in the order of decreasing mobility), all with molecular weight about 32,000–42,000, except ω-gliadins in MW range of 74,000. The ω-gliadins contain very few or no sulfur-containing amino acids, while the other gliadins contain two cysteine and one methionine. The disulfide bonds are all intramolecular.

The detailed conformation of A-gliadins (a particular fraction of α-gliadins from ultracentrifugation separation) has been investigated extensively. It is observed that at neutral pH, A-gliadin molecules associate to form microfibrils 70–80 Å in diameter, and several thousand Å long. Dissociation of the microfibrils into monomers occurs with decreasing pH. Hydrogen bonding and hydrophobic interactions are involved in the association of gliadin monomers (44).

The secondary structure of α-, β-, and γ-gliadins contains 30–35% α-helix and 10–20% β sheet, while ω-gliadins contain β turns, but no α-helix or β sheet. The ω-gliadins are stabilized by strong hydrophobic interactions. The other gliadins are stabilized, in addition, by intramolecular disulfide bonds (45).

Table 2.6. Amino Acid Composition of Whole Gliadin and Glutenin (Ponca)

AMINO ACID	GLIADIN	GLUTENIN
	(Amino acid residues/100,000 g)	
Asp	20	23
Thr	18	26
Ser	38	50
Glu	317	278
Pro	148	114
Gly	25	78
Ala	25	34
Cys	10	10
Val	43	41
Met	12	12
Ile	37	28
Leu	62	57
Tyr	16	25
Phe	38	27
Trp	5	8
Lys	5	13
His	15	13
Arg	15	20
Amide	301	340

Reprinted with permission from Kasarda et al. (*44*), courtesy of American Association of Cereal Chemists.

Glutenins

After the extraction of albumin and globulin from flour with dilute salt solution, and of gliadins with 70% ethanol, the remaining residue is extracted with acid or alkali (usually acetic acid) to obtain the glutenins. The insoluble residue left behind, which accounts for more than 30% of the total protein, is known as the residue protein or acetic-acid-insoluble glutenin. The acetic-acid-insoluble glutenin can be dissolved by reduction and the action of detergent such as SDS.

The molecular weights of glutenins range into millions, with average molecular weight estimated as 300,000 daltons. Glutenins, after reduction and alkylation, can be separated by gel filtration into three fractions: fraction A which is highly aggregated by hydrophobic interaction, consisting of subunits of MW 68,000 or lower; fraction B of species with MW 68,000–133,000; and fraction C consisting of MW 36,000–44,000 polypeptides with composition and gel patterns similar to the ethanol-soluble gliadins.

Application of SDS-PAGE reveals that glutenins are large complexes of about 15 subunits ranging from 11,000 to 133,000 daltons. The glutenin subunits can be classified according to SDS-PAGE patterns into two groups: high-molecular-weight (HMW) and low-molecular-weight (LMW) subunits. The HMW subunits have MW of 95,000 and higher. The low-molecular-weight subunits have amino acid compositions similar to the α-, β-, and γ-gliadins. Glutenin consisting of only LMW subunits (same as fraction C in gel filtration) are commonly known as low-molecular-weight glutenin, high-molecular-weight gliadin, and aggregated gliadin, which has also been obtained from the purification of gliadin fractions. The high-molecular-weight glutenin contains HMW and LMW subunits cross-linked by intermolecular disulfide bonds. Hydrogen bonding and especially hydrophobic interactions play an important role in the association of the subunits.

Different models have been proposed for the structure of glutenin. The following characteristics are generally recognized: (1) disulfides with intramolecular bonding in polypeptide subunits, and intermolecular bonding of the subunits into a functional glutenin structure; (2) noncovalent interactions—hydrogen bonds, and most importantly hydrophobic interaction playing a significant role in maintaining the glutenin structure. The large number of amide side chains and the high proportion of nonpolar amino acid residues support the significant contribution of these secondary interactions (Fig. 2.19).

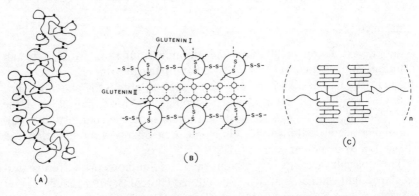

Fig. 2.19. Schematic representation of structures of glutenin: (A) concatenation model, (B) Glutenin I–II complex, (C) cluster model. (A: Reproduced with permission from Ewart (46), courtesy of Society of Chemical Industry. B: Reproduced with permission from Khan and Bushuk (47), courtesy of American Association of Cereal Chemists. C: Reproduced with permission from Graveland et al (48), courtesy of Academic Press, Inc. London.)

The concatenation model (46): In this model, the polypeptide subunits are linked to their neighboring subunits by two intermolecular disulfide bonds to form a linear chain. Subunits linked by one disulfide bond have also been suggested.

The Glutenin I-II complex (47): According to this model, glutenin I (which comprises subunits with MW 68,000 and lower) is held in the complex by strong noncovalent bondings through protein-protein or protein-lipid/carbohydrate-protein interactions. The second type of glutenin, glutenin II, comprises of HMW subunits joined by intermolecular disulfide bonds and associated by strong noncovalent interactions into large aggregates. Glutenin I and II are linked by secondary interactions. The physical interactions between glutenin I and II provide for the viscous flow under stress, and the glutenin II contributes to the characteristic elastic properties.

The cluster model (48): In this model, the glutenins are classified into subunits A (85,000 and above), B (40,000-45,000) and C (33,000-38,000). The high-molecular-weight subunits A are linked head to tail by intermolecular disulfide bonds to form a linear long chain. The glutenin B and C subunits form clusters by intramolecular disulfides and secondary interactions, which are linked via disulfide bonds to the long chain of subunits A.

Transformation of Flour into Dough

A gluten complex can be visualized as a composition of the various flour proteins bound by covalent bonding and noncovalent interactions. The high-molecular-weight glutenins form network with fibrillar extension, in which the gliadins and other proteins are dispersed (49).

Wheat flour contains about 12% protein ($\sim$ 80% gluten), 70% starch (amylose:amylopectin = 1:3), and 2% lipid. Wheat flour lipids can be classified into free (extractable by petroleum ether) and bound (extractable with water-saturated n-butanol). Both free and bound lipids contain glycolipids and phospholipids as major components in their polar fractions. In dough mixing, these flour ingredients are integrated into the gluten network, resulting in a starch-protein-lipid complex matrix.

The reduction of disulfide bonds during the mixing of dough and reoxidation during the resting period have been extensively studied. Disulfide cleavage and reoxidation occur primarily within the high-molecular-weight cross-linked glutenins. The intramolecular disulfides in the gliadins are usually not accessible to the reaction. The initiation of reduction in dough mixing may be due to: (1) thiol-disulfide exchange and/or (2) possible formation of free radicals or superoxide anions, which mediates the reduction of certain metalloproteins and, in turn, the reduction of disulfides. Disulfide reduction physically "loosens" the gluten, and facilitates better interactions with the lipids, starch, and other additives (e.g., emulsifiers, fortified soy proteins), via noncovalent interactions to form a continuous network of starch-protein-lipid complex.

During baking, the bonds between the polar lipids, especially the glycolipids and the gluten proteins, weakens due to the denaturation of the proteins. As the dough temperature increases and the starch gelatinizes above 50°C, a translocation of these lipids occurs in which the lipid molecules are now strongly complexed with the starch. (Refer to Chapter 1 for complexation of emulsifiers with amylose.) Surfactants, such as monoglycerides, ethoxylated monoglycerides, sodium stearoyl-2-lactylate, succinylated monoglycerides (widely added in breadmaking as dough conditioners), softeners, and antistaling agents, are shown also to exhibit translocation from the gluten proteins to complexing with starch. The function of these surfactants depends largely on how well the additive is dispersed in the dough before starch gelatinization occurs (50).

The shortening effect of lipids is exhibited by an increase in loaf volume and an improvement of crumb grain (soft, even-textured). Often, the addition of 0.5–1% shortening or hardened vegetable fat is required, especially in the rapid-breadmaking process. The effectiveness of shortening as an improver depends on the presence of a small amount of solid fat at the temperature of mixing and final proof. Contrary to the polar lipids, little of the shortening becomes bound to the dough proteins during mixing. Free rather than bound fat solids are important to the final quality of the bread. There are a number of hypotheses to explain the shortening effect of nonpolar lipids in dough (51). The fat may form a layer between the gluten proteins and the starch granules, to stabilize and preserve the continuous gas-retaining structure during expansion. The fat may melt during baking and seal the pores and passages in the gluten network so that the escape of carbon dioxide is retarded during the expansion of the dough. There is also some indication that the transition temperature for the induction period during which there is no loss of carbon dioxide is raised by the presence of shortening. Consequently, there is a delay in the release of carbon dioxide in the early stage of baking. It is also likely that the fat forms an ordered structure, such as a liquid crystalline mesophase that facilitates the production of orientated structures in the dough that favors gas retention during baking.

Saturated monoglycerides are known to have a better softening property than the unsaturated ones. There is also evidence that monoglycerides are bound to starch more than to the gluten proteins during dough mixing. It may be that saturated lipids bound to starch, and polar lipids to proteins, form a continuous lipid layer along the structural network of the dough, which allows flexibility in performing various functions as mentioned in the above hypotheses.

Two enzymes present in wheat flour are significant in baking technology. Both of these will be discussed in Chapter 5. Amylases catalyze the degradation of starch to fermentable sugars which are essential for the formation of carbon dioxide. The addition of enzyme-active soy flour in dough often results in an increase in dough relaxation time and improvement of the gluten proteins and mixing tolerance (52). The enzyme lipoxygenase catalyzes the formation of radicals (LOO·) from unsaturated free lipid substrates in dough, which in turn, causes both lipid-protein and protein-protein cross-links. (Refer to the discussion on the reactions of products from lipid oxidation with proteins, Chapter 2.)

COLLAGEN

Collagen is the fibrous protein that contributes to the unique physiological functions of connective tissues in skin, tendons, bones, cartilage, etc. The structural unit of collagen is tropocollagen. It is a rod-shaped protein (15 Å diameter, 3000 Å length) consisting of three polypeptide units (called α chains) interwined to form a triple-helical structure. Each α chain coils in a left-handed helix with three residues per turn, and the three chains are twisted right-handed to form the triple helix (Fig. 2.20).

Triple Helix

The α chains contain ~1000 amino acid residues and vary in amino acid composition. These variations in the α chains constitute at least four major types of collagen (Table 2.7). More than one collagen type is usually present in a particular tissue.

Amino Acid Composition

The amino acid composition of collagen as a whole is unique in that it is exceptionally high in glycine (33%), proline (12%), and alanine (11%), plus two amino acids that are not commonly present in many other proteins, hydroxyproline (12%) and hydroxylysine (1%) (Table 2.8) (53).

Hydroxylysine usually has a carbohydrate residue (glucose or galactose) attached. The amino acid sequence indicates that most of the polypeptide chain consists of glycine-led triplets of the following distribution:

TRIPLET	PROPORTION
Gly-X-X	0.44
Gly-X-I	0.20
Gly-I-X	0.27
Gly-I-I	0.09

Fig. 2.20. Triple-helical structure of collagen.

Table 2.7. Collagens and Their Distribution

TYPE	TRIPLE HELIX	DISTRIBUTION
I	two identical α 1(I) chain + one α 2 chain	skin, tendon, bone
II	three α 1(II) chains	intervertebral disc, cartilage
III	three α 1(III) chains	cardiovascular vessel, uterus
IV	three α 1(IV) chains	basement membrane, kidney glomeruli, lens capsule

Reprinted with permission from *Science* **207**, p. 1316. Copyright 1980 by American Association for the Advancement of Science.

Table 2.8 Amino Acid Composition of Collagen from Rat-tail Tendon and Its Component α Chains

AMINO ACID	WHOLE COLLAGEN	α1 CHAIN	α2 CHAIN
	(Amino acid/100 residues)		
Ala	10.7	11.0	10.3
Arg	5.0	4.9	5.1
Asp (+Apn)	4.5	4.7	4.4
Glu (+Gln)	7.1	7.4	6.8
Gly	33.1	32.9	33.5
His	0.4	0.2	0.7
Hydroxylysine	0.7	0.5	1.0
Hydroxyproline[a]	9.4	9.7	8.6
Ile	1.0	0.6	1.5
Leu	2.4	1.8	3.1
Lys	2.7	3.0	2.1
Met	0.8	0.9	0.7
Phe	1.2	1.2	1.1
Pro	12.2	12.9	11.5
Ser	4.3	4.1	4.3
Thr	2.0	2.0	2.0
Tyr	0.4	0.4	0.4
Val	2.3	1.9	3.0
Gly	33.1	32.9	33.5
Pro + Hyp	21.6	22.6	20.1
Other	45.3	44.5	46.4

[a]Includes both 3- and 4-hydroxyproline.
Reprinted with permission from Fraser and MacRae (*53*), courtesy of Academic Press, Inc.

where I = imino acid residue (proline or hydroxyproline) and X = other amino acid residues. The segment of the polypeptide chain consisting of repeating triplets with imino acid residues are the nonpolar regions, and the segments containing Gly-X-X triplets are mostly polar.

Cross-Linking

Each of the three polypeptide chains in the tropocollagen unit forms its own helix, held together by hydrogen bonding between the NH groups of glycine residues in one chain and the CO group on the other chain. The pyrrolidine rings of proline and hydroxyproline impose restrictions on the conformation of the polypeptide chain and help to strengthen the triple helix. The hydroxyl group of the hydroxyproline plays a part in the stability of the helix by interchain hydrogen bonding via a bridging water molecule as well as direct hydrogen bonding to a carbonyl group (Fig. 2.21) (54).

The arrangement of tropocollagen to form collagen fibrils is determined by the region of basic and acidic amino acids in the polypeptide chain. Each tropocollagen stacks over the other by one-fourth of its length (Fig. 2.22).

The polypeptide chains of the tropocollagens are covalently cross-linked involving side chains of lysine, hydroxylysine, and histidine. The reaction originates from the oxidative deamination of lysine or hydroxylysine by lysyl oxidase to form the aldehydes, allysine or hydroxyallysine. The aldehydes can then condense with lysine or with each other, or with histidine. Some typical cross-linking reactions are outlined in two schemes, one based on allysine (Eq. 2–41.1) and the other on hydroxyallysine (2–41.2). Most of these cross-links are glycosylated, because the hydroxyallysine residues are usually glycosylated.

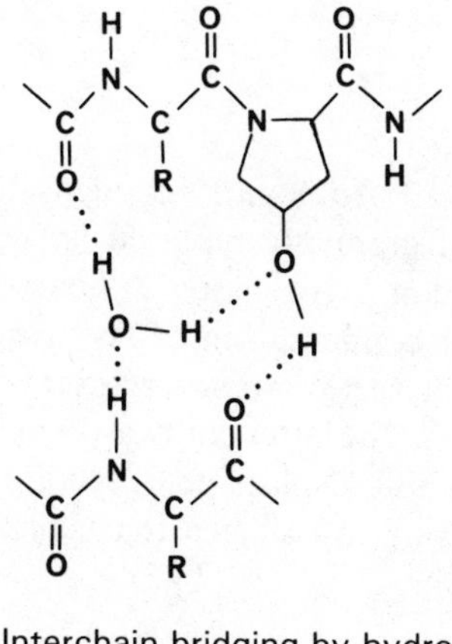

Fig. 2.21. Interchain bridging by hydroxyproline.

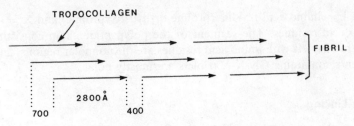

Fig. 2.22. Arrangement of tropocollagen.

2–41.1

Eq. 2-41

These cross-links derived from aldol condensation or Schiff-base reactions are reducible and gradually replaced by nonreducible cross-links during maturation of the connective tissue. A nonreducible cross-link, pyridinoline, is formed by the condensation of one hydroxylysine and two hydroxyallysines or by the interaction of two residues of hydroxylysino-5-ketonorleucine (Eq. 2–42) (55). The latter pathway involves an initial aldol addition followed by a nucleophilic displacement reaction. It is this type of nonreducible cross-links occurring in collagen that causes the tough texture of meat from aged animals.

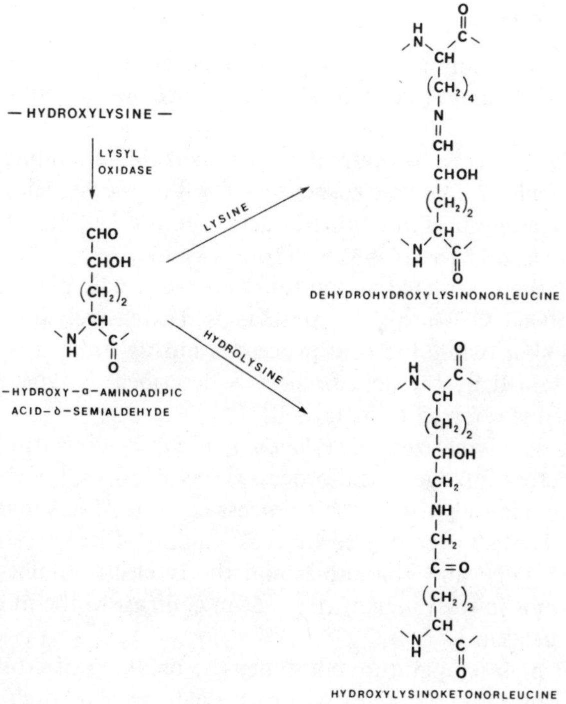

— HYDROXYLYSINE —

LYSYL
OXIDASE

CHO
|
CHOH
|
$(CH_2)_2$
|
—NH—CH—C—
           ‖
           O

δ—HYDROXY—α—AMINOADIPIC
ACID—δ—SEMIALDEHYDE

LYSINE

DEHYDROHYDROXYLYSINONORLEUCINE

HYDROLYSINE

HYDROXYLYSINOKETONORLEUCINE

Eq. 2-41.2

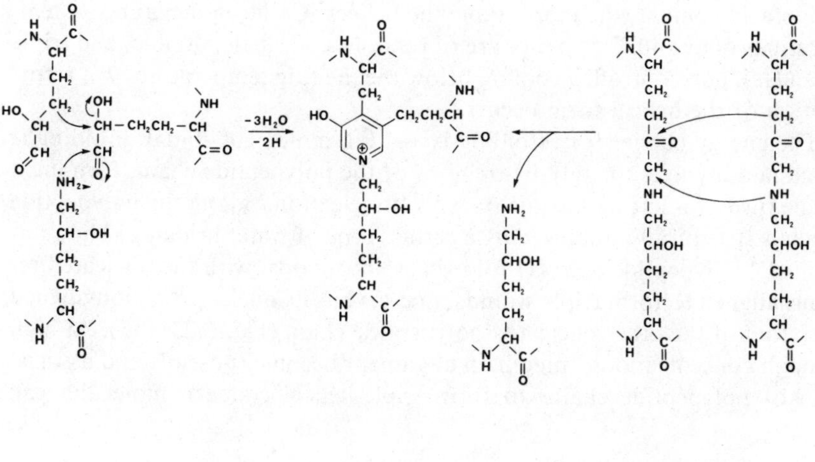

$$\xrightarrow[-2H]{-3H_2O}$$

Eq. 2-42

From Collagen to Gelatin

Commercially, the common sources of collagen are pigskin, cattle hides, and bones. In the United States, pigskin is the chief raw material for edible gelatin production.

The raw material has to be pretreated by (1) soaking at ambient temperature (15–20°C) with 2–5% lime suspension for 8–12 weeks before neutralization, or (2) soaking in dilute mineral acid solution (<5%, pH 3.5–4.5) at ambient temperature for 24–48 hr before washing.

The gelatin is then extracted at neutral or weakly acidic pH at a temperature range of 50–60°C. The alkali process is used extensively for processing cattle hide and skin, while the acid process is mostly utilized for pigskin. The gelatin obtained by the acid process is described as type A. Alkali-processed gelatin is referred to as type B.

Pretreatment depolymerizes the collagen by breaking down the inter- and intramolecular cross-linkages. Acid process is less effective, but the raw material, pigskin, commonly used in this process consists of less mature collagen and hence has a lower degree of cross-linking. Lime treatment also hydrolyzes asparagine and glutamine, and the isoelectric point of alkali-processed gelatin is in the range of pH ~5, in contrast to the pI of ~9 for acid-processed gelatin.

Extraction at high temperature continues the breaking of cross-linkages but, most importantly, serves to break down the hydrogen bonds which are the key stabilizing factor in collagen structure. The gelatin extract is then subjected to clarification and filtration, evaporation, sterilization, and drying to a powdered product.

The Mechanism of Gelation.   The thermal stability of collagen is related to the content of imino acids (proline and hydroxyproline). The higher the imino acid content, the more stable the helices. Collagen denatures at temperatures above 40°C to a mixture of random-coil single, double, and triple strands. Upon controlled cooling below the melting temperature, T_m, reformation of the helical form occurs.

The energy barrier for refolding is ~4 KJ/mole. The initial refolding is rapid and involves the -gly-I-I- regions of the polypeptide chain, forming a single turn of a left-handed helix. This "nucleation" along the polypeptide chain is structurally stabilized by a certain type of water bridging. The "nucleated" polypeptide then (1) folds back into loops, with the nucleated regions aligned to form triple strands, and (2) has its nucleated region aligned with that of the other nucleated polypeptide chain (Fig. 2.23) (56). At high enough concentrations, interchain alignment becomes possible and association of polypeptide chains to form triple-helical collagen molecules can occur.

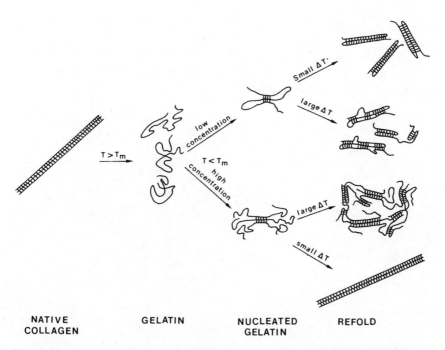

Fig. 2.23. Scheme for the concentration- and temperature-dependent pathways for helix formation in α chains derived from collagen. (Reprinted with permission from Harrington, W. F., and Venkateswara, R. Collagen structure in solution. I. Kinetics of helix regeneration in single chain gelatins. *Biochemistry* **9**, 3723. Copyright 1970 American Chemical Society.)

In both cases, once the nucleated regions are aligned, the remainder of the chain (or chains) start renaturation, the rate of which depends on the cooling temperature. Rapid cooling with large ΔT would cause rapid renaturation, resulting in areas unavailable for the formation of helical structures. Renatured collagen with various degrees of perfection is obtained.

The setting of gelatin solution corresponds to the re-formation of disordered gelatin molecules to the collagen structure. However, commercial gelatin is heterogeneous in structure and composition, due to the processing conditions and the source of raw material. Gelatin may contain polypeptide chains ranging from MW of 30,000 to 300,000. Amino acid compositions differ; alkali-processed gelatin contains more free carboxyl groups, for example. Therefore, in a gelatin solution, the process of reformation is limited. Various processes contribute to the formation of a gelatin gel network. Nucleation involving imino-rich sections provides junction zones for the gel network. Upon further cooling, additional ordering is achieved by intra-

and interchain association. At certain regions, recoiling and refolding give rise to re-formation of the collagen structure, which provides reinforcement and hence rigidity of the gel. The resulting gel is essentially an open network through chain association with imino-rich junctions and reinforced by regions of reformed collagen helices.

3
CARBOHYDRATES

Studies of the chemistry of monosaccharides and oligosaccharides is largely directed to the degradation of these carbohydrates and the subsequent rearrangements and reactions. The many flavor and color compounds in food are the results of these changes and the various interactions among the breakdown products and with other food constituents in the system. Among the many reaction mechanisms involved, the Maillard reaction remains the most important and extensively studied. The basic chemistry of these reactions is discussed and many more secondary reactions are covered in Chapter 6.

Polysaccharides are important due to their special physical and chemical characteristics in solution. The study of the unique flow properties and the mechanism of gelation of a particular polysaccharide requires an understanding of the orientation of its molecular structure in crystal and in solution. Quite often, gelling of polysaccharide molecules requires complexation with metal ions, and the degree and nature of the substituent groups play a significant role. A number of polysaccharides that are of industrial use are presented, with particular emphasis on the relationship between the molecular arrangement and gelling characteristics.

GLYCOSIDIC LINKAGE

Many carbohydrates are found in nature linked to various phenolic compounds as glycosides. The monosaccharide and the glycoside in Eq. 3-1 are hemiacetal and acetal, respectively. Glycosides are fairly stable in aqueous alkali solution but hydrolyzed when heated in aqueous acid.

Acid hydrolysis proceeds via a carbonium ion intermediate, the exocyclic oxygen atom being protonated, followed by the C1-oxygen bond cleavage,

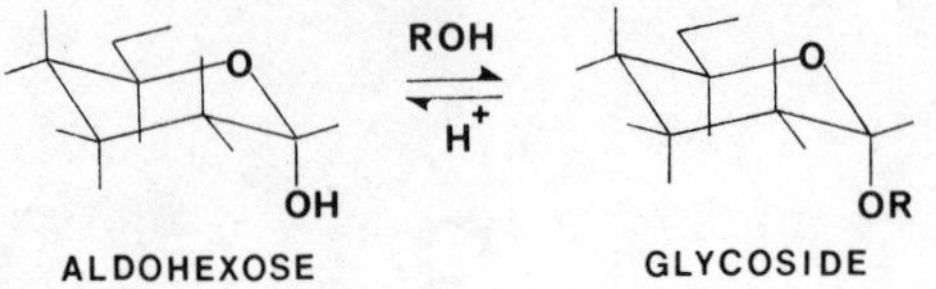

ALDOHEXOSE   GLYCOSIDE

Eq. 3-1

as shown in Eq. 3-2 for the hydrolysis of D-glucopyranoside. The carbonium ion is stabilized by conjugation with the ring oxygen atom, under which conditions the molecule assumes a half-chair conformation.

GLUCOPYRANOSIDE   CARBONIUM ION

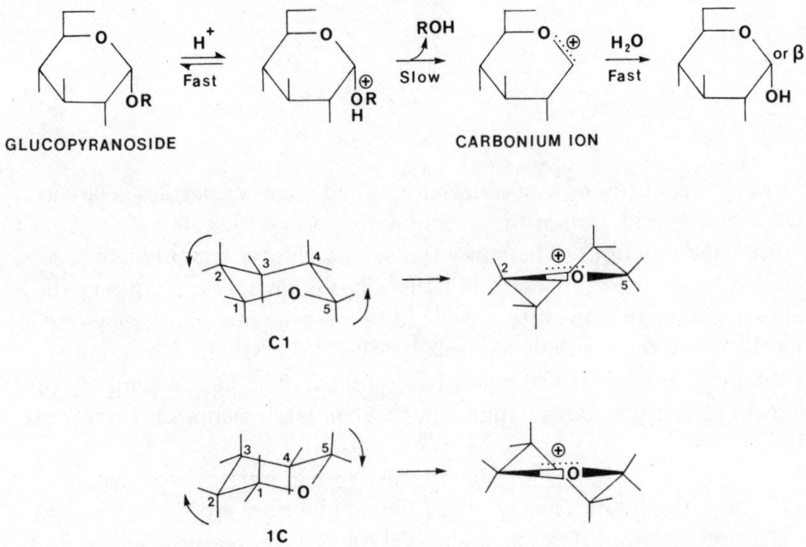

Eq. 3-2

The conversion of the chair form to the half-chair conformation, and hence the stability of the glycoside to acid depends on the steric interaction of the substituents. The formation of the half-chair form, in which the $-C1(OH, Me)-O-$ bridge is planar, requires rotation about the $C2-C3$ and $C4-C5$ bonds as illustrated in Eq. 3-2 for both the C1 and 1C conformers. Therefore, bulky substituents at these positions tend to increase steric hindrance to the rotation. Deoxyglycosides such as 2- and 3-deoxyglycosides are more labile to acid hydrolysis as compared with the corresponding normal glycosides.

The conversion of the chair to the half-chair form is also assisted by the changes of the C2 and C5 axial substituents away from the C4 and C3 axial substituents, respectively. Therefore, in their C1 conformation the order of stability for the methylpyranosides will be: glucose (no axial OH) > mannose (one axial OH) > gulose (two axial OH (Eq. 3-3). Similarly, for methyl pentopyranosides, xylose > arabinose > lyxose.

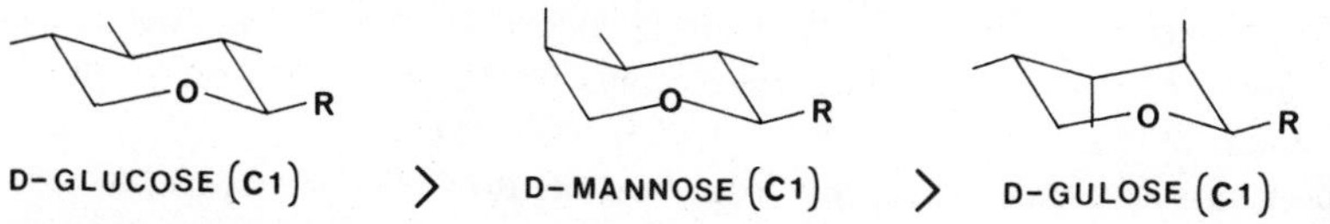

D-GLUCOSE (C1) > D-MANNOSE (C1) > D-GULOSE (C1)   Eq. 3-3

Only substituents at C2, C3, C4, C5 shown.

—— = OH

—— R = ——CH₂OH

Base-catalyzed hydrolysis of glycosides occurs only under extreme conditions with highly concentrated hydroxide and at high temperatures. For example, experiments have been done using 10% sodium hydroxide and heating at 170°C for the hydrolysis of methylpyranosides. *Trans* 1,2-glycosides are suggested to undergo hydrolysis via intermediate 1,2 anhydro formation by inversion. This 1,2-anhydro sugar reacts by a second inversion with the C6 alkoxide ion to form the 1,6-anhydride (Eq. 3-4) (*1*).

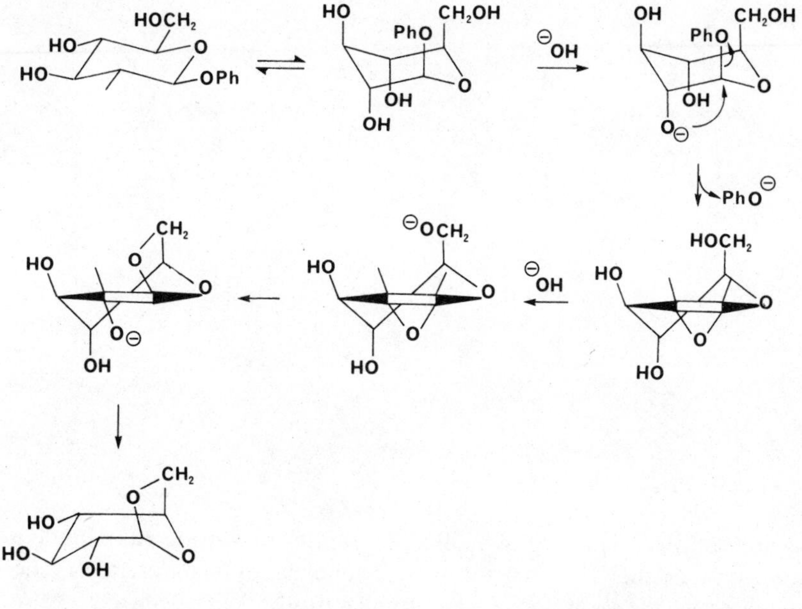

Eq. 3-4

Alkyl glycosides are very resistant to alkaline degradation, but aryl glycosides, especially with electron-withdrawing substituents, are found to react readily. *Trans* 1,2-glycosides generally react faster than the *cis* isomers, since formation of the 1,2-anhydride intermediate involves a *trans* configuration for the nucleophilic attack of the C2 alkoxide at C1. From Eq. 3-4, it is also conceivable that hydrolysis may proceed via a bimolecular nucleophilic substitution in which a back side attack of the hydroxide ion displaces the phenoxide (*1*).

The monosaccharides resulting from both acid or alkaline hydrolysis enter into further degradation by a number of pathways.

ACTION OF ALKALI ON MONOSACCHARIDES

Aldose-Ketose Rearrangement

In the presence of dilute base, D-glucose enolizes to form a mixture of anomers of D-glucose, D-fructose, and D-mannose. The interconversion proceeds via a 1,2-enediol intermediate in a rearrangement commonly known as the Lobry de Bruyn-Alberda van Ekenstein reaction (Eq. 3-5).

Eq. 3-5

The formation of 1,2-enediol in alkaline solution is considered to involve the ionization of the C1 − OH, followed by enolization via a pseudo-cyclic intermediate which allows an intramolecular proton shift from C2 to the C5 oxygen (Eq. 3-6) (*2*). The resulting enediol anion rearranges in the reverse process of enolization to form the isomers.

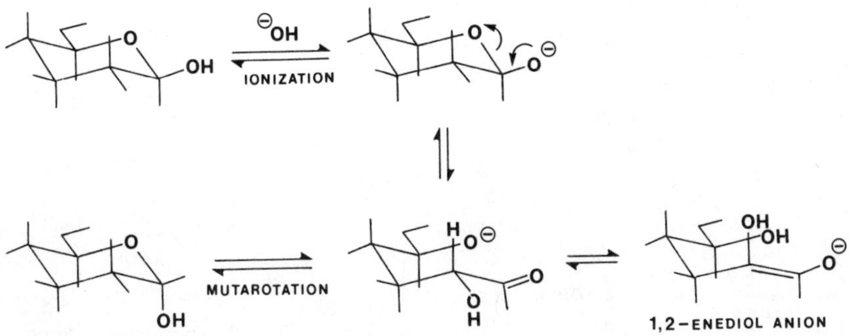

Eq. 3-6

1,2−ENEDIOL ANION

Formation of Saccharinic Acid

In strong alkaline solution, the 1,2-enediol undergoes elimination at C3 to yield 3-deoxyglycosulose, which undergoes a benzilic acid type of rearrangement to metasaccharinic acid (for simplification, glucose is the aldohexose used in the reaction schemes) (Eq. 3-7.1). Formation of the 2,3-enediol, which then undergoes similar reaction sequences, leads to the formation of saccharinic and isosaccharinic acids (Eq. 3-7.2) (3). For saccharinic acid, elimination occurs at C1, while elimination at C4 of the 2,3-enediol, followed by benzilic rearrangement, yields the isosaccharinic acid.

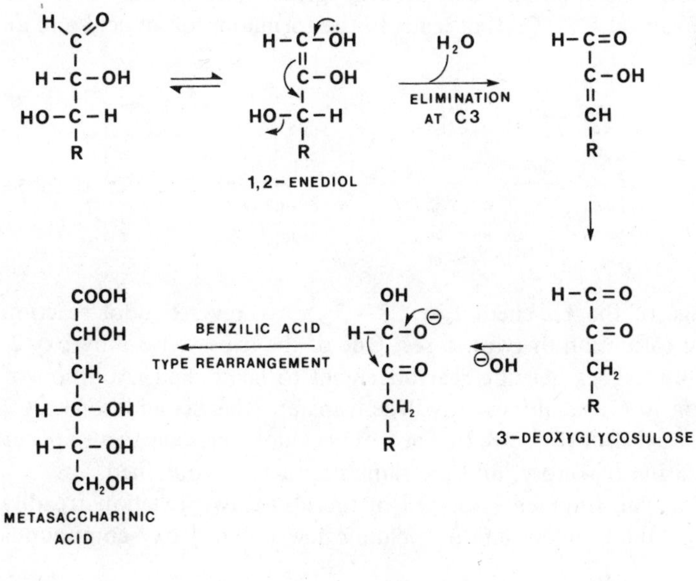

3 − 7.1

1,2−ENEDIOL

Eq. 3-7

METASACCHARINIC ACID

3−DEOXYGLYCOSULOSE

3 – 7.2

$$
\begin{array}{c}
\text{H} \\
\text{H}-\overset{|}{\text{C}}-\text{OH} \\
\overset{|}{\text{C}}=\text{O} \\
\text{HO}-\overset{|}{\text{C}}-\text{H} \\
\overset{|}{\text{R}} \\
\text{KETOSE}
\end{array}
\rightleftharpoons
\begin{array}{c}
\text{H}-\text{C}=\text{O} \\
\text{H}-\overset{|}{\text{C}}-\text{OH} \\
\text{HO}-\overset{|}{\text{C}}-\text{H} \\
\overset{|}{\text{R}} \\
\text{GLUCOSE}
\end{array}
$$

$$
\begin{array}{c}
\text{H}_2\text{C}-\text{OH} \\
\overset{|}{\text{C}}-\text{OH} \\
\overset{\|}{\text{C}}-\text{OH} \\
\overset{|}{\text{R}} \\
\text{2,3-ENEDIOL}
\end{array}
\xrightarrow[\text{at C1}]{\overset{\text{H}_2\text{O}}{\text{Elimination}}}
\begin{array}{c}
\text{CH}_2 \\
\overset{\|}{\text{C}}-\text{OH} \\
\overset{|}{\text{C}}=\text{O} \\
\overset{|}{\text{R}}
\end{array}
\rightleftharpoons
\begin{array}{c}
\text{CH}_3 \\
\overset{|}{\text{C}}=\text{O} \\
\overset{|}{\text{C}}=\text{O} \\
\overset{|}{\text{R}}
\end{array}
\xrightarrow[\text{Rearrangement}]{\overset{\ominus}{\text{OH}}\ \text{Benzilic}}
\begin{array}{c}
\text{COOH} \\
\text{CH}_3-\overset{|}{\text{C}}-\text{OH} \\
\text{H}-\overset{|}{\text{C}}-\text{OH} \\
\text{H}-\overset{|}{\text{C}}-\text{OH} \\
\overset{|}{\text{CH}_2\text{OH}} \\
\text{SACCHARINIC ACID}
\end{array}
$$

Eq. 3-7 (cont.)

$$
\xrightarrow[\text{at C4}]{\overset{\text{H}_2\text{O}}{\text{Elimination}}}
$$

$$
\begin{array}{c}
\text{H}_2\text{C}-\text{OH} \\
\overset{|}{\text{C}}=\text{O} \\
\overset{\|}{\text{C}}-\text{OH} \\
\overset{|}{\text{CH}} \\
\overset{|}{\text{R}}
\end{array}
\rightleftharpoons
\begin{array}{c}
\text{H}_2\text{C}-\text{OH} \\
\overset{|}{\text{C}}=\text{O} \\
\overset{|}{\text{C}}=\text{O} \\
\overset{|}{\text{CH}_2} \\
\overset{|}{\text{R}}
\end{array}
\xrightarrow[\text{Rearrangement}]{\overset{\ominus}{\text{OH}}\ \text{Benzilic}}
\begin{array}{c}
\text{COOH} \\
\text{HOCH}_2-\overset{|}{\text{C}}-\text{OH} \\
\overset{|}{\text{CH}_2} \\
\overset{|}{\text{CHOH}} \\
\overset{|}{\text{CH}_2\text{OH}} \\
\text{ISOSACCHARINIC ACID}
\end{array}
$$

Fragmentation

1. Dicarbonyl compounds such as deoxyglycosulose in Eq. 3-7.1 can undergo cleavage at C1 − C2 that leads to the formation of an acid and an aldehyde (Eq. 3-8.1).

$$
\begin{array}{c}
\text{H}_2\text{C}=\text{O} \\
\overset{|}{\text{C}}=\text{O} \\
\overset{|}{\text{CH}_2} \\
\text{H}-\overset{|}{\text{C}}-\text{OH} \\
\text{H}-\overset{|}{\text{C}}-\text{OH} \\
\overset{|}{\text{CH}_2}
\end{array}
\xrightarrow{\overset{\ominus}{\text{OH}}}
\begin{array}{c}
\overset{\text{OH}}{\text{H}-\text{C}-\text{O}^{\ominus}} \\
\overset{|}{\text{C}}=\text{O} \\
\overset{|}{\text{CH}_2} \\
\text{H}-\overset{|}{\text{C}}-\text{OH} \\
\text{H}-\overset{|}{\text{C}}-\text{OH} \\
\overset{|}{\text{CH}_2\text{OH}}
\end{array}
\longrightarrow
\begin{array}{c}
\text{CHO} \\
\overset{|}{\text{CH}_2} \\
\text{H}-\overset{|}{\text{C}}-\text{OH} \\
\overset{|}{\text{CH}_2\text{OH}}
\end{array}
+
\begin{array}{c}
\overset{\text{O}}{\overset{\|}{\text{H}-\text{C}-\text{OH}}}
\end{array}
$$

Eq. 3-8

2. Cleavage of the 1,2-enediol at C3 − C4 via a reverse aldol reaction results in the formation of two trioses. One of the trioses, 1,3-dihyroxy-2-propanone, undergoes benzilic rearrangement to lactic acid. Addition of hydroxide ion to C1 results in a hydride transfer, whereas addition to C2 involves a methyl shift (Eq. 3-8.2). The reverse aldol reaction can also occur via the formation of ketose, and the same products are obtained.

3. The aldehyde fragments formed in the above two reactions readily undergo cross-aldol condensation, yielding new polyhydroxy compounds (4).

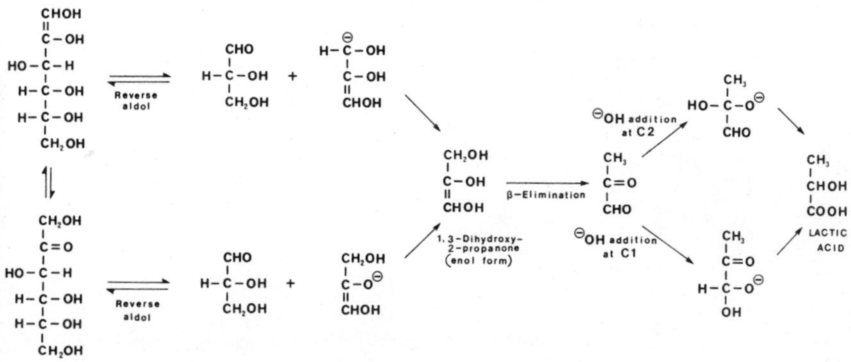

Eq. 3-8 (cont.)

ACTION OF ACID ON MONOSACCHARIDES

Formation of Furan Derivatives

When a monosaccharide is heated in a strong acidic solution, dehydration occurs, resulting in the formation of furfural compounds (3, 5).

1. The first step is the protonation of the carbonyl oxygen, followed by 1,2 enolization to form 1,2 enediol (Eq. 3-9.1). (Glucose is the aldohexose used in the reaction scheme.) The 1,2-enolization is the rate-determinating step. And the enediol is formed from both aldose and ketose, since it is the common intermediate in the aldose-ketose rearrangement (Eq. 3-5).

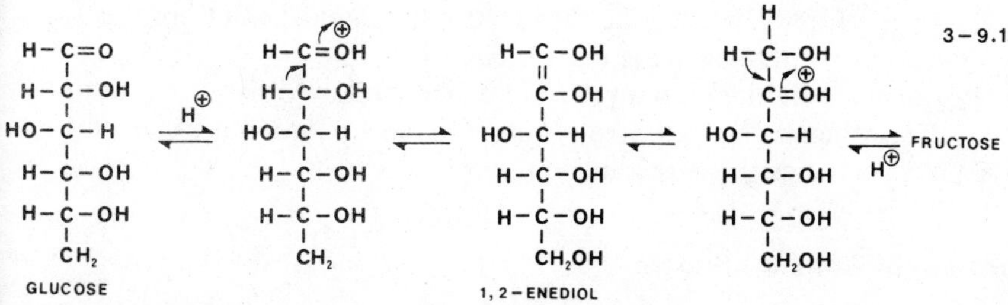

Eq. 3-9

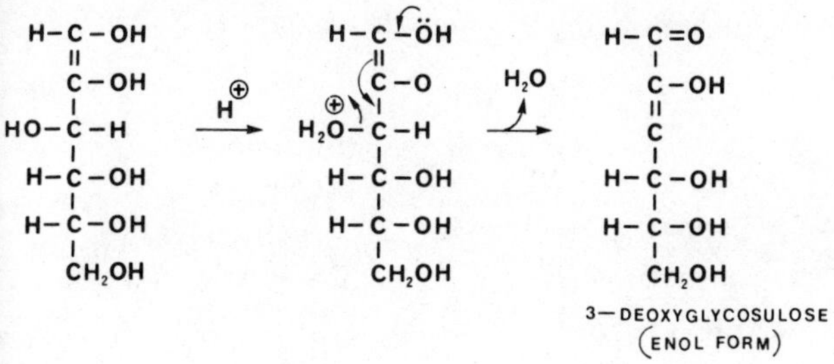

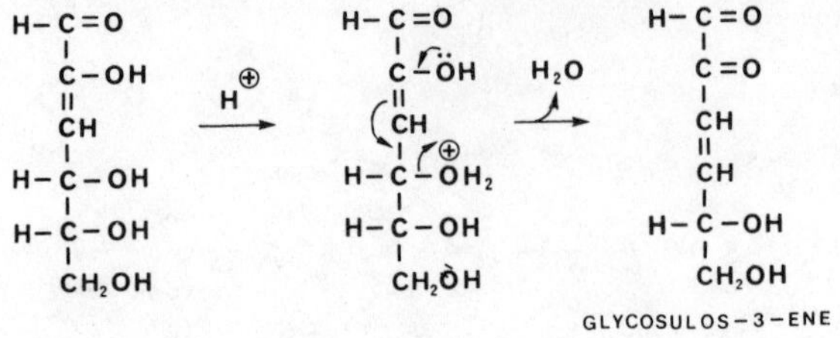

GLYCOSULOS-3-ENE

Eq. 3-9
3-9.4 (cont.)

HYDROXYMETHYLFURFURAL

LEVULINIC ACID

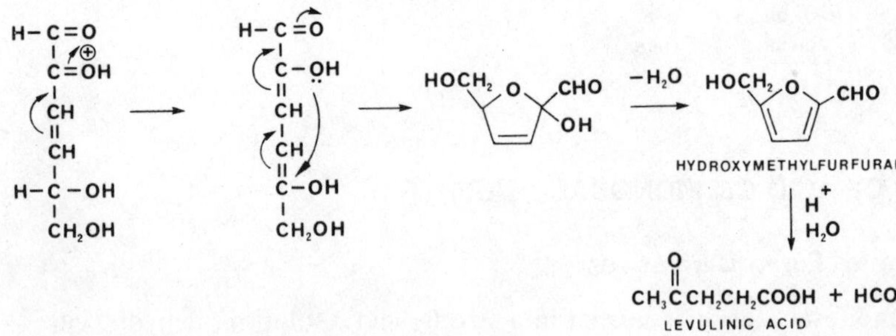

2. Elimination at C3, assisted by the protonation of the hydroxyl group at C3 and nucleophilic addition at C1, leads to the formation of the enol form of 3-deoxyglycosulose (Eq. 3-9.2). This elimination is an allylic shift or oxotropic rearrangement.

3. Elimination at C4, assisted by the protonation of the oxygen at C4 and nucleophilic addition at C2, yields the 3,4-unsaturated glycosulose (glycosulos-3-ene) (Eq. 3-9.3). Unlike the reaction in alkaline solution, the 3-deoxyglycosulose intermediate is not formed (6).

4. Protonation of the carbonyl group at C2, followed by enolization, extends the conjugated unsaturated system. Cyclodehydration involving C5 and C2 oxygen yields the furfural (Eq. 3-9.4).

Formation of Anhydro Sugars

In dilute acid solution, aldohexoses form anhydro products with the loss of a water molecule, the most common being 1,6-anhydro sugars (Eq. 3-10) (7).

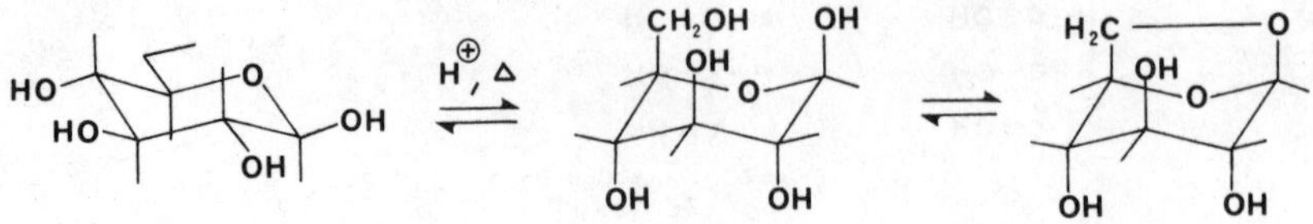

Eq. 3-10

Table 3.1. Percentage of 1,6-Anhydro Sugars in Equilibrium Mixture

SUGAR SOLUTION (0.25–0.50%, 0.25 M H$_2$SO$_4$, 100°C)	% 1,6-ANHYDROHEXOSE
D-Glucose	0.2
D-Mannose	0.8
D-Altrose	65.5
D-Gulose	65.0
D-Idose	86.0

Reprinted with permission from Cěrný and Stanek (7), courtesy of Academic Press, Inc.

Note that in β-D-glucopyranose, in order for the C1 − OH and C6 − OH to be in the axial position, the other hydroxyl groups will also be in axial positions, which constitute the unstable conformation due to steric interaction. In contrast, with D-altrose, D-gulose, and D-idose, the molecule assumes a more stable conformation when the C1 − OH and C6 − OH are oriented axially. This conformation factor affects the proportion of 1,6-anhydro products found in a particular sugar solution, as indicated in Table 3.1.

NONENZYMATIC BROWNING (MAILLARD REACTION)

The Maillard reaction is one of the most important reactions that occur in food. The reaction is critical in the production of the many flavor and color compounds in processed food products, both desirable and undesirable (Chapter 6). Reduction of nutritional quality and possible generation of toxic and mutagenic compounds, via the Maillard reaction, are causes for concern (Chapter 8).

The Maillard reaction comprises a series of reactions that are far from being clearly elucidated. The following reactions are generally recognized (5, 8, 9). (Glucose is the aldohexose used in the reaction scheme).

1. Formation of glycosylamine, via a Schiff-base formation between the carbonyl group of a reducing sugar and the amino group of amine. The reaction is reversible. Formation of the glycosylamine is acid-catalyzed and favored at low water content (Eq. 3-11.1).

2. Amadori rearrangement in which the glycosylamine is transformed to the ketoseamine. The nitrogen of the glycosylamine accepts a proton to form the amine salt, which is in equilibrium with the cation of the Schiff base (iminium ion). Rearrangement of the latter gives the enol form (1,2-enaminol), which then tautomerizes to yield the keto form of the Amadori compound (1-amino-1-deoxy-2-ketose) (Eq. 3-11.2). The tautomerization

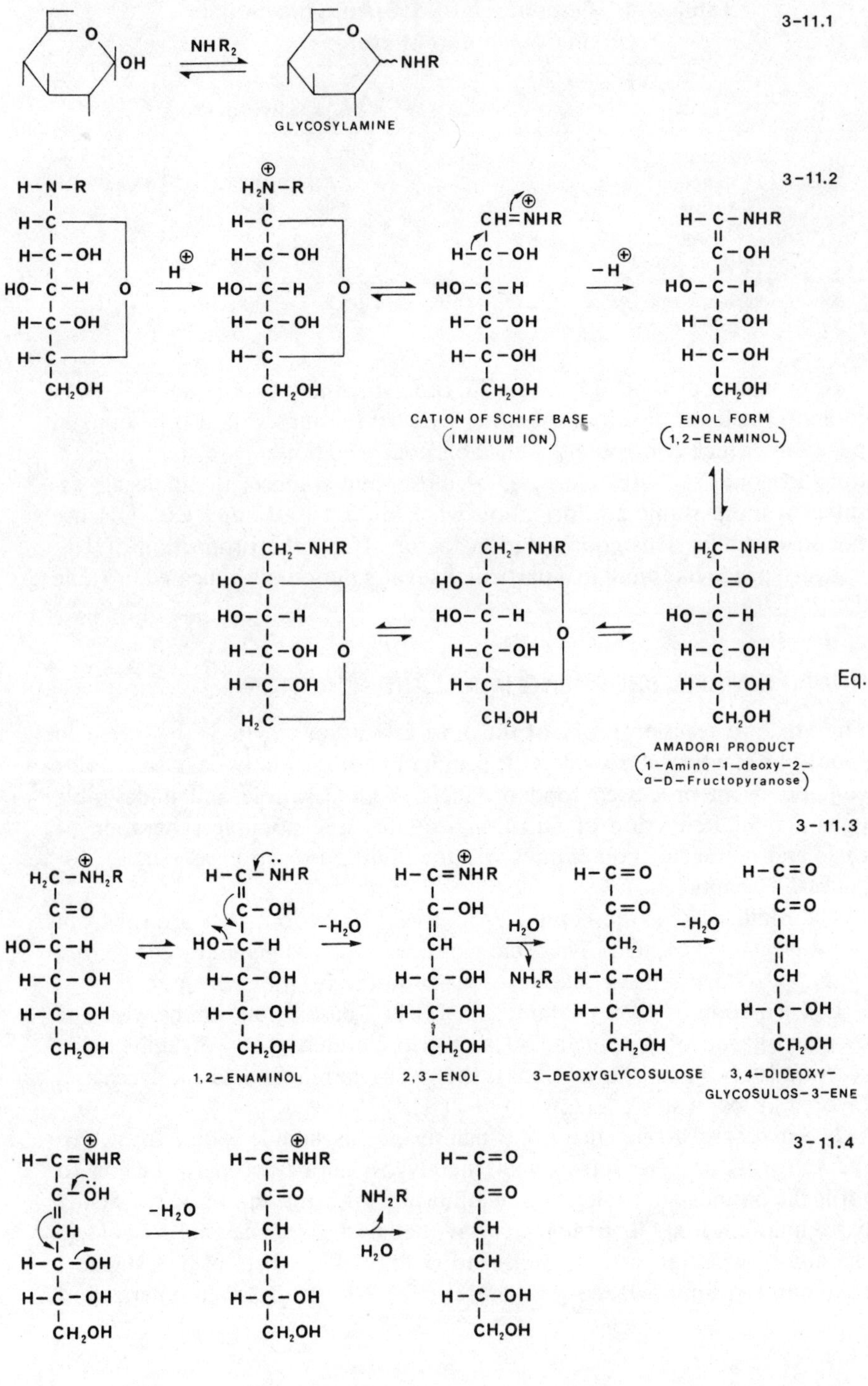

3-11.1

GLYCOSYLAMINE

3-11.2

CATION OF SCHIFF BASE
(IMINIUM ION)

ENOL FORM
(1,2-ENAMINOL)

Eq. 3-11

AMADORI PRODUCT
(1-Amino-1-deoxy-2-
α-D-Fructopyranose)

3-11.3

1,2-ENAMINOL   2,3-ENOL   3-DEOXYGLYCOSULOSE   3,4-DIDEOXY-
                                                          GLYCOSULOS-3-ENE

3-11.4

3-11.5

5-HYDROXYMETHY-2-
FURALDEHYDE
(Hydroxymethyl furfural)

Eq 3-11
(cont.)

to the keto form is driven by the formation of the cyclic structure (*10*). The mechanism of Amadori rearrangement is catalyzed by a weak acid, and in the reaction involving amino acids, the carboxyl group of the amino acid may act as the catalyst by furnishing the necessary proton.

3. Under acidic conditions, the Amadori compound exists in its salt form. The presence of the positively charged amino group assists in shifting the equilibrium to the enol form, which undergoes elimination of the hydroxyl group from C3 to yield the 2,3-enol, which is readily hydrolyzed at the C1 Schiff base to the glycosulose. Further elimination of the hydroxyl group at C4 yields an unsaturated glycosulos-3-ene (Eq. 3-11.3).

4. Alternately, eliminations at C4 before hydrolysis of the Schiff base also give the glycosulos-3-ene. Once the hydroxyl group has been eliminated at C3, the course of reaction is not affected by the removal of the amino group (Eq. 3-11.4).

5. The glycosulos-3-ene, which is an unsaturated dicarbonyl compound, undergoes cyclodehydration to form furaldehyde (Eq. 3-11.5).

6. While in the above pathway the 3-deoxyglycosulose is formed from the 1,2-enaminol resulting by 1,2-enolization, a second pathway may also proceed via irreversible 2,3-enolization of the Amadori compound by β-elimination of the C1 amino group to form a methyl α-dicarbonyl intermediate. Rapid enolization of the unstable intermediate yields the α, β-unsaturated carbonyl compound, which may form reductones and carbonyl products via hydrolytic fission or O-heterocyclic compounds by cyclization (Eq. 3-12).

Chemistry of the Reactions

The formation of glycosylamine involves the addition of amine to the acyclic aldehyde or the cation generated from the sugar by mutarotation. Most ketoses and hexoses exist in less than 0.05% as the open aldehydic form, the stability of which depends on the conformational orientation of the hydroxyl groups. The formation of acyclic cation $[R-CH=OH]^+$ by general acid catalysts enhances the reactivity of the carbonyl function. Nucleophilic attack of the amine at C1, followed by the elimination of the

$$
\begin{array}{ccc}
\text{H}_2\text{C}-\text{NHR} & \text{H}-\text{C}-\overset{\oplus}{\text{NH}_2}\text{R} & \text{CH}_3 \\
\text{C}=\text{O} & \text{C}-\text{OH} & \text{C}=\text{O} \\
\text{HO}-\text{C}-\text{H} \xrightarrow{\text{2,3 ENOLIZATION}} & \text{C}-\overset{..}{\text{OH}} \xrightarrow{\text{NH}_2\text{R}} & \text{C}=\text{O} \\
\text{H}-\text{C}-\text{OH} & \text{H}-\text{C}-\text{OH} & \text{H}-\text{C}-\text{OH} \\
\text{H}-\text{C}-\text{OH} & \text{H}-\text{C}-\text{OH} & \text{H}-\text{C}-\text{OH} \\
\text{CH}_2\text{OH} & \text{CH}_2\text{OH} & \text{CH}_2\text{OH}
\end{array}
$$

METHYL α–DICARBONYL
INTERMEDIATE (UNSTABLE)

Eq. 3-12

$$
\begin{array}{c}
\overset{\text{OH}}{\underset{}{}}\quad\overset{\text{OH}}{\underset{}{}} \\
\text{CH}_3\text{C}=\text{CHCH}=\text{CCH}_3 \\
\overset{\text{O}}{\underset{}{||}} \\
\text{CH}_3\text{CCHO} \\
\overset{\text{O}}{\underset{}{||}}\ \overset{\text{O}}{\underset{}{||}} \\
\text{CH}_3\text{C}-\text{CCH}_3 \\
\overset{\text{O}}{\underset{}{||}}\ \overset{\text{O}}{\underset{}{||}} \\
\text{HOCH}_2\text{C}-\text{CCH}_3
\end{array}
\xleftarrow{\text{Hydrolytic fission}}
\begin{array}{c}
\text{CH}_3 \\
\text{C}=\text{O} \\
\text{C}-\text{OH} \\
\text{C}-\text{OH} \\
\text{H}-\text{C}- \\
\text{CH}_2\text{OH}
\end{array}
\xrightarrow{\text{Cyclization}}
$$

α,β–UNSATURATED
COMPOUND

hydroxyl ion, results in forming the iminium ion of glycosylamine (Eq. 3-13) (*11*).

$$
\begin{array}{ccc}
\text{H}-\text{C}-\text{OH} & \overset{\oplus}{\text{HO}}=\text{C}-\text{H} & \overset{\text{H}}{\underset{}{}}\ \overset{\oplus}{} \\
\text{H}-\text{C}-\text{OH} & \text{H}-\text{C}-\text{OH} & \text{HO}-\text{C}-\text{NH}_2\text{R} \\
\text{HO}-\text{C}-\text{H} \xrightarrow{\text{H}^{\oplus}} & \text{HO}-\text{C}-\text{H} \xrightarrow{\text{NH}_2\text{R}} & \text{H}-\text{C}-\text{OH} \\
\text{H}-\text{C}-\text{OH} & \text{H}-\text{C}-\text{OH} & \text{HO}-\text{C}-\text{H} \\
\text{H}-\text{C}-\text{O} & \text{H}-\text{C}-\text{OH} & \text{H}-\text{C}-\text{OH} \\
\text{CH}_2\text{OH} & \text{CH}_2\text{OH} & \text{H}-\text{C}-\text{OH} \\
& & \text{CH}_2\text{OH}
\end{array}
$$

$$\downarrow -\text{H}^{\oplus}$$

Eq. 3-13

$$
\begin{array}{ccc}
\text{H}-\text{C}-\text{NHR} & \text{H}-\text{C}=\overset{\oplus}{\text{NH}} & \overset{\text{H}}{\underset{}{}} \\
\text{H}-\text{C}-\text{OH} & \text{H}-\text{C}-\text{OH}\ \ \text{H}_2\text{O} & \text{HO}-\text{C}-\text{NHR} \\
\text{HO}-\text{C}-\text{H}\quad\text{O} \xleftarrow{} & \text{HO}-\text{C}-\text{H} \xleftarrow{} & \text{HO}-\text{C}-\text{H} \\
\text{H}-\text{C}-\text{OH} & \text{H}-\text{C}-\text{OH} & \text{H}-\text{C}-\text{OH} \\
\text{H}-\text{C} & \text{H}-\text{C}-\text{OH} & \text{H}-\text{C}-\text{OH} \\
\text{CH}_2\text{OH} & \text{CH}_2\text{OH} & \text{CH}_2\text{OH}
\end{array}
$$

GLYCOSYLAMINE   SCHIFF BASE
CATION

In the Amadori rearrangement as outlined in Eq. 3-11.2, the initial step is suggested as *N*-protonation. The mechanism has also been interpreted to involve the addition of the proton to the ring oxygen rather than the nitrogen atom, since the ammonium ion is stable and unreactive. Enolization of the resulting immonium ion by elimination of the hydrogen atom at C2 yields the enol form, 1,2-enaminol. Tautomerization of the enolic glycosylamine leads to same end product, 1-amino-1-deoxy-2-ketose (Eq. 3-14).

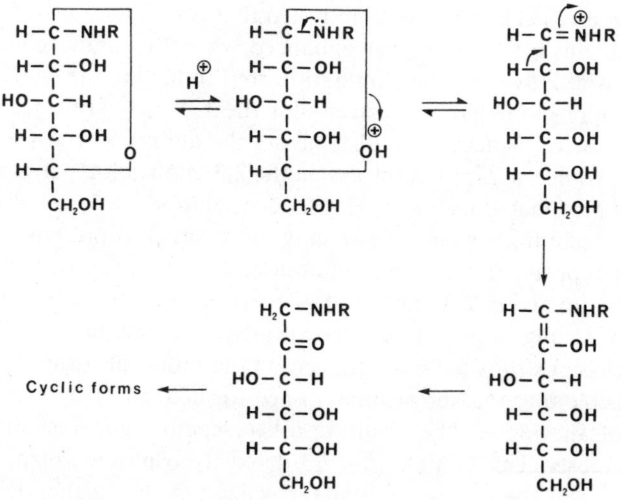

Eq. 3-14

The mechanism of the Amadori rearrangement is dependent on the N-substituent of the amino component. The formation of the iminium ion is favored by strongly basic amine. However, the following enolization step by the elimination of the hydrogen atom at C2 is facilitated by a weakly basic amino component. In reactions involving amino acids, the enolization reaction is assisted by the carboxyl groups. Condensation of the iminium ion with the carboxylate anion, followed by intramolecular decomposition, leads to the formation of the glycosylamine (Eq. 3-15).

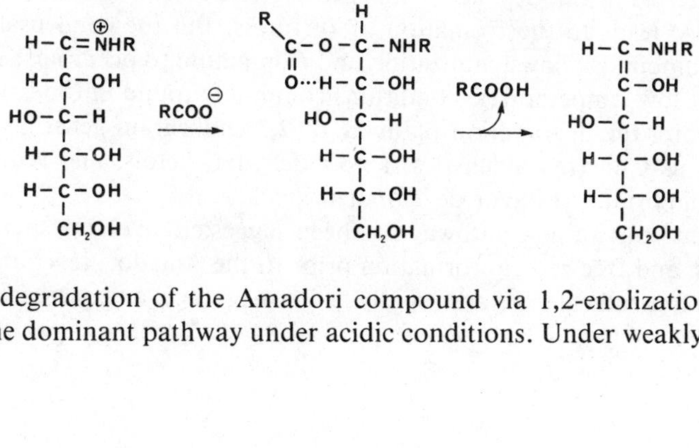

Eq. 3-15

The degradation of the Amadori compound via 1,2-enolization constitutes the dominant pathway under acidic conditions. Under weakly alkaline conditions, and with a strongly basic secondary amino component, the 2,3-enolization pathway is favored. In the former pathway, the Amadori compound has the nitrogen protonated and 1,2-enolization is assisted by the positively charged nitrogen acting as an electron sink. The effect is strongest with the Amadori compound of weak bases. In the free-base form, as present in an alkaline condition, the Amadori compound has the electron density at C1 increased by the amino nitrogen that is unfavorable for 1,2-enolization, and this effect is further enhanced when the amino component is strongly basic. However, the elimination step following the enolization exhibits the opposite enhancing effect. For the 1,2-enolization pathway, the elimination at C3 is accelerated by having the amino nitrogen in the free-base form, while the elimination step in the 2,3-enolization pathway is facilitated by a protonated nitrogen. Hence, degradation via 1,2-enolization occurs under conditions where the weakly basic amine is present in approximately equal proportions of the salt and the free-base forms. Similarly, the degradation involving 2,3-enolization occurs at optimum conditions in the presence of strong basic amine salts with excess amine (5).

The discussions thus far have concerned the monosubstituted amines. It is also possible that the ketoseamine, once formed, can react with another molecule of an aldose. The resulting diketoseamine undergoes rearrangement and subsequent elimination to 3-deoxyglycosulose, which, on dehydradation, yields the glycosulos-3-ene. Cyclization and further dehydration of the latter yields the furfural (Eq. 3-16) (12). This reaction pathway is probably more important in systems having high sugar-to-nitrogen substrate ratios and low pH conditions.

The Amadori rearrangement explains the conversion of an aldosylamine to a ketoseamine. However, the reverse rearrangement in which a ketosylamine is rearranged to give the corresponding aldosamine also occurs in a similar mechanism, known as Heynes rearrangement, in the conversion of D-fructosylamine to 2-amino-2-deoxy-D-glucose (D-glucoseamine) (Eq. 3-17).

As suggested in Eq. 3-9, acid-catalyzed 1,2-enolization in the absence of amines also leads to the formation of furfurals. But the condensation of amino compounds allows enolization and elimination to occur near neutral pH and at low temperatures, conditions commonly found in food systems. Furthermore, the degradation pathway of 2,3-enolization generates products such as C-methyl aldehydes, keto-aldehydes, ketols, and reductones that are important to flavor development.

In recent years, a new pathway has been suggested involving sugar fragmentation and free-radical formation prior to the Amadori rearrangement (13). The free radical develops rapidly at an early stage, and decreases while

Eq. 3-16

the formation of the Amadori compound and 3-deoxyglycosulose remain increasing. Carbonyl compounds with enediol groups and amino compounds with primary amino groups are most effective in the formation of free radicals. Free-radical formation occurs at neutral pH, increases with pH, and disappears above pH 11. It has been shown that the radical is N, N'-disubstituted pyrazine cation radical formed by dimerization of a two-carbon enaminol product from fragmentation of the glycosylamine. The mechanism, as shown in Eq. 3-18, involves (1) a reverse-aldol reaction of the glycosylamine to give the 2-carbon enaminol, glycoaldehyde alkylimine, and (2) condensation of the latter to form the unstable dialkyldihydropyrazine, which is readily oxidized to the dialkylpyrazinium product via the cation radical intermediate. It has been shown that the dialkylpyrazinium product is likely the active intermediate for the polymerization in browning.

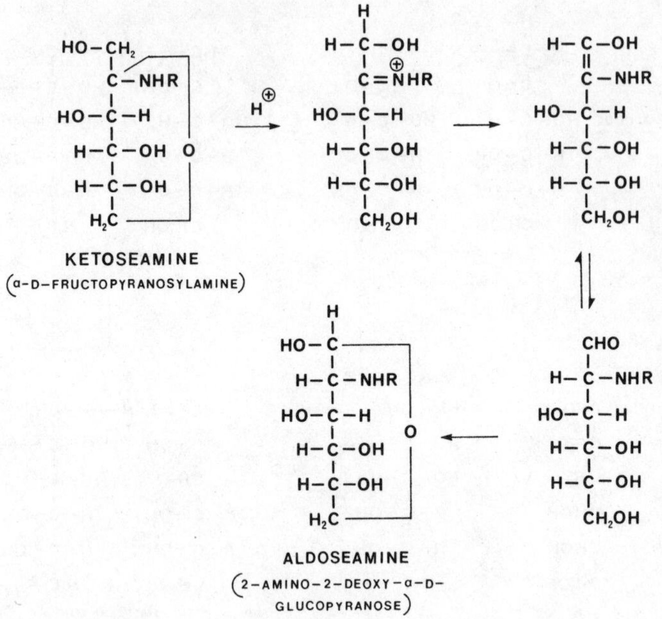

KETOSEAMINE
(α-D-FRUCTOPYRANOSYLAMINE)

ALDOSEAMINE
(2-AMINO-2-DEOXY-α-D-
GLUCOPYRANOSE)

Eq. 3-17

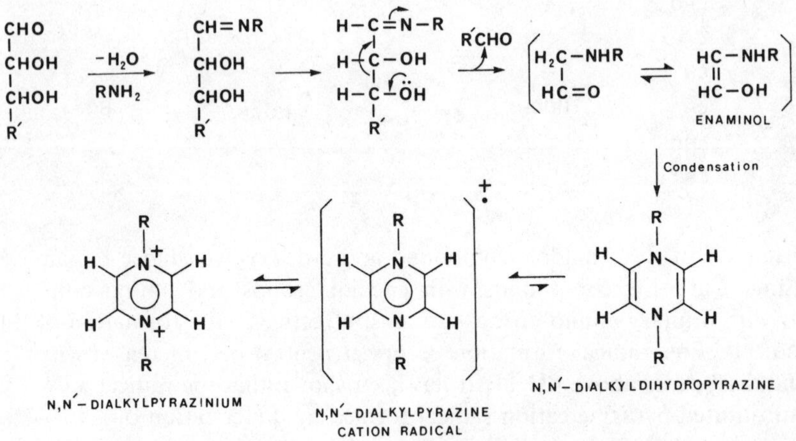

ENAMINOL

Condensation

Eq. 3-18

N,N′-DIALKYLPYRAZINIUM

N,N′-DIALKYLPYRAZINE
CATION RADICAL

N,N′-DIALKYLDIHYDROPYRAZINE

Secondary Reactions

Both degradation pathways via 1,2- or 1,3-enolization provide dicarbonyl compounds that are the key intermediate for subsequent reactions and degradations that are of most significance in changing the characteristic attributes in food systems. (Refer to Chapter 6 for more discussion on secondary reactions.)

Reaction with Sulfite and Bisulfite. Inhibition of nonenzymatic browning reaction involves the reaction between browning intermediates such as α, β-unsaturated carbonyls with sulfite to form stable sulfonic acids. Sulfur dioxide has been used extensively to retard the Maillard reaction in food.

At pH 6 or above, the 3-deoxyglycosulose in its enol form may have the C4 hydroxyl group replaced by sulfite ion, with possible inversion at C4 (Eq. 3-19). The α, β-unsaturated ketone, such as glycosulos-3-ene, reacts with sulfite by addition, and the same product, 3,4-dideoxy-4-sulfo-D-glycosulose, has been isolated and identified (5).

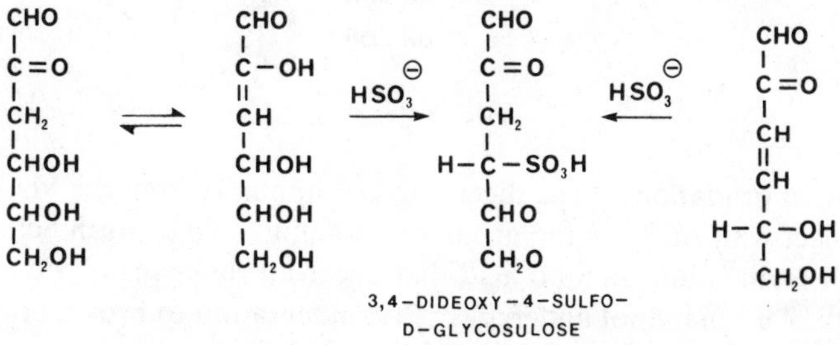

Eq. 3-19

3,4-DIDEOXY-4-SULFO-D-GLYCOSULOSE

Formation of Melanoidins. Melanoidin is the result of polymerization of the unsaturated carbonyl compounds formed in the Maillard reaction with amines. One possible structure may be composed of repeating units of Schiff base formed bewween 3-deoxyglycosulose and its enamine as shown in Eq. 3-20.1 (14), although polymers of furan (3-20.2) and pyrrole (3-20.3) structures have also been suggested. The proposed structure consists of ether bonds and reductone systems.

3 - 20.1

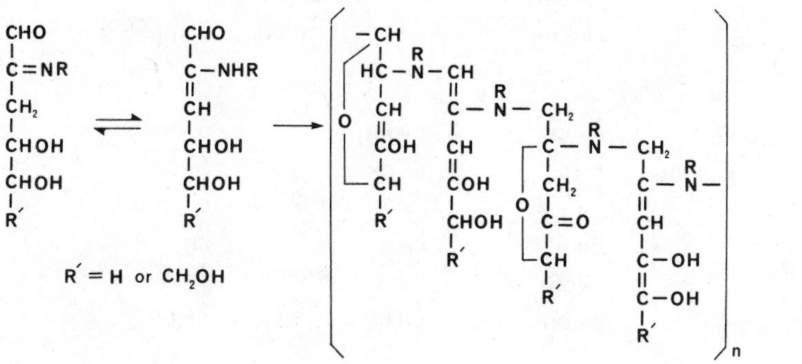

Eq. 3-20

$R' = H$ or CH_2OH

3-20.2

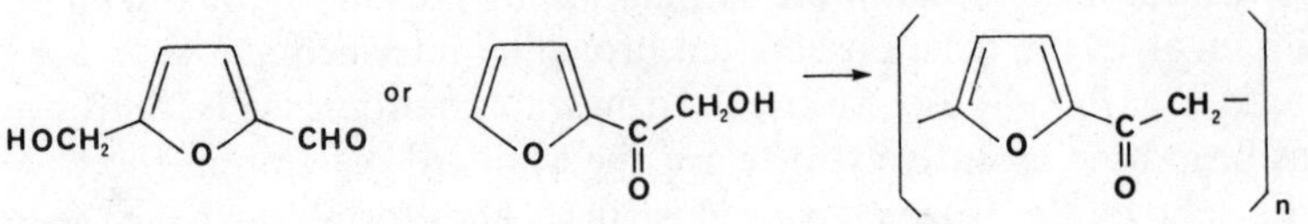

Eq. 3-20
(cont.)

3-20.3

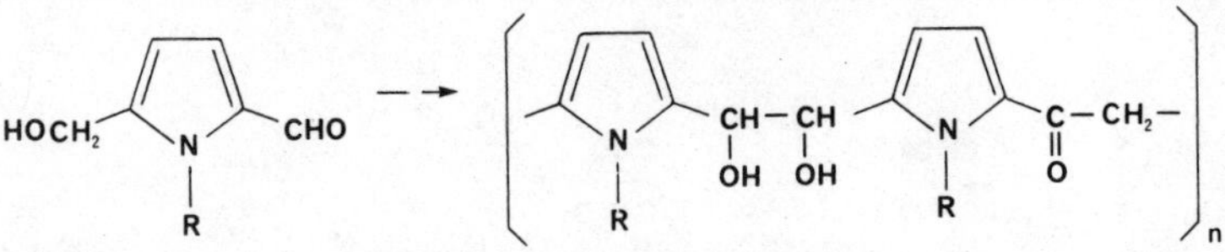

Strecker Degradation. The dicarbonyl compounds from the Maillard reaction react with the α-amino group of an amino acid to form Schiff base.
The enol form is an α-amino acid that decarboxylates readily to yield the
enaminol. The enaminol undergoes self-condensation to brown polymer or
hydrolysis to the amine and the aldehyde, with the latter corresponding to
the original amino acid with one less carbon (Eq. 3-21) (8). The aldehydes
derived from Streker degradation constitute many of the important flavor
compounds in food systems.

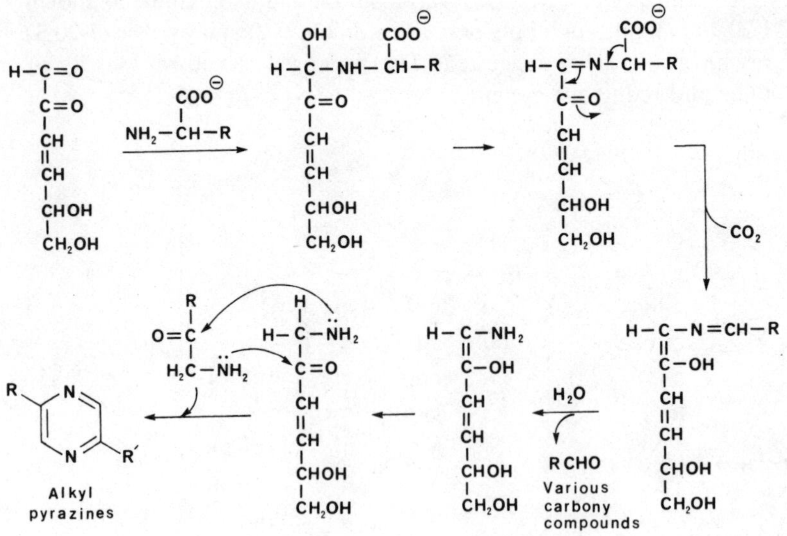

Eq. 3-21

COMPLEXES OF SUGARS WITH METAL IONS

In Neutral Solution

Sugar molecules containing axial-equatorial-axial hydroxyl groups in a six-member ring or a *cis-cis* sequence in a five-member ring form complexes with metal cations (Fig. 3.1). Only cations with an ionic radius >0.8 Å complex readily. Alkaline earth metals such as Ca^{2+} and Ba^{2+} usually form strong complexes. Univalent cations complex only slightly.

Cis-inositol, with three axial-equatorial-axial sequences, possesses four sites for complexing, and subsequently the inositol-metal complex is very stable (*15*).

Reducing sugars, as hemiacetals, are readily hydrolyzed by water and the α- and β-anomers are interconverted via mutarotation to an equilibrium mixture. If either anomer is combined with metal cation to form stable complexes, the equilibrium is shifted in favor of that anomeric form. For example, the α-anomer of allopyranose contains the axial-equatorial-axial hydroxy groups at C1, C2, and C3. Addition of $CaCl_2$ increases the percent proportion of the α-anomer as indicated in Eq. 3-22.

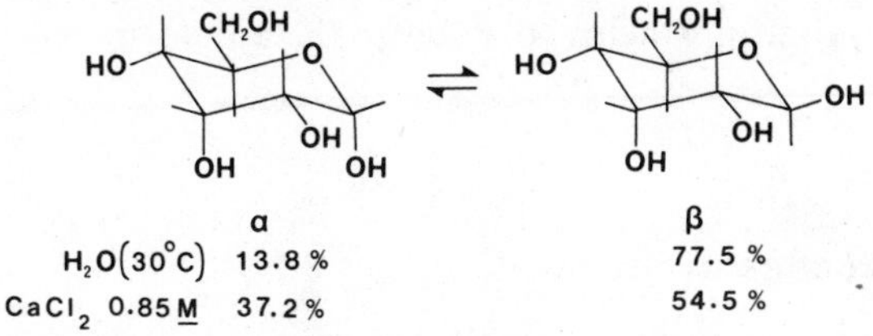

	α	β
$H_2O(30°C)$	13.8 %	77.5 %
$CaCl_2$ 0.85 M	37.2 %	54.5 %

Eq. 3-22

Similar reasoning can be applied to the equilibrium shift in favor of the conformation that complexes with metal ions. In the equilibrium mixture of α-D-ribopyranoside, only the 1C conformation has the axial-equatorial-axial sequence to complex metal ions (Eq. 3-23). Addition of $CaCl_2$ shifts the equilibrium in favor of the 1C form.

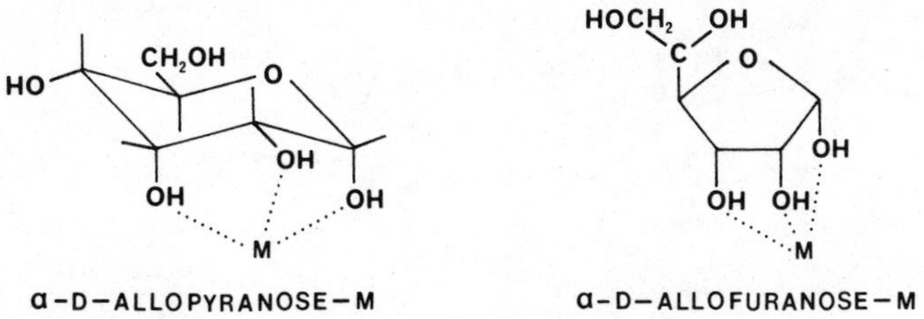

α−D−ALLOPYRANOSE−M   α−D−ALLOFURANOSE−M

Fig. 3.1. Complexation of allopyranose and allofuranose with metal.

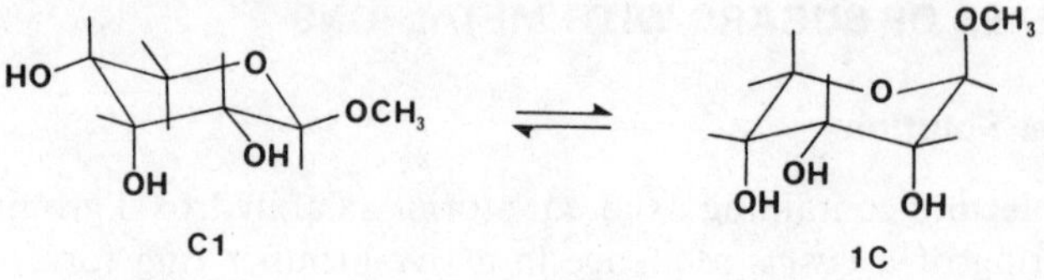

$$\text{Eq. 3-23}$$

C1   1C

In Alkaline Solution

In alkaline pH, the sugar molecule readily loses protons to form alcoholates with metal hydroxides and oxides (Fig. 3.2A) (*16*). Both alkaline metal and alkali-earth metal alcoholates can be prepared.

Polyhydroxy compounds are comparatively more acidic than monohydric alcohols. Dissociation constants for carbohydrates usually are in the range of 10^{-12}–10^{-14}, while ethanol $\sim 10^{-16}$. The higher acidity of carbohydrates is due to the inductive effect of neighboring substituents and intramolecular hydrogen bonding.

Electron-withdrawing substituents usually tend to cause the neighboring hydroxyl group to be more acidic. For example, in a methyl-α-pyranoside, the hydroxyl group at C2 is more acidic. Hydrogen bonding between neighboring hydroxyl groups helps to stabilize the anion oxygen (Fig. 3.2B).

STARCH

Chemical Structure of Starch

Starch contains two polysaccharide fractions: amylose and amylopectin. Amylose is a linear chain consisting of up to 4000 glucosyl residues connected by α-1,4 glucosidic linkages. Amylopectin is a branched polymer of repeating glucose units connected by α-1,4 linkages and branched with α-1,6 linkages (Fig. 3.3).

Amylose exists in polymorphic forms, A (in cereal starch) and B (in tuber starch). The V form occurs as collapsed or compact helices when starch

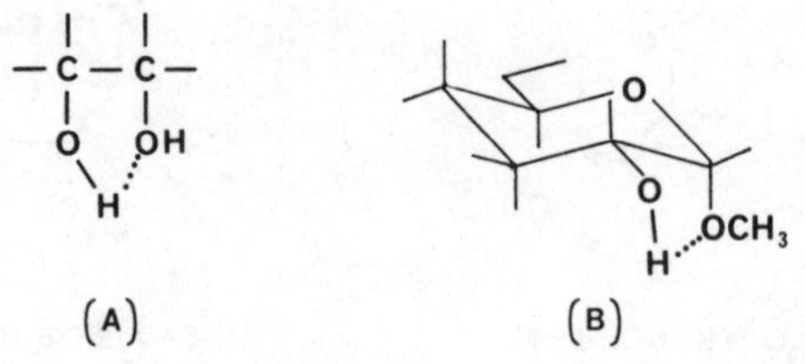

(A)   (B)

Fig. 3.2. Alcoholate (A) general structure, and (B) methyl-α-pyranoside.

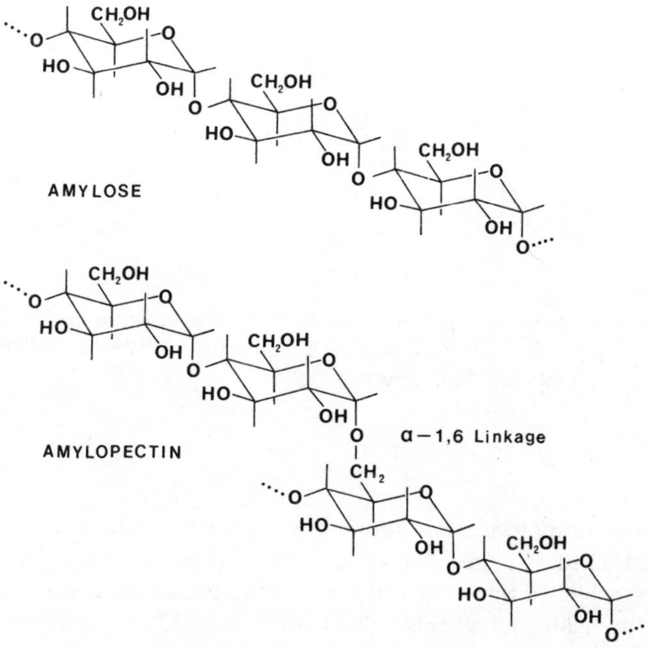

CH₂OH

HO

OH

CH₂OH

HO

OH

CH₂OH

HO

OH

AMYLOSE

CH₂OH

HO

OH

CH₂OH

HO

OH

α−1,6 Linkage

AMYLOPECTIN

CH₂

HO

OH

CH₂OH

HO

OH

Fig. 3.3. Structure of amylose and amylopectin.

complexes with small molecules such as iodine, alcohols, or fatty acids, and single helices with six, seven, or eight glucose units per turn (8 Å) have been suggested. Both the A and B forms assume a double helix with six glucose residues per turn (Fig. 3.4) (*17*). Recently, an antiparallel left-handed double helix structure has been suggested.

In aqueous solution, the configuration of amylose largely exists as a random coil, with a very small amount of left-handed psuedohelical backbone. In the presence of complexing agents, the amylose assumes a helical structure.

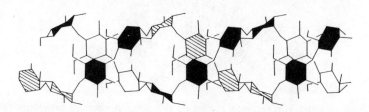

Fig. 3.4. Native starch double helix. (Reprinted with permission from French and Murphy (*17*), courtesy of American Association of Cereal Chemists.)

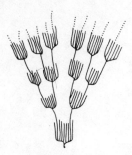

Fig. 3.5A. Tassel-on-a-string arrangement of amylopectin molecules. (Reprinted with permission from Osaka (*18*), copyright 1978, by VCH Verlagsgesellschaft, Weinheim, Germany.)

Amylopectin molecule has commonly been suggested to assume a randomly branched structure. However, a tassel-on-a-string arrangement is equally possible. In this model, the branching points are collected together towards the reducing end (Fig. 3.5A) (*18*).

Gelatinization

In plants, starch exists in granules (2–100 μ), composed of radially arranged starch molecules of linear and branched chains (Fig. 3.5B). The amylopectin associates with straight-chain amylose to form regions of crystalline micelles. As a result of this association through hydrogen bonding, the starch

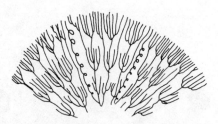

Fig. 3.5B. Schematic representation of the arrangement of amylopectin and amylose in a starch granule. (Reprinted with permission from Osaka (*18*), copyright 1978, VCH Verlagsgesellschaft, Weinheim, Germany.)

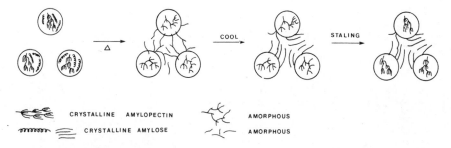

CRYSTALLINE AMYLOPECTIN   AMORPHOUS

CRYSTALLINE AMYLOSE   AMORPHOUS

Fig. 3.6. Changes in a starch granule during heating and cooling.

granule is insoluble in cold water. When heating provides sufficient energy to disrupt the weak bonding between crystalline micelles, the granule starts to hydrate and swell. The starch (largely the amylose) is solubilized into solution, forming an intergranular matrix which causes an increase in viscosity (Fig. 3.6). The granules lose their birefringence. These changes in gelatinization take place between 52 and 60°C for wheat starch. A continuous increase in temperature eventually causes the granule to collapse and rupture. The degree of hydration depends on temperature, pH, shear, and concentration (*19*).

Retrogradation

Upon cooling in a dilute starch solution, the linear molecules realign themselves by hydrogen bonding into an insoluble precipitate. In concentrated starch solution (5–10%), realignment is rapid and disordered and association of the molecules occurs at limited locations, with water entrapped in the interstices. This association of the linear molecules, and the resulting decrease in solubility, is commonly referred to as retrogradation. The amylopectin, being branched, associates relatively slowly after cooling, forming tighter amylopectin micelles. Such a crystallization process continues slowly and is chiefly responsible for the staling of bread.

Chemical Modification

Chemically, the gelling characteristics of starch can be changed by modifying the functional groups in the starch molecule (*20*).

1. *Cross-linking:* It involves the reaction between the hydroxyl groups with di- or poly-functional reagents. Phosphorus oxychloride (Eq. 3-24.1), adipic acid anhydride (Eq. 3-24.2), and epichlorohydrin (Eq. 3-24.3) are commonly used to form ester linkages with the hydroxyl group.

3-24.1

$$POCl + 3 \boxed{ST}\!-\!OH \longrightarrow \boxed{ST}\!-\!O\!-\!\overset{\overset{\displaystyle O}{\|}}{P}\!-\!O\!-\!\boxed{ST} + 3HCl$$

PHOSPHORUS
OXYCHLORIDE

$$O\!-\!\boxed{ST}$$

3-24.2

$$CH_3\!-\!\overset{\overset{\displaystyle O}{\|}}{C}\!-\!O\!-\!\overset{\overset{\displaystyle O}{\|}}{C}\!-\!\left(CH_2\right)_4\!-\!\overset{\overset{\displaystyle O}{\|}}{C}\!-\!O\!-\!\overset{\overset{\displaystyle O}{\|}}{C}\!-\!CH_3 + \boxed{ST}\!-\!OH \longrightarrow \boxed{ST}\!-\!O\!-\!\overset{\overset{\displaystyle O}{\|}}{C}\!-\!\left(CH_2\right)_4\!-\!\overset{\overset{\displaystyle O}{\|}}{C}\!-\!O\!-\!\boxed{ST} + CH_3\!-\!\overset{\overset{\displaystyle O}{\|}}{C}\!-\!OH$$

ADIPIC ACID ANHYDRIDE

3-24.3

$$\overset{O}{\overset{/\backslash}{CH_2\!-\!CH\!-\!CH_2Cl}} + \boxed{ST}\!-\!OH \longrightarrow \boxed{ST}\!-\!O\!-\!CH_2\!-\!\overset{\overset{\displaystyle OH}{|}}{CH}\!-\!CH_2\!-\!O\!-\!\boxed{ST}$$

EPICHLOROHYDRIN

Eq. 3-24

The cross-linked starch exhibits more resistance to acid hydrolysis, increased shear resistance, greater heat tolerance, and higher temperature requirement for hydration of the granules. The modified starch is used extensively in acid food products.

2. *Substitution with stabilizing functional groups:* Substituted starch is utilized in frozen or refrigerated food products. The reaction involves the addition of monofunctional groups, such as hydroxylpropyl (Eq. 3-25.1), phosphate (Eq. 3-25.2), and acetyl (Eqs. 3-25.3, 3-25.4). Attachment of these functional groups interferes with the association of the amylose molecules by hydrogen bonding and allows the starch to remain hydrated, clear, and stable. The modified starch has reduced temperature required for hydration, increased viscosity in solution, reduced syneresis, and increased stability to freezing and freeze-thaw.

$$CH_3\!-\!\overset{O}{\overset{/\backslash}{CH\!-\!CH_2}} + \boxed{ST}\!-\!OH \longrightarrow \boxed{ST}\!-\!O\!-\!CH_2CH_2CH_2OH$$

3-25.1

PROPYLENE OXIDE

$$\overset{\overset{\displaystyle OH}{|}}{\underset{\underset{\displaystyle OH}{|}}{HO\!-\!P\!=\!O}} + \boxed{ST}\!-\!OH \overset{H_2O}{\longrightarrow} \boxed{ST}\!-\!O\!-\!\overset{\overset{\displaystyle OH}{|}}{\underset{\underset{\displaystyle OH}{|}}{P\!=\!O}} \longrightarrow \boxed{ST}\!-\!O\!-\!\overset{\overset{\displaystyle O-\boxed{ST}}{|}}{\underset{\underset{\displaystyle O-\boxed{ST}}{|}}{P\!=\!O}}$$

3-25.2

MONOPHOSPHORIC
ACID

TRIALKYL PHOSPHATE
ESTER

$$CH_3\!-\!\overset{\overset{\displaystyle O}{\|}}{C}\!-\!O\!-\!\overset{\overset{\displaystyle O}{\|}}{C}\!-\!CH_3 + \boxed{ST}\!-\!OH \longrightarrow CH_3\!-\!\overset{\overset{\displaystyle O}{\|}}{C}\!-\!O\!-\!\boxed{ST} + CH_3COOH$$

3-25.3

ACETIC ANHYDRIDE

Eq. 3-25

$$CH_3-\overset{\overset{O}{\|}}{C}-O-CH=CH_2 \; + \; \boxed{ST}-OH \; \longrightarrow \; CH_3-\overset{\overset{O}{\|}}{C}-O-\boxed{ST} \; + \; CH_3CHO$$

3 – 25.4

Eq. 3-25
(cont.)

3. *Cleavage:* Acid hydrolysis can be used to produce limited, random cleaving of the starch molecules to obtain a modified starch with less viscosity, low-temperature hydration, low resistance to heat, and greater retention of gel structure. Oxidative cleavage with sodium hypochlorite produces modified starch with excellent flow properties that will not cake during processing.

ALGINATE

Chemical Structure of Alginate

Alginates are extracted from a number of brown seaweeds (Class *Phaeophyceae,* species *Fucus serratus, Ascophyllum nodosum, Laminaria digitata, Laminaria cloustonii, Ecklonia maxima,* and *Macrocystis pyrifera*).

Alginic acid is a linear polymer of 1,4 linked β-D-mannuronic acid and α-L-guluronic acid (Fig. 3.7). Commercial alginates have a degree of polymerization in the range of 100–1000.

Alginic acid is insoluble in water. Salts (alginates) of alkali metals (e.g., sodium, potassium) are soluble and salts of di- and trivalent metals (e.g., calcium) are insoluble. In practice, the seaweed is acidified to remove undesirable impurities and the aliginic acid, then neutralized to a soluble salt for extraction (*21*).

The D-mannuronate residues and the L-guluronate residues are arranged in the following sequences:

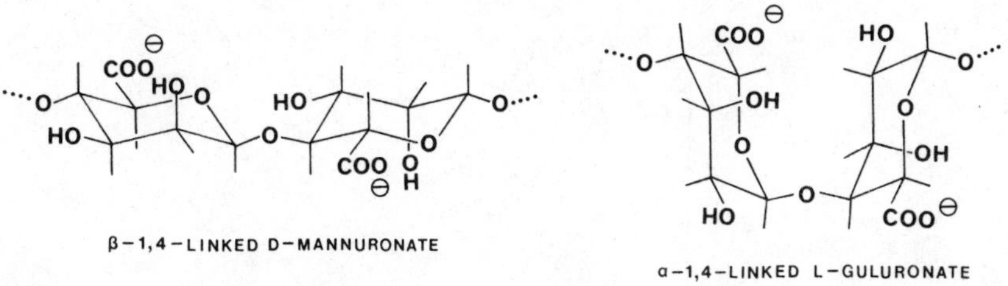

β–1,4–LINKED D–MANNURONATE

α–1,4–LINKED L–GULURONATE

Fig. 3.7. Structure of alginic acid of (A) D-mannuronate, and (B) L-guluronate residues.

1. Mannuronic acid blocks: -M-M-M-M-M-
2. Guluronic acid blocks: -G-G-G-G-G-
3. Alternating blocks: -M-G-M-G-M-

Each block type contains about 20 residues. The homogeneous block types (1 and 2) are comparatively more resistent to hydrolysis than the alternating block.

Gelling

Alginate solution gels quickly with calcium ions. In order to produce a uniform gel structure, soluble calcium salts are not used. Instead, slightly soluble calcium salts such as calcium citrate, calcium tartrate, or dicalcium phosphate dihydrate is used. Calcium salts that are insoluble in neutral solution (e.g., calcium carbonate, dicalcium phosphate anhydrous), but become soluble with increasing acidity, are also used. In this case, the pH of the system can be controlled by using glucono-δ-lactone, which hydrolyzes slowly in solution to gluconic acid. Gels can also be formed by controlled diffusion of soluble calcium salts through the alginate solution. This last method is used primarily in the production of fabricated foods.

At low concentrations of calcium (less than 35% of the sodium replaced in a 1% sodium alginate solution), the alginate solution is thixotropic, breaking down to a liquid on agitation and reverting back to a thick solution or weak gel on standing. The apparent viscosity decreases with increasing shear rate, and also is related to the duration of shear. With the same shear rate, the apparent viscosity decreases with time. The resulting plot of apparent viscosity against shear rate shows a hysteresis loop—a deformation where the thinning and thickening do not coincide (Fig. 3.8).

Aliginate gels are not thermally reversible. The mechanism of gelatin has

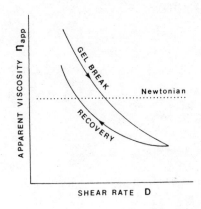

Fig. 3.8. Viscosity vs. shear rate for thixotropic flow.

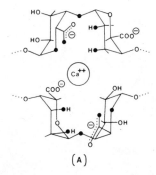

Fig. 3.9A. Interaction of polyguluronate with Ca^{2+} to form an "egg box."

been shown to be the interaction between calcium ions and polyguluronate blocks. Polymannuronic acid exists as a flat, ribbonlike chain with the sugar units in the C1 chair form equatorially 1,4-linked. However, polyguluronic acid has sugar units in the 1C chair form, diaxially 1,4-linked to form chains of a rodlike conformation favorable for the binding of calcium ions. The polyguluronate is shown to form chains of twofold symmetry, which allows four-oxygen coordination involving C6-O, C5-O, C2-O and C3-O (Fig. 3.9A). The association between polyguluronic acid and calcium ions provides junctions which cross-link the alignate polymers into a three-dimensional network. This arrangement has been termed the "egg box" model (Fig. 3.9B) (22).

Application

Alginate has been found to be most useful as a gelling agent in the production of fabricated foods, such as artificial cherries, gelled confectionary powder with dairy desserts, and fabricated onion rings. It is also used as a thickener for sauces and condiments, sausages, and bakery fillings; as a stabilizer for ice cream to prevent ice crystal formation; and for whipped

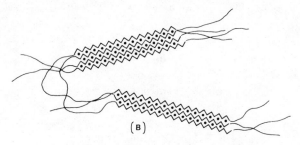

Fig. 3.9B. The "egg box" model.

cream and toppings, yogurt, and milk drinks. The concentration of alignate used in food gel ranges from 0.5 to 1.5%.

At low pH, alginate is converted to alginic acid, which becomes increasingly insoluble and finally precipitates. For the alginate to be used as a stabilizer in acidic foods, the carboxylic groups are esterified partially with propylene oxide to form propylene/glycol alignate ester (Eq. 3-26).

$$CH_3-CH-CH_2 + RCOO^{\ominus} \xrightarrow{H^{\oplus}} CH_3-\overset{OH}{\underset{|}{CH}}-CH_2-O-\overset{O}{\overset{\parallel}{C}}-R \qquad \text{Eq. 3-26}$$

PECTIN

Chemical Structure of Pectin

Commercial pectin is obtained from apple pomace and citrus fruit pulp. The native pectin (protopectin) in the pulp is made soluble during extraction with hot, acidic solution. Pectin is a polysaccharide with 150–500 galacturonic acid units (molecular weight 30,000–100,000), partially esterified with a methoxy group (Fig. 3.10). The backbone chain also contains residues of L-rhamnose and is branched with side chains composed mainly of β-D-galactopyranose and α-L-arabinofuranose.

The percentage of galacturonic acid units esterified is called the degree of esterification (DE). Amidation occurs when ammonia is used to deesterify pectin having a high percentage of methoxy groups (Eq. 3-27).

$$R-\overset{O}{\overset{\parallel}{C}}-OCH_3 + NH_3 \longrightarrow R-\overset{O}{\overset{\parallel}{C}}-NH_2 + CH_3OH \qquad \text{Eq. 3-27}$$

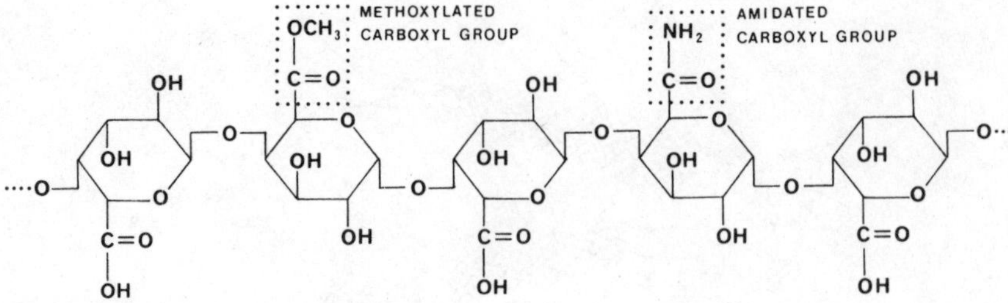

Fig. 3.10. Structure of pectin.

Classification of Pectin

Pectin can be classified according to the degree of esterification into:

1. *High-methoxy pectin (HM):* High-methoxy pectin has over 50% DE and gels in a medium with soluble solid content (usually sugar) greater than 55%, at a pH range of 2.0–3.5.

2. *Low-methoxy pectin (LM):* LM pectins have a DE lower than 50%. Gelation is controlled by introducing calcium ions and occurs in a medium with 10–20% soluble solids at pH between 2.5 and 6.5. These pectins make suitable gels when present at 0.5–1.5%.

Gelling

The mechanism of gelling for HM and LM pectins involves stacking of the polysaccharide chains to form junction zones similar to that in the alginate gel. The HM pectin gel is stabilized by hydrophobic binding of methyl ester groups as well as by intermolecular hydrogen bonding. The free energy for the formation of a junction zone for HM pectin is represented by Eq. 3-28 *(23)*.

$$\Delta G^{\circ}_{\substack{\text{JUNCTION}\\\text{ZONE}}} = \Delta G^{\circ}_{\substack{\text{HYDROPHOBIC}\\\text{INTERACTION}}} + \Delta G^{\circ}_{\substack{\text{HYDROGEN}\\\text{BONDING}}} - T\Delta S_{\substack{\text{CONFIGURATIONAL}\\\text{ENTROPY}}} \qquad \text{Eq. 3-28}$$

For DE 70% HM   −18.6   −37.5   +41.1 KJ/mol

The low pH used in HM gelling causes protonation of the carboxylate groups, a decrease in electrostatic repulsion between pectic chains, and an increase in intermolecular hydrogen bonding.

However, hydrogen bonding alone is insufficient to overcome the energy of entropy loss (loss of freedom of motion of the polymer during formation of the junction zone). Addition of a cosolute, such as sucrose, lowers the water activity; and the water is less "free" to solvate the polysaccharide molecule. Added sucrose, therefore, increases hydrophobic interactions between the methyl ester groups. Without sucrose, the contribution of hydrophobic interaction is too small for forming a stable junction zone.

In LM gel, the junction zone is stabilized by interchain bridging by Ca^{2+} involving five polyanion oxygens: C5-O, C6-O, and C2-O from one chain with C5-O and C6-O from the same galacturonic acid residue, and C2-O and C6-O from the adjacent chain (Fig. 3.11) *(24)*.

Rhamnose in the backbone and side chains of neutral sugars also interferes with the stacking. At sites with rhamnose, the backbone is distorted. At alkali pH, pectin is unstable and starts to depolymerize (Eq. 3-29).

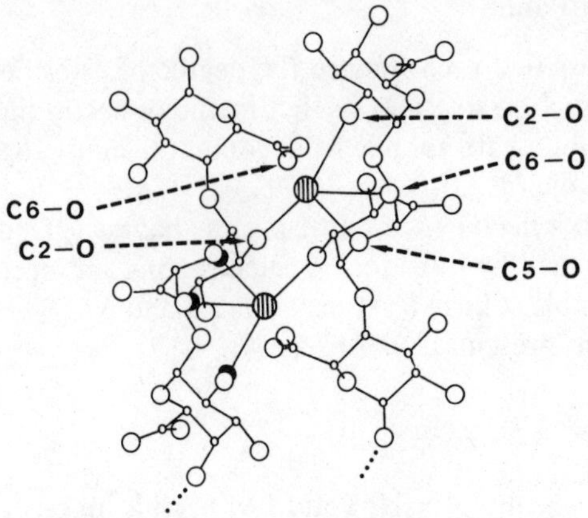

Fig. 3.11. The calcium pectate unit cell. (Reprinted with permission from Walkinshaw and Arnott (24), courtesy of Academic Press, Inc.)

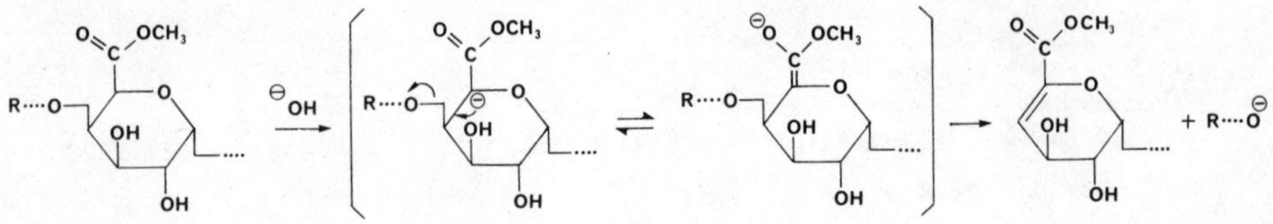

Eq. 3-29

CARRAGEENAN

Chemical Structure of Carrageenan

Carrageenan is isolated from a number of closely related species of red seaweed (Class *Rhodophyceae; Chondrus, Eucheuma,* and *Gigartina* species) by hot alkali extraction.

Carrageenan is a polysaccharide chain consisting of galactose units and 3,6-anhydro galactose, both sulfated and nonsulfated, joined by alternating α-1,3, β-1,4 glycosidic linkages.

The number of sulfate groups as well as their location in the polysaccharide chain contribute to the gelling characteristics of carrageenan. Accordingly, there are three types of carrageenan, κ, ι, and λ (Fig. 3.12). Some of their properties are listed in Table 3.2 (25).

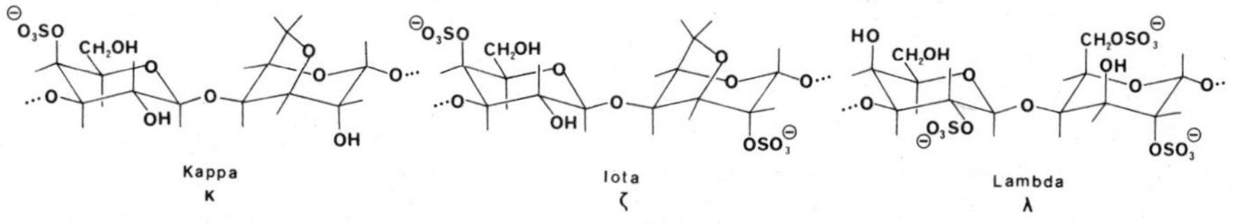

Fig. 3.12. Structure of kappa, iota, and lambda carrageenan.

Gelling

Carrageenan (κ and ι) forms thermally reversible gels with frameworks that are cross-linked by double helices. In solution, the polysaccharide chains are dispersed randomly. On cooling, gelation occurs when sufficient helices provide cross-linking junctions to build a continuous network (Fig. 3.13A) (26).

Table 3.2. Structures and Functional Properties of Carrageenan

KAPPA (κ)	IOTA (ι)	LAMBDA (λ)
REPEATING UNIT		
galactose-4-sulfate-3, 6-anhydro-galactose	galactose-4-sulfate-3, 6-anhydro-galactose-2-sulfate	galactose-2-sulfate-galactose-2,6-disulfate
DEGREE OF ESTERIFICATION		
25% ester sulfate	32% ester sulfate	35% ester sulfate
34% 33,6-AG	30% 3,6-AG	small amount
ION FOR GELLING		
K$^+$	Ca^{2+}	no gel
TEXTURE		
strong, rigid, opaque	elastic, clear	no gel
SYNERESIS		
+	−	−
FREEZE/THAW STABILITY		
−	+	+
ACID STABILITY		
+	+	+
SHEAR STABILITY		
−	+	−

Reprinted with permission from *Marine Colloids, Introductory Bulletin A-1 (25)*, courtesy of FMC Corporation.

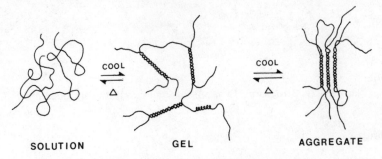

Fig. 3.13A. Gelling mechanism of carrageenan.

As more helices continues to form, they start to associate into aggregates and the gel turns opaque. Finally, enough aggregation will cause contraction of the network with subsequent exclusion of the liquid from the interstices, and the gel becomes brittle.

The function of the sulfate anion is to keep the carrageenan in solution. Transformation of the polysaccharide chains from random coil to helice is favored with decreasing sulfate content. Therefore, the least sulfated κ-carrageenan forms gel that is opaque. With higher percentages of ester sulfate, the ι-carrageenan tends to stay in solution as a random coil, and the gel is clear and elastic and does not aggregate.

The helices are held by specific hydrogen bonding between the O-2 and O-6 of the galactose residues in different strands of the same double helix, and every unsubstituted hydroxyl group is then hydrogen bonded within the

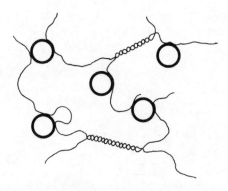

Fig. 3.13B. Interaction between κ-carrageenan and κ-casein in gelling.

double helix, making the conformation very stable. Sulfate residues located on the outside reinforce the backbone of the helices. Association of the chain, and stacking of helices to form junctions, is reinforced by the electrostatic attraction between cations (K^+, Ca^{2+}) and the sulfate anions. The gelling of carrageenan can be regulated by adjusting the cation concentration. The addition of cations increases the gelling temperature.

Interaction with Protein

Carrageenan has the ability to form stable complexes with milk protein through the electrostatic interaction between the sulfate anions with the many positive charges on the surface of casein micelles (27).

The interaction between κ-carrageenan and κ-casein has been extensively studied. At neutral pH, κ-casein has an extensive region of positively charged amino acid residue (between residues 20 and 115). Since κ-casein is the only milk protein to interact with carrageenan, it is suggested that this region is responsible for the electrostatic interaction. The interaction of a carrageenan molecule with a micelle leaves the unadsorbed segments of the polysaccharide chain free in solution, in the form of loops and tails. The free segments then associate to form the gel network (Fig. 3.13B).

The milk protein–carrageenan interaction serves to reinforce the double helices that normally occur in aqueous systems, increasing the gel strength tenfold. Carrageenan has found wide use in dairy products, whipped cream, pie filling with milk, imitation milk, coffee creamer, etc. The concentration is usually in the range of 0.1–0.5%.

Dilute concentration (0.01–0.04%) of κ-carrageenan and milk protein form a weak thixotropic gel. This special property is utilized to suspend cocoa in making chocolate milk.

Synergism with Locust Bean Gum

Locust bean gum and κ-carrageenan interact to increase gel strength, making the gel more elastic and resistant to syneresis. Locust bean gum consists of a backbone chain of mannose residue with side chains composed of essentially galactose units. The unbranched region of the backbone can assume a ribbonlike conformation stabilized by the carrageenan helices (Fig. 3.14). The branched regions (with galactose side chains) cannot bind, but have the following important functions: (1) They serve to connecting the cross-linked junctions in the framework; (2) because they are randomly dispersed, they confer elasticity to the gel; and (3) because they cannot associate, they resist aggregation and hence syneresis (28).

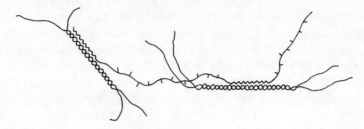

Fig. 3.14. Synergistic interaction between κ-carrageenan and locust bean gum.

CELLULOSE

Chemical Structure of Cellulose

Cellulose is the structural material of plant cell walls. Chemically, it is a polysaccharide chain of a repeating cellobiose unit (which consists of two glucose units linked by β-1,4 linkage) (Fig. 3.15A).

The hydroxyl groups can be substituted with methyl, hydroxylpropyl-methyl, or carboxymethyl groups (Fig. 3.15B). These cellulose ether derivatives are "cellulose gums."

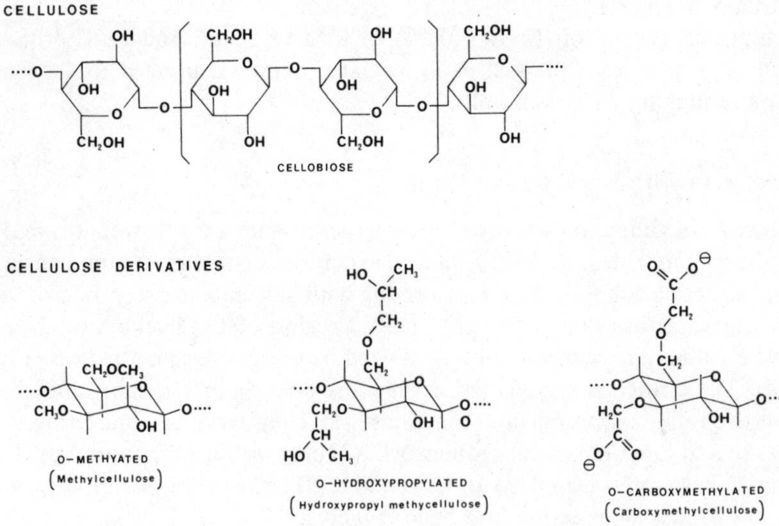

CELLULOSE

OH	CH₂OH	OH	CH₂OH

CELLOBIOSE

CELLULOSE DERIVATIVES

O—METHYATED
(Methylcellulose)

O—HYDROXYPROPYLATED
(Hydroxypropyl methycellulose)

O—CARBOXYMETHYLATED
(Carboxymethylcellulose)

Fig. 3.15. Structure of (A) cellulose, and (B) cellulose derivatives.

Gelling

The gelling characteristics depend largely on: (1) degree of polymerization (DP)—usually in the range > 100,000, and (2) degree of substitution (DS). The maximum theoretical degree of substitution is 3.0 when all the hydroxyls are derivatized. Usually the ether substituents are unevenly distributed, with some segments of the polysaccharide chain more densely substituted.

As the DP increases, the viscosity of the solution increases. The degree of substitution may increase or decrease the viscosity depending on the nature of the substituents.

Methylcellulose (*29*). O-methylated cellulose is stable over a wide pH range of 3.0—11.0. It exhibits an unusual property of thermal gelation—it gels when heated and melts upon cooling, in contrast to the common gelling characteristics of pectin, carrageenan, and alginate.

The gelling temperature ranges from 50 to 70°C for most methylcellulose products. The phenomenon of thermal gelation can be explained in terms of the structural effect of the polysaccharide molecules in water. In solution, the polysaccharide molecules are hydrated, but the molecules contain some segments that are more densely substituted than the others and are less water soluble. An increase in temperature weakens the many hydrogen bonds maintained by water molecules and helps facilitate phase separation of the less polar segments. The relatively nonpolar densely substituted segments are excluded from the liquid phase to form clusters analogous to lipid micelles, while the less densely substituted segments remain in solution, cross-linking the clusters of less polar segments into an extended network.

Formation of clusters of less polar segments depends on the degree of hydrophobicity along the chain, and not the cooperative binding of chains through electrostatic interactions among substituted groups or with cations as described in the cross-linking in other gel types.

Hydropropylmethyl Cellulose (*30*). O-hydroxypropylated cellulose gels when heated and precipitates at high temperatures. During the substitution reaction, the hydroxypropyl groups are not only attached to the hydroxyls of the glucose residues, but some may condense with each other to form propylene glycol side chains.

These hydrophobic side chains are capable of forming clusters in a way similar to the O-methylated segments, reinforcing cross-linking, and end up with a tighter network.

Carboxymethyl Cellulose (*31*). Carboxymethylated cellulose has properties quite different from the two derivatives described. Here, the substitution group contains a hydrophilic carboxyl group which tends to make the polysaccharide more soluble in water.

In general, all carboxymethyl cellulose solutions (and in fact, many polysaccharide solutions) are pseudoplastic, in which the apparent viscosity decreases with increasing shearing rate and is independent of duration of shear. When the shear stress is removed, it instantly reverts to its original viscous state. Pseudoplasticity is usually referred to as time-independent steady-state flow (Fig. 3.16). However, solutions of carboxymethyl cellulose types of high DP and low DS show thixotropic behavior.

The viscosity of cellulose gum solution decreases with high temperature and increases with acidic pH. High temperature, especially long periods of heating, degrades the cellulose. Under acidic pH, the less soluble carboxylic acid predominates and viscosity increases. Carboxymethyl cellulose solutions exhibit maximum stability at a pH of 7–9.

The polysaccharide chain is dispersed in solution in a loose network. Maximum viscosity is obtained when the polysaccharide chains dissolve in water while partially retaining the loose network. Increasing the degree of substitution makes the cellulose more hydrophilic, and therefore more dissolving, and it finally reaches a point when the polysaccharide network is completely disaggregated and the viscosity drops.

Gels can be produced by the addition of trivalent metal salts, such as aluminum acetate, to the cellulose gum solution. The polysaccharide molecules are cross-linked by the metal ions between the carboxymethyl groups. The gel texture depends on the concentration of the gum, the ratio of metal ions to carboxylate anions, pH, and degree of polymerization. High concentration, low DP, and high DS yield more elastic gels. High concentration

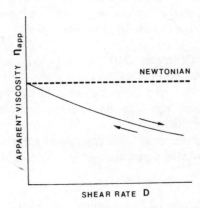

Fig. 3.16. Viscosity vs. shear rate for pseudoplastic flow.

of metal ions gives brittle gels. High temperature and low pH tend to increase the solubility of aluminum acetate and speed up its release into the solution, and hence the rate of gel formation.

XANTHAN GUM

Chemical Structure of Xanthan Gum

Xanthan gum is commonly produced by aerobic fermentation of *Xanthomonas campestris* culture in a medium of carbohydrates and nutrients. The gum is precipitated with isopropanol and then dried and milled.

The polymer backbone consists of 1,4-linked and β-D-glucose, which is similar to cellulose in this respect. A trisaccharide side chain of one glucuronic acid and two mannose units branches from the 3-position of alternate glucose unit. About half of the terminal D-mannose residues contain a pyruvic acid residue linked via the keto group to the 4- and 6-position (Fig. 3.17).

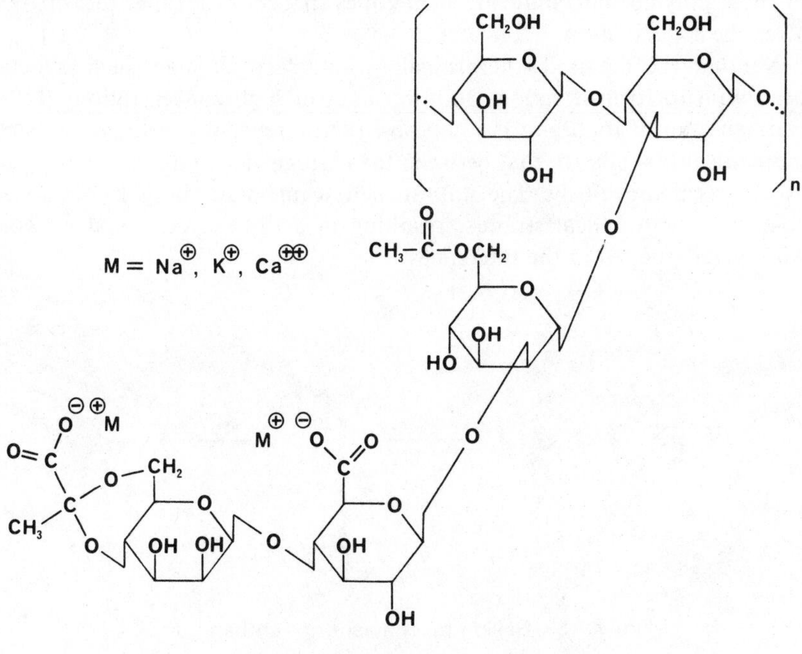

Fig. 3.17. Structure of xanthan gum.

Gelling

In solution, the trisaccharide side chains align with the backbone, forming a stiff polymer that is stable to a temperature in excess of 100°C before transforming into the coil form. The stiff chains can also exist as a double helix or multiple-stranded assembly through intramolecular association forming a network of entangled, stiff, rod-shaped polymers (Fig. 3.18) (*32, 33*).

Metal ions or a low concentration of electrolyte enhances the stability of the structure by reducing the electrostatic repulsion between the carboxyl groups. The stable helical conformation is resistant to temperature denaturation and accounts for the properties of (1) uniform viscosity over a temperature range of −18°–80°C, (2) exceptional stability to changes in pH from 1 to 11, (3) compatibility with high salt concentration. The trisaccharide side chains tend to protect the glucosidic linkage on the backbone against hydrolytic cleavage chemically or enzymatically, making the structure exceptionally stable.

Xanthan gum solution exhibits a high degree of pseudoplastic flow over a broad shear rate and concentration range. Shear thinning and recovery are instantaneous. This unique flow property is attributed to the stiffness of the native conformation of the xanthan polymer and to the aggregation of these polymer into multistranded zones that allow a rapid reformation after shear dissipation.

Xanthan is often used synergistically with galactomannans such as locust bean gum to form a viscous solution and, at high concentrations (0.5% xanthan gum of total gum), a cohesive thermoreversible gel. The possible mechanism is similar to that between the carrageenan and locust bean gum as discussed previously. The unbranched segment of the galactomannan complexes with helical strands, resulting in a three-dimensional network with water trapped in the interstices.

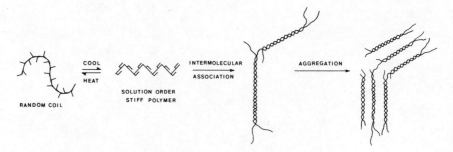

Fig. 3.18. Gelling mechanism of xanthan gum.

HEMICELLULOSE

Hemicellulose is found together with cellulose, lignin, and pectin in plant cell walls. The name hemicellulose is designated for those cell wall polysaccharides extractable by aqueous alkali after delignification, and its composition varies with the plant source.

Unlike cellulose, which has a degree of polymerization (DP) in the range of 6,000–10,000, hemicellulose is commonly composed of 50–100 units. Most hemicelluloses are heteropolysaccharides containing two to four different types of sugar residues, and can be classified based on the backbone chains into (1) xylans, (2) mannans and glucomannans, and (3) galactans and arabinogalactans (34). The partial structures of some common hemicelluloses are listed in Fig. 3.19.

The most interesting hemicelluloses are the arabinoxylans, which are also found in cereal flour. These xylans, obtained by cold-water extraction of flour and subsequent fractionations of the extract, are commonly included as water-soluble pentosans in the literature.

Pentosans

The water-soluble pentosans of wheat flour are composed of two fractions.

1. Linear arabinoxylan with single α-L-arabinofuranosyl side groups attached at the C2 or C3 position (Fig. 3.20A).
2. Arabinogalactans with branching at C3, and covalently linked to proteins or polypeptides consisting of a high amount (16–20%) of 4-hydroxyproline (Fig. 3.20B).

The α-L-arabinose side groups prevent the association and precipitation of the arabinoxylan polymer. Removal of the side chains causes precipitation of the pentosan.

One of the most unique properties of pentosans is their ability to form gels under oxidative conditions (e.g., H_2O_2-peroxidase present in flour). This ability of "oxidative gelation" is due to the small amount of ferulic acid present in the arabinoxylan (Fig. 3.21A). The ferulic acid residue self-cross-links and is in part responsible for the formation of the three-dimensional network (Fig. 3.21B) (35), or it can cross-link with tyrosine in proteins or polypeptides (Fig. 3.21C). The cross-linking in these cases involves the aromatic ring. Another possible reaction may involve 1,4-addition of the α,β-unsaturated bond of the ferulic acid as indicated in Fig. 3.21D (36).

These types of cross-linking are believed to be the causes of the increase

(1) XYLAN
(A) D-xylan
-(1,4)-β-D-Xylp-(1,4)-β-D-Xylp-(1,4)-
3
1
-(1,4)-β-D-Xylp-(1,4)-β-D-Xylp-(1,4)-
(B) L-arabino-D-xylan
D-xylan------β-D-Xylp----------
3
1
L-Araf
(C) L-arabino-D-glucurono-D-xylan
D-xylan-----β-D-Xylp------β-D-Xylp--------
3
2 or 3
1
1
L-Araf
4-Me-α-D-GpA
(2) MANNANS and GLUCOMANNANS
(A) D-mannan
β-D-Manp-(1,4)-(β-D-Manp)n-(1,4)-D-Manp
(B) D-gluco-D-mannan
-(1,4)-(β-D-Manp)4-(1,4)-(β-D-Gp)2-(1,4)-(β-D-Manp)4-
(3) GALACTANS and ARABINOGALACTANS
(A) D-galactan
-(1,3)-β-D-Galp-(1,3)----------------β-D-Galp-(1,3)-
6
6
1
1
β-D-Galp-(1,6)-β-D-Galp
β-D-Galp-(1,6)-β-D-Galp
(B) L-arabino-D-galactan

Fig. 3.19. Partial structures of common hemicellulose.

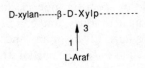

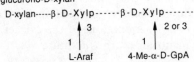

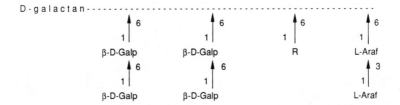

```
D-galactan- - - - - - - - - - - - - - - - - - - - - - - - - - - - - - - - - - - - - - - - - - - - - - - - -
                ▲6            ▲6           ▲6           ▲6
              1 │           1 │          1 │          1 │
              β-D-Galp      β-D-Galp        R          L-Araf
                ▲6            ▲6                        ▲3
              1 │           1 │                       1 │
              β-D-Galp      β-D-Galp                   L-Araf
```

Xylp = xylopyranosyl
Araf = arabinofuranosyl
GpA = Glucuronic acid
Manp = Mannopyranosyl
Gp = Glucopyranosyl
Galp = Galactopyranosyl
R = Galp or Araf

Fig. 3-19. (cont.)

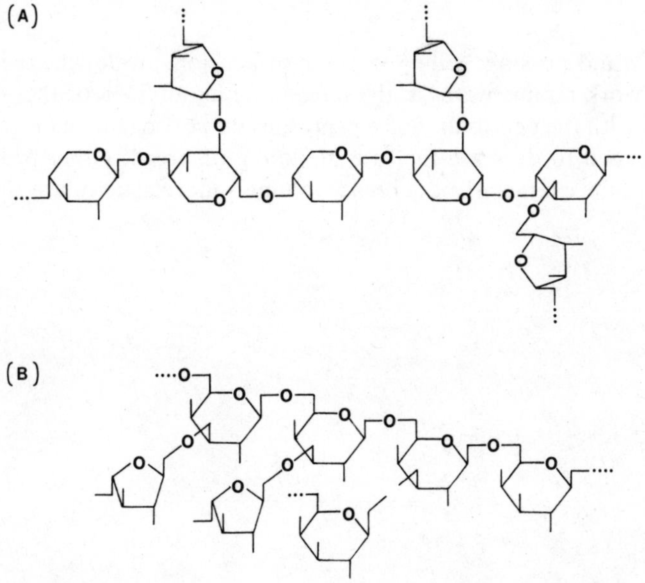

(A)

(B)

Fig. 3.20. Structure of (A) arabinoxylan, and (B) arabinogalactan.

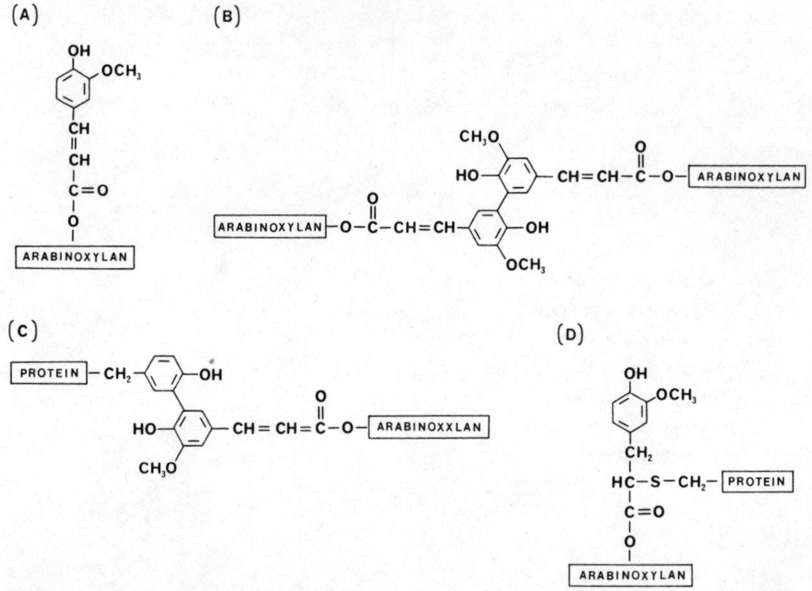

Fig. 3.21. Structure of (A) ferulic acid, (B) self-cross-linked, (C) cross-linked with tyrosine, and (D) cross-linked with thiol.

in viscosity and possibly gelling in flour suspension. In dough, the cross-linked network retains water in the interstices; about 23% of the water is associated with the pentosan. Since pentosans do not coagulate or retrograde, they tend to decrease the rate of staling of bread. Other properties such as crumb characteristics, bread volume, and elasticity are also improved.

4
COLORANTS

Color compounds constitute a unique class in that they are structurally diverse, and their chemical and physical properties are extremely complex. For many color compounds, their chromophoretic properties can only be adequately explained by referring to their molecular orbital structures.

Based on theoretical grounds, color compounds can be classified into two groups: One has chromophores with conjugated systems, and the other has metal-coordinated porphyrins. The former group includes carotenoids, anthocyanins, betanains, caramel, dyes, and lakes. The latter is comprised of myoglobins, chlorophylls, and their derivatives.

From the standpoint of application, an understanding of the basic chemistry of these compounds provide important information that enables development and formulation of color additives. It is also important for the proper processing, in retaining the natural color, or eliminating undesirable color changes in food products.

LIGHT ABSORPTION

The quantum light energy is related to the wavelength (λ) and frequency (ν) by Eq. 4-1, where h = Planck's constant and c = speed of light in a vacuum. A molecule absorbs a quantum of light if the energy of the quantum is equal to the energy of transition: E(light $= E$(excited) $-$ E(ground).

$$\varepsilon \;=\; hc\,/\,\lambda \;=\; h\nu \qquad\qquad \text{Eq. 4-1}$$

A number of molecular transitions can occur depending on the wavelength. The present discussion will be limited to the visible region where the molecule is electronically excited. The electrons may undergo several types

147

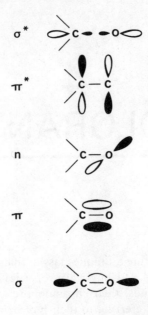

Fig. 4.1. Electronic energy transitions.

of transitions of different energies, schematically presented in Fig. 4.1. Not all of the above transitions are equally probable; some are forbidden transitions. Experimentally, transitions are measured by the molar extinction coefficient (Eq. 4-2).

$$\log\left(I_0 \,/\, I\right) = \varepsilon\, c\, l \qquad\qquad \text{Eq. 4-2}$$

where

I_o = intensity of incident light
I = intensity of light transmitted
c = molar concentration of solute
l = length of pathway through solution

The position and intensity of an absorption band is affected by substituent group and conjugation, as well as the conformation of the molecule. The effects may be hypsochromic, bathochromic, hypochromic, or hyperchromic (Fig. 4.2).

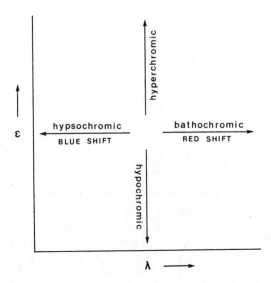

Fig. 4.2. Various effects on position and intensity of absorption.

CONJUGATION

The effect of conjugation on transition energy can be visualized from considering the π molecular orbitals of 1,3-butadiene (Fig. 4.3).

$$\text{Total } \pi\text{-electron energy} = 2(\alpha + 1.62\beta) + 2(\alpha + 0.62\beta)$$
$$= 4\alpha + 4.48\beta$$

If the orbitals are localized in the 1,3-butadiene, the total π electron energy will be equal to $4(\alpha + \beta)$. The energy of delocalization, therefore, is $4\alpha + 4.48\beta - 4\alpha - 4\beta$, which equals 9 kcal/mol ($\beta = 19$ kcal/mol) (Fig. 4.4).

The higher the degree of conjugation, the lower energy it will require for the π to π^* transition from the highest occupied molecular orbital (HOMO) to the lowest unoccupied molecular orbital (LUMO). Increased conjugation in the molecule shifts the absorption maximum to a higher wavelength (red shift).

A simple set of rules (Fieser-Kuhn rules) has been used for estimating the λ_{max} and hence the ϵ_{max} for long-chain polyenes like β-carotene.

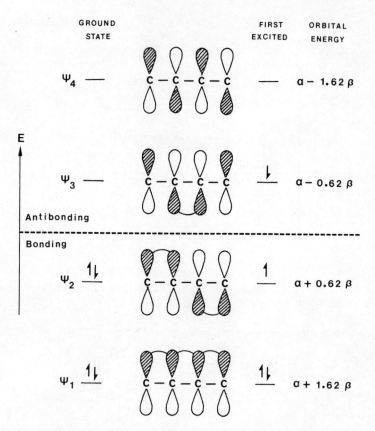

Fig. 4.3. Configuration of π-electrons and orbital energy in ground state and first excited state.

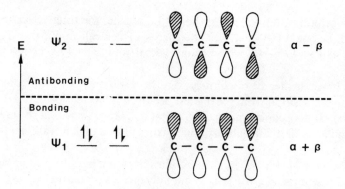

Fig. 4.4. Configuration of π-electrons and energy in localized orbitals.

λ_{max} (in hexane) $= 114 + 5M + n(48.0 - 1.7n) - 16.5R_{endo} - 10R_{exo}$
ϵ_{max} (in hexane) $= 1.74 \times 10^4 n$

where

M = number of alkyl substituents
n = number of conjugated double bonds
R_{endo} = number of rings with endocyclic double bonds
R_{exo} = number of rings with exocyclic double bonds

For β-carotene, $\lambda_{max} = 452$ nm in hexane, and the calculated $\lambda_{max} = 114 +$ $(5 \times 10) + 11(48.0 - 1.7 \times 11) - (16.5 \times 2) - (10 \times 0) = 453$ nm.

The same red shift effect of increasing conjugation is observed in poly-nuclear aromatic compounds. However, in these asymmetric compounds, the nodal planes intersect along either of the x or y axes, producing different energy level orbitals. Transitions along the x or y directions give separate distinct absorption bands (1). Take for example, the molecular orbital of anthracene: The π to π^* transition along the x and y directions, as indicated in Fig. 4.5, results in absorption bands at λ_{max} 250 and 370, respectively.

SUBSTITUENT EFFECTS

Substituent groups with lone pairs of electrons tend to increase π conjugation through resonance (Eq. 4-3). Similar resonance structures can be written for $-OR$ ($R = CH_3, C_2H_5$), $-X$ (Br, Cl), $-NH_2$ (or $-NRH$), etc. Electron-withdrawing groups, such as $-COR$, $-CN$, $-COOH$, $-CHO$, $-NO_2$, $-COOR$, $-SO_3H$, are also capable of π-conjugation (Eq. 4-4).

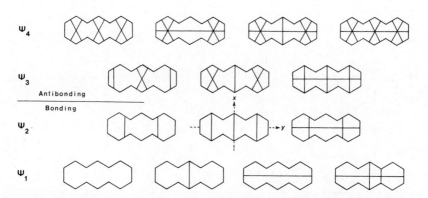

Fig. 4.5. The molecular orbitals of anthracene. (Reprinted with permission from Sinsheimer (1), copyright 1955 by McGraw-Hill Book Co.)

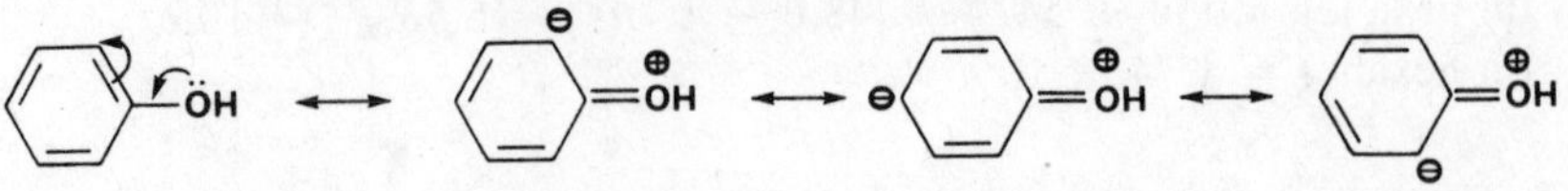

Eq. 4-3

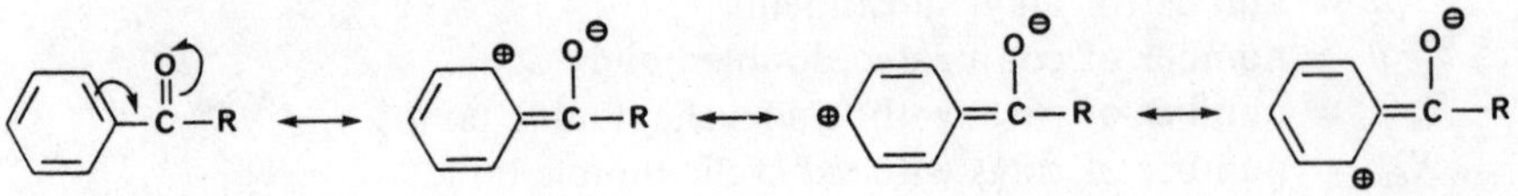

Eq. 4-4

In all the above examples, the conjugated π systems are the benzenoid rings, which constitute the absorbing chromophore. Substituents that cause a red shift in the λ_{max} and increase the ϵ are often called auxochromes.

As expected, disubstitution also shows the effect of extended π conjugation. Para-substitution with both π-electron donor and acceptor groups is particularly effective in enhancing a bathochromic shift and increasing absorption intensity due to resonance stability (Eq. 4-5).

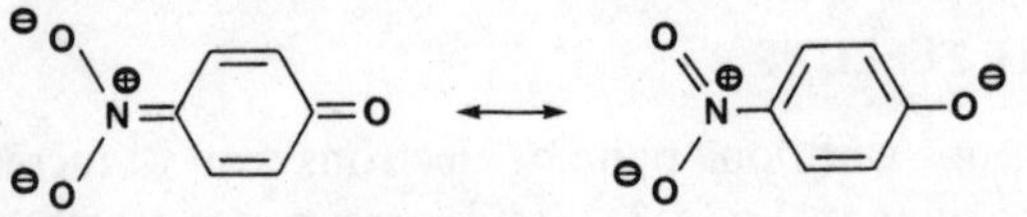

Eq. 4-5

With the substituent effect in mind, it is then not so hard to understand that the anthocyanin with the structural formula in Eq. 4-6 is intensely colored.

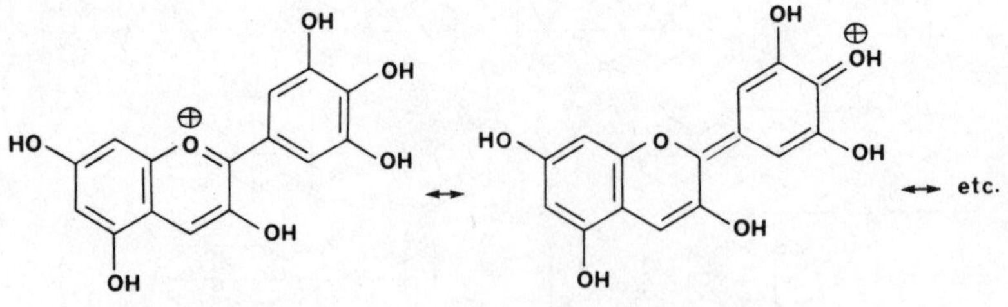

etc. Eq. 4-6

DELPHINIDIN

For simple reference, the relationship between the absorption wavelength and the observed color (transmitted light) is tabulated in Fig. 4.6.

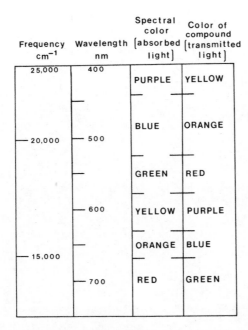

Fig. 4.6. Visible absorption spectrum.

CAROTENOIDS

Carotenoids represent one of the most widespread classes of naturally oc-curring pigments in higher plants. The basic carbon skeleton of carotenoid consists of repeating isoprene units (Fig. 4.7).

There are more than 300 identified carotenoids, and they can be classified into two main groups (Fig. 4.8) (2):

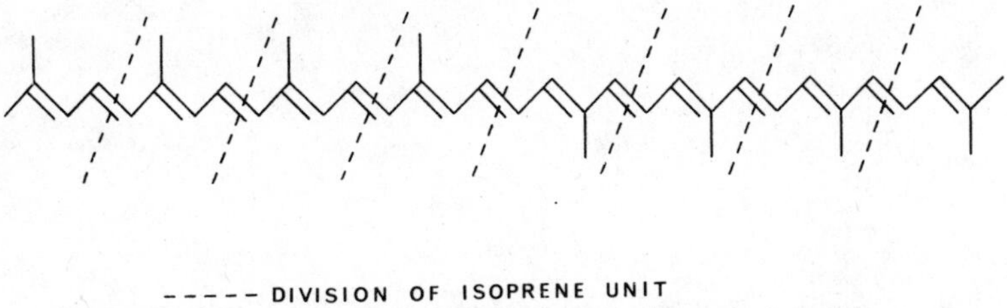

---- DIVISION OF ISOPRENE UNIT

Fig. 4.7. Repeating isoprene units.

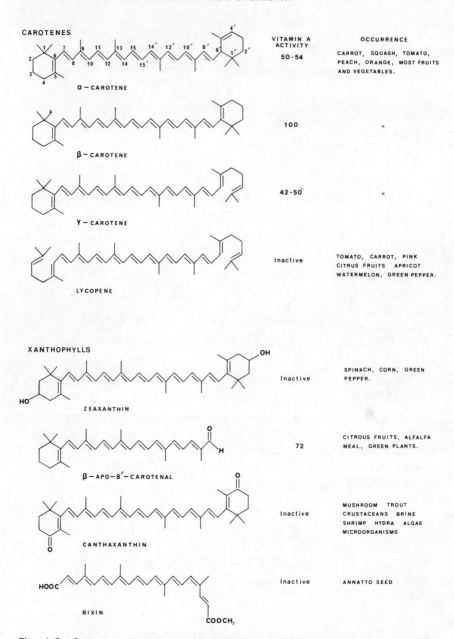

Fig. 4.8. Structure, activity, and occurrence of some selected carotenoids. (Reprinted with permission from Bauernfeind, J. C. Carotenoid, vitamin A precursors and analogs in foods and feeds. *J. Agric. Food Chem.* **20**, 458. Copyright 1972 American Chemical Society.)

1. Carotenes—carotenoid hydrocarbons.
2. Xanthophylls (or oxycarotenoids)—oxygen-containing derivatives.

Isomerization

Naturally occurring carotenoids exist in the all-*trans* form. When carotenoids are exposed to light, acid, or heat, isomerization occurs resulting in a mixture of stereoisomers. One example is the isomerization of β-carotene to 15, 15′-mono-*cis*-β-carotene (Eq. 4-7) (*3*).

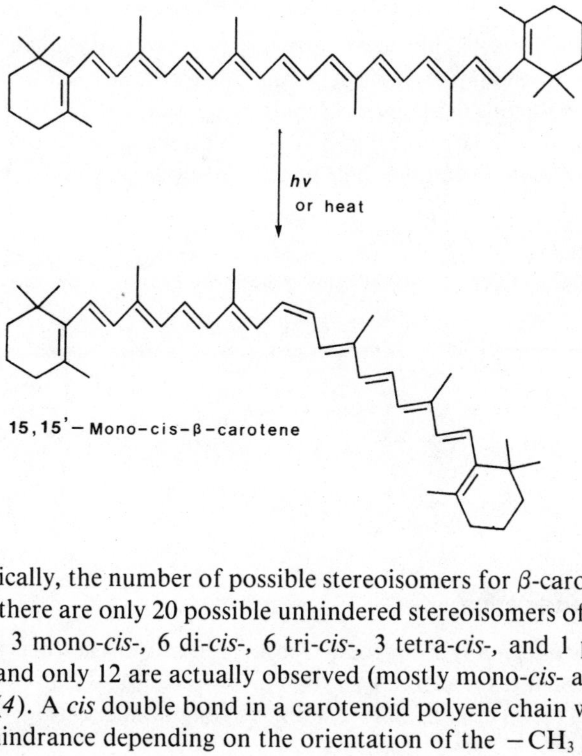

hv
or heat

Eq. 4-7

15, 15′ — Mono–cis–β–carotene

Theoretically, the number of possible stereoisomers for β-carotene is 272. However, there are only 20 possible unhindered stereoisomers of β-carotene (all-*trans*-, 3 mono-*cis*-, 6 di-*cis*-, 6 tri-*cis*-, 3 tetra-*cis*-, and 1 penta-*cis*-β-carotene) and only 12 are actually observed (mostly mono-*cis*- and di-*cis*-β-carotene) (*4*). A *cis* double bond in a carotenoid polyene chain will encounter steric hindrance depending on the orientation of the $-CH_3$ group (Fig. 4.9).

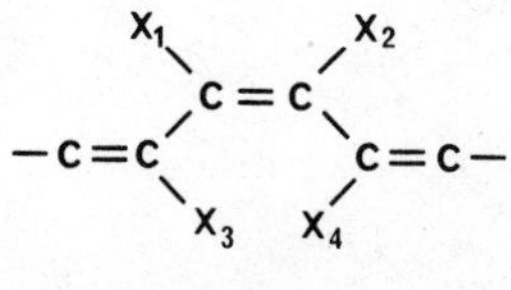

Fig. 4.9. Effect of orientation of methyl substituents on *trans-cis* isomerization.

If it happens that X_3 or X_4 are methyl groups, *trans-cis* isomerization becomes unfavorable. Another restricting factor on the *trans-cis* rearrangement is the fact that the probability of a given configuration decreases with increasing number of *cis* double bonds. The equilibrium ratio of mono-*cis*/all-*trans* is 1/10, while that of tetra-*cis*/all-*trans* is 1/1000.

The *cis* isomers absorb light at shorter wavelengths and exhibit lower extinction coefficients than the parent all-*trans* carotenoids, and the *cis* peak appears in the near-ultraviolet region at 320–380 nm.

Epoxides

Controlled chemical oxidation of carotenoids gives mono- or di- epoxy derivatives (*3*). In the presence of acid, the 5,6-epoxide is isomerized to 5,8-furanoxide, reducing the length of conjugation (Eq. 4-8). The 5,6-epoxide occurs in potatoes, red pepper, paprika, and orange peel. Epoxide isomerization often occurs in canned fruit juice, causing appreciable loss of color.

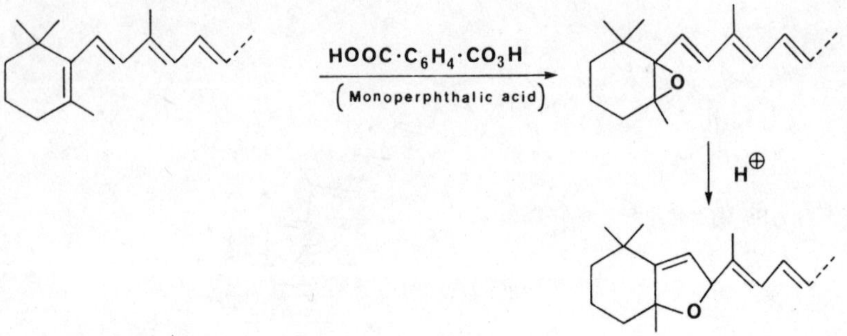

$$HOOC \cdot C_6H_4 \cdot CO_3H$$
$$(\text{Monoperphthalic acid})$$

$$H^{\oplus} \qquad \text{Eq. 4-8}$$

Thermal Degradation

Carotenoids and many long-chain conjugated polyenes, when heated at a sufficiently high temperature ($\sim 190°C$), forms degradation products of ionene, toluene, *m*-xylene, and 2,6-dimethylnaphthalene (Fig. 4.10). The

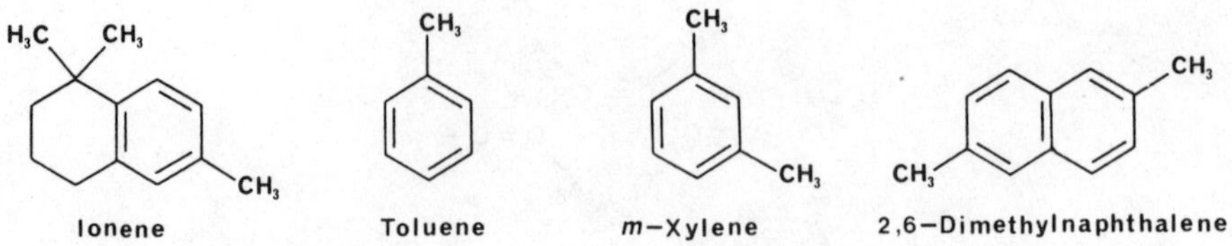

Ionene   Toluene   *m*—Xylene   2,6—Dimethylnaphthalene

Fig. 4.10. Degradation products of thermal degradation of β-carotene.

mechanism involves cyclization and elimination reactions, involving a four-membered ring intermediate. Rearrangement of the eight-electron system results in the formation of tolune and the corresponding shortened polyene (Eq. 4-9) (5).

$$\text{TOLUENE}$$
$$\text{SHORTENED POLYENE} \qquad \text{Eq. 4-9}$$

Photochemical Reactions

(A) Quenching of Triplet Sensitizers. Carotenoids can be involved in triplet energy transfer with other photosensitizers. One of the most studied systems is the quenching of triplet chlorophyll by carotenoids. The quenching reaction explains the protective role played by carotenoids against photobleaching of chlorophylls (Eq. 4-10).

$$^3\text{Chl} + {}^1\text{Car} \longrightarrow {}^1\text{Chl} + {}^3\text{Car} \qquad \text{Eq. 4-10}$$

(B) Quenching of Singlet Oxygen (6). Beta-carotene has been shown to quench singlet oxygen to the triplet (ground) state by electronic energy transfer. The quenching rate is close to that expected for diffusion-controlled reactions. (Refer to Chapter 10, vitamins A and E.)

Carotenoids as Color Additives

Three naturally occurring carotenoids are currently used as color additives in most countries, including the United States. These are the β-carotene (orange-red), β-apo-8'-carotenal (red), and canthaxanthin (purplish red). All three are commercially synthesized from β-ionone (7). The two most important steps involved in the chemical synthesis are the Grignard and Wittig reactions. In the Grignard reaction, two β-C_{19}-aldehydes are condensed with acetylene to form a C_{40} compound. In the Wittig reaction, two C_{15} Wittig salts condense with the C_{10}-dialdehyde (Eq. 4-11).

Beta-apo-8'-carotenal and canthaxanthin are synthesized by similar processes. All these color additives are commercially marketed in two main forms. (1) The oil-soluble forms: These are suspensions of the color compound in a modified vegetable oil carrier, usually partially hydrogenated cottonseed or soybean oils, fractionated triglycerides, and monoglycerides.

Eq. 4-11

Antioxidants such as BHA or α-tocopherol are incorporated to provide stability of the product. (2) The water-dispersible forms: The color additives are dissolved in oil which is emulsified into an aqueous matrix before drying. Various matrixes are employed, the common ones being gelatin/sucrose/modified food starch and gum/acacia/dextrin. Antioxidants such as BHT, BHA, and ascorbyl palmitate are used.

ANNATTO

Annatto is the pigmented extract found in the pericarp of the fruit of *Bixa orellana* L., a large shrub 2–5 m high, native to tropical America. The orange pigment in the annatto is primarily a carotenoid, *cis*-bixin (9'-*cis*-6, 6'-diapocarotene-6,6'-dioate) (*8*) (Fig. 4.11). The *cis*-bixin is insoluble in oil. Heat treatment used in extraction usually converts the *cis*-bixin to *trans*-bixin, which is red and oil soluble.

Oil-soluble annatto is produced by mechanically abrading (a process known as "raspeeling") the pericarp immersed in heated vegetable oil (70°C). Commercially produced annatto contains 0.2–0.25% bixin. Alternatively, the pigment can be extracted by suitable solvents, such as acetone. The solvent is removed to prepare a high-strength bixin powder. The powder is then suspended into oil to produce 3.5–5.2% bixin.

Water-soluble annatto is prepared by abrading the pericarp of annatto seed in aqueous alkali at 70°C. The product is salts of norbixin (both *cis* and *trans*) (Eq. 4-12).

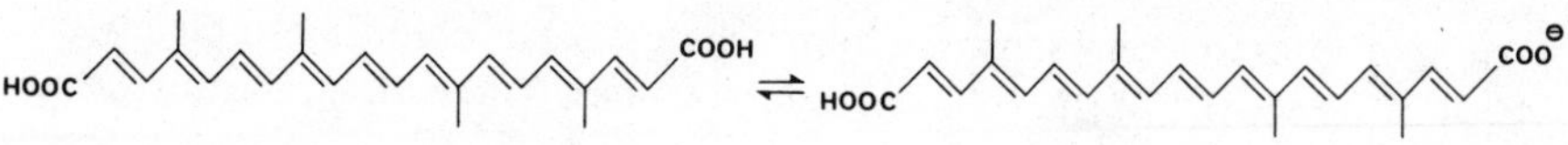

Eq. 4-12

Norbixin is orange, and being a dicarboxylic acid, is not soluble in water. However, the salt formed in alkaline media is readily soluble in water. Most food products have pH's in the acid range. When a solution of norbixin salt is added, the pigment is dispersed into the food system and turns insoluble due to the lowering of the pH. This unique characteristic of norbixin enables the preparation of uniformly colored food products that will not leach any color. For this reason, norbixin is valuable for use in breakfast cereals where leaching of color into milk is undesirable.

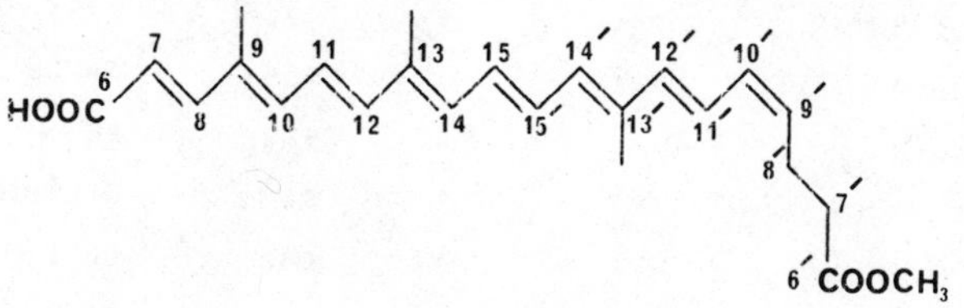

Fig. 4.11. *cis*-Bixin.

ANTHOCYANINS

The anthocyanins are the water-soluble pigments responsible for the brilliant orange red through deep purple colors in flowers and fruits. Anthocyanins are glycosides of anthocyanidins. The anthocyanidins are thus the aglycones of the glycoside anthocyanins. The basic structure of anthocyanidin is flavylium (2-phenyl-benzopyrylium). Some of the common anthocyanidins are listed in Fig. 4.12.

In nature, the anthocyanidins always occur as glycosides, most often the 3-monoglycosides and sometimes 3,5-diglycosides. Different monosaccharides (glucose, galactose, rhamnose, arabinose), disaccharides, and trisaccharides may glycosylate the same group of anthocyanidins. In some cases, the sugar is acylated with p-coumaric, caffeic, or ferulic acids. For illustration, Fig. 4.13 shows the structural formula of the major anthocyanin in eggplant, delphinidin-3-[4-(p-coumaroyl)-L-rhamnosyl(1,6)glucosideo]-5-glucoside (9).

Effect of pH on the Color of Anthocyanins

Anthocyanins are very sensitive to pH changes. The color is completely lost by shifting the pH of the solution to high values. The structural transformation is shown for malvidin 3-glucoside (Eq. 4-13).

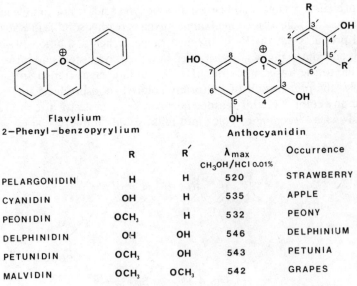

Flavylium
2—Phenyl—benzopyrylium

Anthocyanidin

	R	R'	λ_{max} $CH_3OH/HCl\ 0.01\%$	Occurrence
PELARGONIDIN	H	H	520	STRAWBERRY
CYANIDIN	OH	H	535	APPLE
PEONIDIN	OCH₃	H	532	PEONY
DELPHINIDIN	OH	OH	546	DELPHINIUM
PETUNIDIN	OCH₃	OH	543	PETUNIA
MALVIDIN	OCH₃	OCH₃	542	GRAPES

Fig. 4.12. Chemical structure of common anthocyanidins.

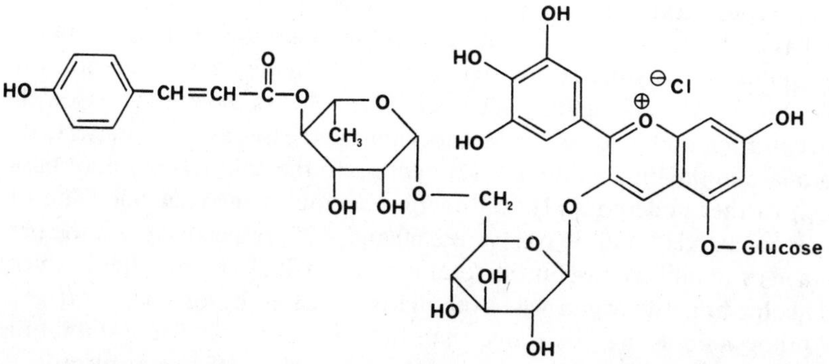

Fig. 4.13. Delphinidin-3-[4-(*p*-coumaroyl)-L-rhamnosyl(1,6)glucosideo]-5-glucoside.

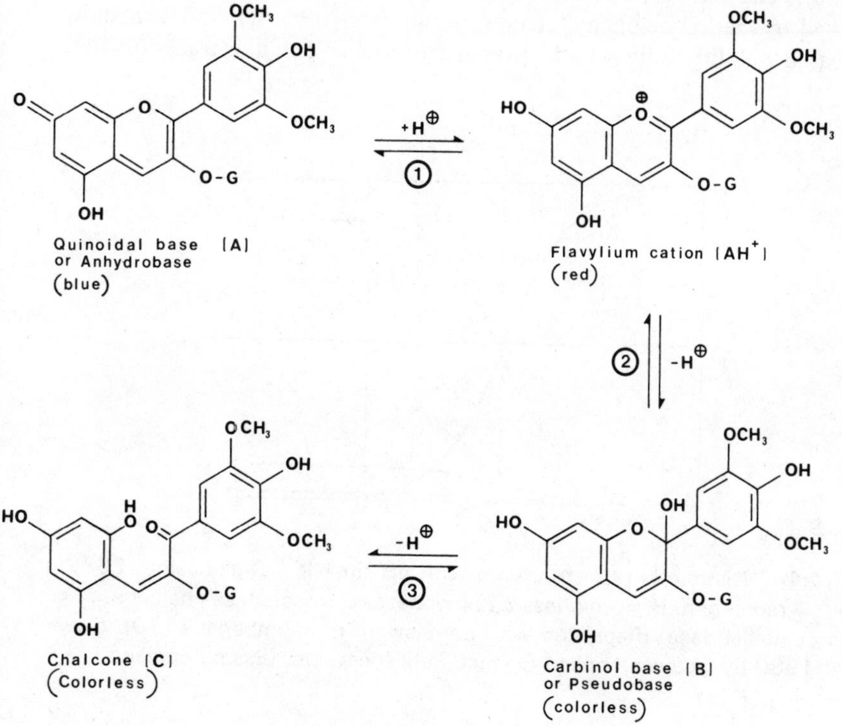

OCH₃
OH
O
OCH₃
O-G
OH
+H⊕
①
Quinoidal base [A]
or Anhydrobase
(blue)
HO
⊕
Flavylium cation [AH⁺]
(red)
②
-H⊕
HO
O
OH
-H⊕
③
Chalcone [C]
(Colorless)
Carbinol base [B]
or Pseudobase
(colorless)

Eq. 4-13

The proton transfer reaction of A to AH^+ is very fast (microseconds) and has a pK of 4.25. The percentage of quinoidal base (A) is very small in the equilibrium mixture at any pH (Fig. 4.14). In very acidic solution (pH 0.5), the red AH^+ is the only species in solution. As the pH increases, the concentration and color of the anthocyanin decreases as AH^+ may (1) deprotonate to the blue quinoidal, (2) hydrate to the colorless carbinol base, which further undergoes (3) tautomeric reaction to the chalcone. The pK values for B/AH^+ and A/AH^+ are 2.60 and 4.25, respectively. As the percentage of quinoidal base in the total is very small at any pH, there is very little color in anthocyanin when the pH is increased beyond pH 4 (10).

Anthocyanidins are less stable than the anthocyanins. The instability is due to the lack of substituents at position 3, and the chalcone form is an unstable α-diketone which is readily hydrolyzed irreversibly to give protocatechuic acid (Eq. 4-14).

Decoloration by Sulfur Dioxide

Sulfur dioxide adds to the 4-position of the anthocyanin to form a bisulfite addition compound (11) (Fig. 4.15). The addition does not occur at the 2-position, and the reaction is reversible. Anthocyanins with the 4-position blocked by methyl or phenyl groups are unaffected by SO_2 and also show increased stability to light in the presence of ascorbic acid or traces of iron.

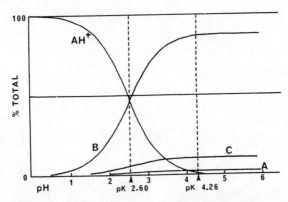

Fig. 4.14. Distribution of structures with pH (malvidin 3-glucoside; 25°C). AH^+ = red cation, B = colorless carbinol base, C = colorless chalcone, A = blue quinoidal base. (Reprinted with permission from Timberlake (10), copyright 1980 by Elsevier Applied Science Publishers, Ltd., Essex, England.)

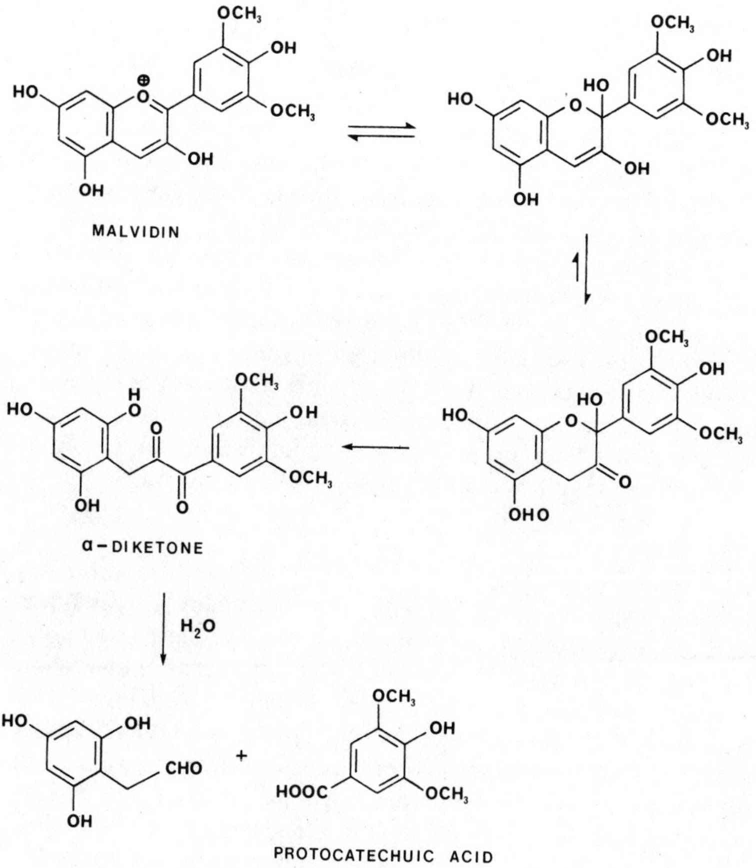

MALVIDIN

Eq. 4-14

α−DIKETONE

H_2O

PROTOCATECHUIC ACID

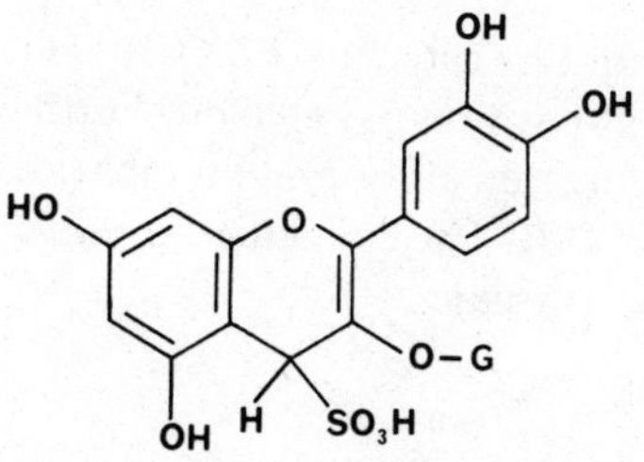

Fig. 4.15. Bisulfite addition product of anthocyanin.

Self-association and Copigmentation

Plant tissues have pH's typically in the range of 3.5–5.5. How can the anthocyanins retain the vivid color without being transformed to the colorless forms as in solution? The anthocyanins are shown to self-associate to form helical stacks through the hydrophobic attraction and hydrogen bonding between the flavylium nuclei. Stacking tends to shelter the chromophores behind the sugar groups from the hydration reaction (*12*).

Another protective mechanism is copigmentation, probably through hydrogen bonding of the phenolic groups between anthocyanin and flavone molecules. The flavone molecules are interleaved between anthocyanin molecules to form alternate stacking. Addition of flavonols, aurone, tannins, and polypeptides also has been shown to stabilize anthocyanins. In some anthocyanins, the acylating groups, caffeic acid, are tucked between and under the stacked flavylium nuclei. Copigmentation usually causes bathochromic shifts in the visible λ_{max} and a large increase in intensity.

Condensation

Anthocyanins condense at position 4 with other flavonoids such as flavanols (e.g., catechins) via electrophilic substitution to give dimers and polymers. The product, a flavene, is oxidizable to a 4-"phenyl" anthocyanin (*13*). The oxidation of flavene can be accomplished by the oxidation-reduction reaction involving intermolecular hydride transfer from the flavene to the flavylium ion. The progressive loss of anthocyanins during wine aging is ascribed to these types of condensation reactions. Generally, about half of the anthocyanins is lost in forming flavene. Condensed pigments are less sensitive to changes in pH than are the parent anthocyanins, and are stable to sulfur dioxide decoloration. One example of these condensations is shown between malvidin-3-glucoside and catechin (Eq. 4-15).

Another mechanism of condensation is observed in which the $-\overset{|}{C}H(CH_3)$ bridges from acetaldehyde link between anthocyanins and other phenolic compounds like catechin (Eq. 4-16). The acetaldehyde reacts as an electrophile at the 8-position of catechin. Further electrophilic substitution of the product into an anthocyanin such as malvidin-3-glucoside yields a condensed dimer (*14*). The condensed product shows bathochromic shift and increased hyperchromic effect.

Oxidation

Hydrogen peroxide oxidizes 3-substituted anthocyanins (e.g., malvidin-3-5-diglucoside) to benzoyloxyphenylacetic acid esters of the malvone type. The

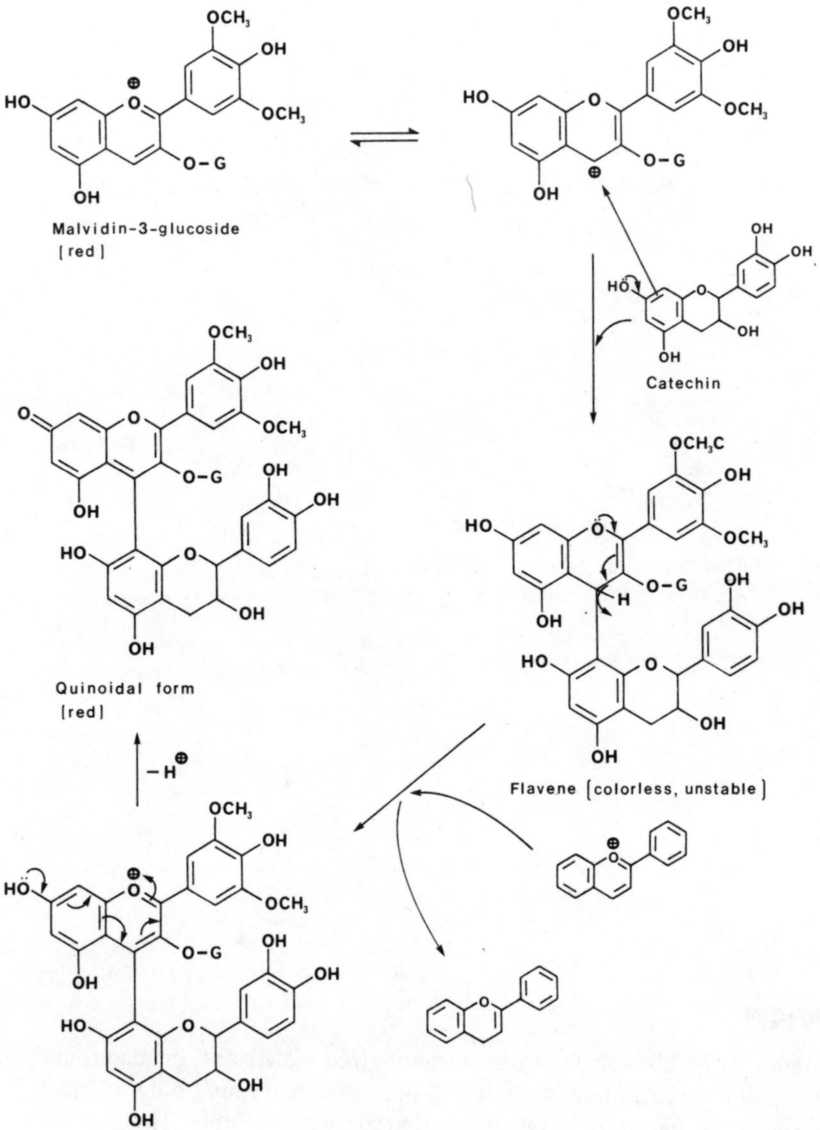

Malvidin-3-glucoside
[red]

Catechin

Quinoidal form
[red]

Eq. 4-15

Flavene [colorless, unstable]

reaction follows a Bayer-Villiger type of oxidation. After the nucleophilic attack of the H_2O at C2, the hydrogen atom migrates to the adjacent oxygen and causes the cleavage between C2 and C3. The esters formed are easily hydrolyzed to various breakdown products (Eq. 4-17) (15).

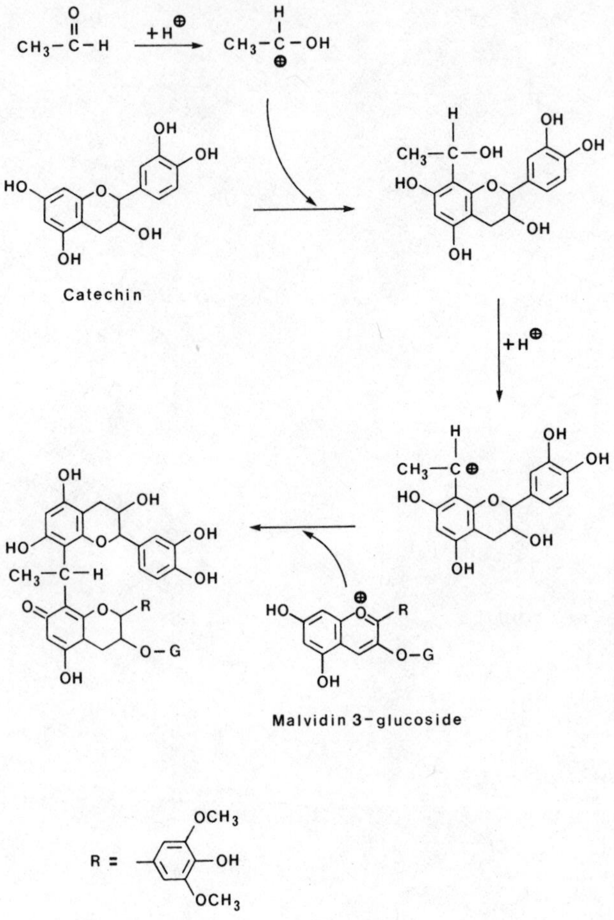

Eq. 4-16

BETANAIN

Betanain can be classified into betacyanins (red-violet) and betaxanthins (yellow), which occur abundantly in red beet, pokeberry, and other plants. Betanains occur naturally in salt forms that are water-soluble. The general structure of betanain is represented in Fig. 4.16, indicating that it is the conjugated cation that constitutes the chromophore.

In betaxanthins, the R or R′ does not extend the 1,7-diazaheptamethin conjugation system (Fig. 4.16), and they are yellow with λ_{max} near 480 nm. An example of betaxanthin is indicaxanthin (Fig. 4.17) isolated from cactus fruit (*Opuntia ficus-indica*).

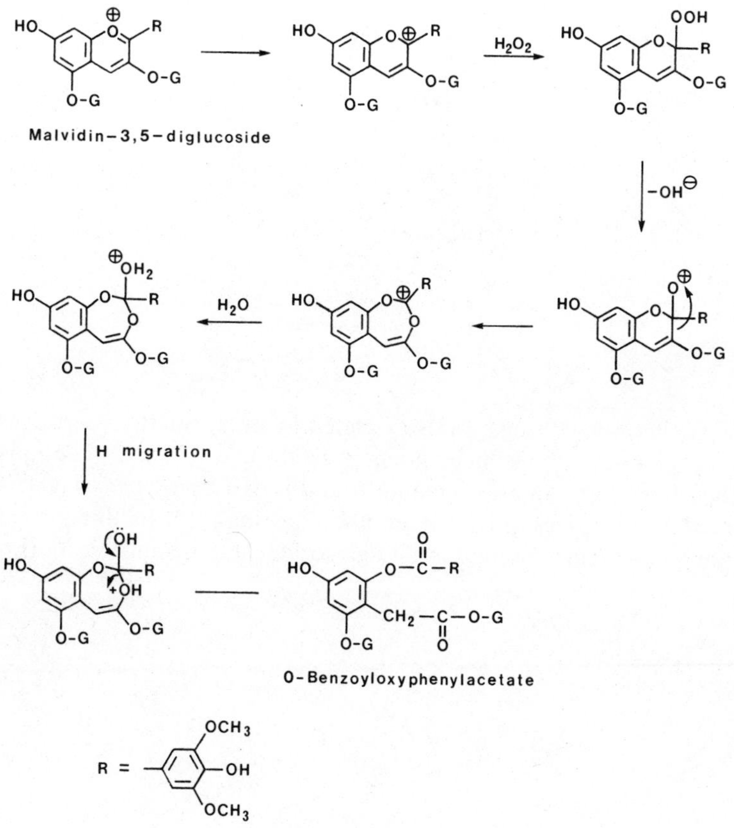

Malvidin-3,5-diglucoside

Eq. 4-17

O-Benzoyloxyphenylacetate

$$R = \begin{array}{c} OCH_3 \\ -OH \\ OCH_3 \end{array}$$

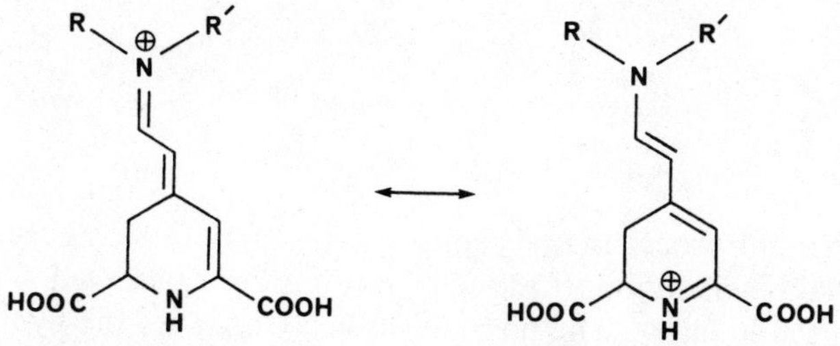

Fig. 4.16. Betanain: resonance of conjugated cation.

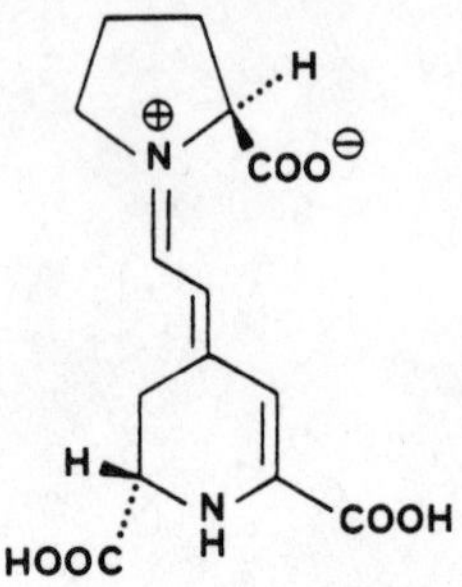

Fig. 4.17. Indicaxanthin.

In betacyanins, the conjugation is extended by an aromatic substituent, and the chromophore shows a bathochromic shift to 540 nm. The most commonly known betacyanin is betanin found in red beet (*Beta vulgaris*). Betanin readily isomerizes to isobetanin upon heating (*16*). In alkaline medium, hydrolysis yields cyclodopa-5-O-glycoside and betamic acid (Eq. 4-18).

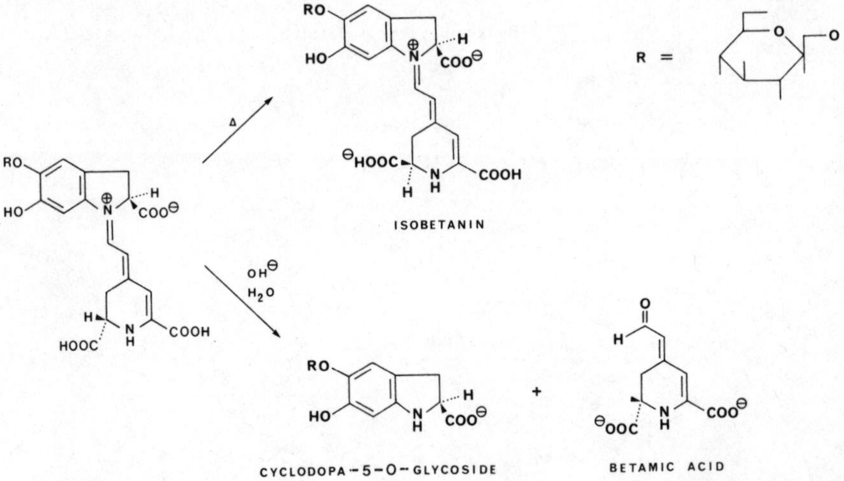

Eq. 4-18

During thermal processing of canned beets, loss of color occurs. However, a partial regeneration of the red color has been observed and known for years in the industry. The mechanism likely involves the hydrolysis of the Schiff base and, in the reverse reaction, the condensation of the amine of cyclodopa-5-O-glycoside and the aldehyde of the betamic acid.

CARAMEL

Caramel is the amorphous, dark brown product resulting from controlled heat treatment of food grade carbohydrates, usually corn syrup with 75% dextrose content. Acids, alkalies, or salts are added in small amounts to increase the caramelization rate and to obtain the desirable characteristics for various food uses.

Caramel consists of a complex mixture of molecular polymers of indefinite compositions. The exact chemical structure is uncertain and caramel is generally classified into three groups:

Caramelan	$C_{24}H_{36}O_{18}$
Caramelen	$C_{38}H_{50}O_{25}$
Caramelin	$C_{125}H_{188}O_{80}$

The net charge of caramel is critical to the choice of application in a particular food product (17). Caramel in aqueous solution can be positively or negatively charged depending on the manufacturing process and the pH of the medium. A positively charged caramel can interact with other negatively charged molecules in the food system and cause precipitation. Caramel for soft drinks and beverages is required to carry strong negative charges. For bakery products, beer, and gravies, the caramel mixture used should carry positive charges. A variety of caramels is commercially available to match particular requirements. The "acid-proof" caramel is compatible with phosphoric acid and is used in soft drinks such as colas and root beers. Baker's caramel is made for breads, biscuits, and other baked goods. Powdered caramel is for cake mixes and species.

DYES AND LAKES

The FDA regulation specifies two types of color additives (18).

1. *Uncertified* color additives are natural colors that may be obtained from natural sources or synthetic duplicates of the natural colors. Many of the uncertified color additives have already been discussed and include the chlorophyll derivatives described in later sections.

2. *Certified* color additives are synthetic colors that do not occur in nature. These are the dyes and lakes which include the most important and most used food colorants. Compared to natural colors, the certified color additives have (a) higher tinctorial (coloring) strength, (b) higher stability, and (c) uniform standardization of hue and color strength.

The dyes are commercially marketed in liquid, powder, dispersion, and paste forms. A complete list of the certified color dyes is presented in Fig. 4.18. Structurally, all these dyes are extensively conjugated and several of

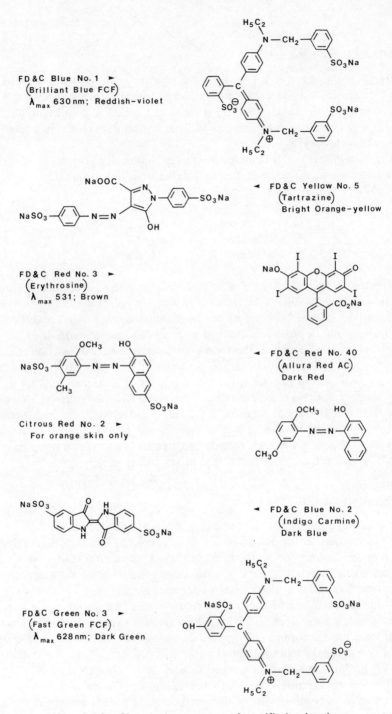

FD & C Blue No. 1 ►
(Brilliant Blue FCF)
λ_{max} 630 nm; Reddish–violet

◄ FD & C Yellow No. 5
(Tartrazine)
Bright Orange–yellow

FD & C Red No. 3 ►
(Erythrosine)
λ_{max} 531; Brown

◄ FD & C Red No. 40
(Allura Red AC)
Dark Red

Citrous Red No. 2 ►
For orange skin only

◄ FD & C Blue No. 2
(Indigo Carmine)
Dark Blue

FD & C Green No. 3 ►
(Fast Green FCF)
λ_{max} 628 nm; Dark Green

Fig. 4.18. Chemical structures of certified color dyes.

them are azo compounds. The azo dyes are susceptible to reduction by reducing agents, resulting in color fading.

Lakes are common names for the aluminum salts of FD&C color dyes. These are formed by the precipitation and adsorption of a water-soluble dye on an insoluble base or substrate, alumina hydrate [Al(OH)$_3$]. The alumina hydrate base is water insoluble, so the product is an insoluble form of dye, that is, a pigment. Lakes, therefore, are basically dyes modified for application in nonaqueous systems. Icings, filling, frostings, confectionary coatings, and gum products are some of the examples. Lakes are applied to foods by dispersion, and lakes are stable in the pH range of 4.0–9.0 with very little "leaching." Lake particles are typically 5 μ in size.

COORDINATION CHEMISTRY

Thus far, we have concentrated on colored compounds that are chiefly conjugated polyenes. There is another unique class of compounds that are highly colored—the metalloporphyrins. In order to understand the chemistry and chromophoretic properties of the metalloporhyrins, a brief presentation of coordination chemistry is needed (19).

A simple approach will be to consider what would happen to the five d orbitals of a transition metal (Fe, Co, etc.) when it is placed in an octahedral array of six ligands along the three cartesian coordinate systems of the d orbitals. Here, the ligands are treated as point negative charges. The $d_{z^2-y^2}$ and the d_{z^2} orbitals have each of their lobes of electron density pointing towards the ligand, while the d_{xy}, d_{xz}, and d_{yz} orbitals have each of their lobes directed between the ligands. The $d_{x^2-y^2}$ and d_{z^2} orbitals thus experience a higher electrostatic potential energy compared to the d_{xy}, d_{xz}, and d_{yz} orbitals. The net result is that the five d orbitals are split into two energy levels (e_g and t_{2g}) (Fig. 4.19).

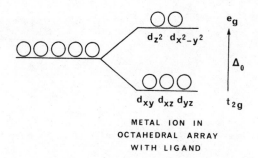

Fig. 4.19. Energy-level diagram for the five d orbitals of a metal ion in an oactahedral field.

Now, taking into account the orbitals of the ligands bonding to the metal ion, the qualitative molecular orbital energy-level diagram for an octahedral complex between a metal ion and the six ligands that do not possess π orbitals will appear as in Fig. 4.20. The d_{xy}, d_{xz}, and d_{yz} orbitals of the metal ion are nonbonding, since they have the wrong symmetry for binding with the ligands. The energy level therefore remains the same. The d_z^2, $d_{x^2-y^2}$ can overlap with the ligand orbitals forming the σ_d^* and σ_d bonds. The nonbonding and the lowest antibonding levels are analogous to the t_{2g} and e_g in the previously discussed point model. Most ligands, however, have π orbitals which also interact with the nonbonding orbitals (d_{xy}, d_{xz}, d_{yz}).

The ligand π orbitals may be simple $p\,\pi$ orbitals or molecular orbitals of a polyatomic ligand (e.g., O_2, CO, NO). The interaction of d_{xy}, d_{xz}, and d_{yz} orbitals of the metal ion with the π orbitals of the ligands affects the magnitude of the splitting energy, Δ_0, negatively or positively. Ligands with π orbital symmetry that destabilize the metal t_{2g} orbitals tend to cause the Δ_0 value to decrease. Likewise, ligands that stabilize the metal t_{2g} orbitals tend to increase the splitting energy, Δ_0.

For a simple $p\pi$ system of a single atom (e.g., Cl^-) overlapping with $d\pi$ orbitals of the metal ion, the lone-pair electrons in the π orbitals of the Cl^- repel the electrons in the d_{xy}, d_{yz}, and d_{xz} orbitals. Such ligand-to-metal

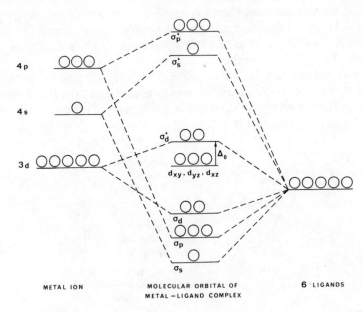

METAL ION   MOLECULAR ORBITAL OF   6 LIGANDS
METAL –LIGAND COMPLEX

Fig. 4.20. Molecular orbital energy-level diagram for octahedral coordination between a metal ion and six ligands.

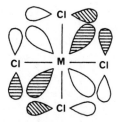

Fig. 4.21. Interaction between $p\pi$ orbital of Cl^- with $d\pi$ orbital of metal.

interaction, $p\pi$-$d\pi$ or L-M(π), is illustrated in Fig. 4.21. In this case, the energy of the t_{2g} level is increased, and Δ_0 is decreased. Such ligands (including Cl^-, Br^-, and OH^-) are called weak-field ligands.

The effect of π interaction via polyatomic ligands is more complicated. Consider the π and π^* orbitals of the ligand CO. The π-bonding orbital destabilizes the t_{2g} orbitals of the metal ion by L-M(π) interaction similar to the Cl^- example discussed above. The antibonding π^* orbital also interacts with the metal $d\pi$ orbitals in a M-L(π) interaction (Fig. 4.22).

In the π-bonding orbital of the ligand, most of the electron density stays in the CO π bond. However, the π^* antibonding orbital of the ligand is unfilled. Therefore, overlapping of the empty π^* orbital with the metal ion $d\pi$ orbital results in the delocalization of the electrons from the metal to the ligand. The interaction is therefore often called π back bonding, indicating the donation of a π electron from metal ion to ligand. Delocalization of electron density stabilizes and lowers the energy of the t_{2g} orbitals of the metal ion, resulting in increased Δ_0. Ligands such as CO, O_2, and NO that increase the splitting energy, Δ_0, this way are called strong-field ligands.

In light absorption, there are two types of electron transitions possible.
1. *d-d* Transition: If ν is equal to Δ_0/h, the metal-ligand can then transfer

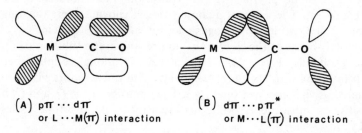

(A) $p\pi \cdots d\pi$
or L$\cdots$M(π) interaction

(B) $d\pi \cdots p\pi^*$
or M$\cdots$L(π) interaction

Fig. 4.22. Interaction between (A) $p\pi$ orbital, (B) $p\pi^*$ orbital of polyatomic ligand and $d\pi$ orbital of metal.

Table 4.1. L-M Transitions

LIGAND	METAL (d ORBITAL)
π	$\pi^*\ (t^*_{2g})$
π	$\sigma^*\ (e_g)$
σ	$\pi^*\ (t_{2g})$
σ	$\sigma^*\ (e_g)$

the light energy to energy of excitation of electrons from t_{2g} to the e_g metal ion centered orbital.

2. Charge-transfer transition: This is the excitation of electrons from the molecular orbital centered in the ligand to the metal ion centered orbital. These L-M transitions are listed in Table 4.1. The process requires higher energy than the d-d transition, and the resulting spectrum lies in the range of shorter wavelengths (less than 400 nm) (19).

METALLOPORPHYRIN—ELECTRONIC STRUCTURE

In metalloporphyrins, the situation is complicated by the fact that the porphorin that complexes with metal ion through the four nitrogen atoms, possesses 18 π-electrons in an 18-atom ring (20, 21). It is no surprise that all porphyrins are highly colored due to the conjugated π-electron system (π-π^* transition). All free-base porphyrins show similar four-banded visible spectra of moderate intensity between 500 and 660 nm and a strong band at 400 nm. The visible bands and the UV bands are labeled Q and B, respectively. The UV band is also known as the Soret band. The highest filled π orbital of a porphyrin has A_{1u} and A_{2u} symmetry and the lowest unfilled orbitals are of E_g symmetry, according to MO calculations which are beyond the scope of this book. The excited states correspond to the π-π^* transition of one of the four electrons from HOMO to LUMO.

In a free-base porphyrin, the π-π^* transition is polarized along the x, y axes (Fig. 4.23). The π-electron conjugation is shown, where the lone-pair electrons in the two imino nitrogens do not enter the π system.

The polarization of the π-π^* transition results in the splitting of Q_o into Q_x and Q_y bands (Fig. 4.24A, B). Further vibrational coupling between these transitions with the Soret bands give rise to two more visible bands designated as vibronic overtones (Q_v). The intensity of the Q_x and Q_y bands is sensitive to substituents, while the intensity of the overtones is less affected.

In the metal-porphyrin complex, the molecule assumes a D_{4h} (full square) symmetry, partly due to the resonance of the complex (Fig. 4.25). Transi-

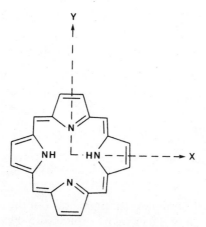

Fig. 4.23. Polarization of the π-π* transition along x, y axes of a free-base porphyrin.

tions in the x and y directions are equivalent. The two LUMOs (e_g) are strictly degenerate, and the HOMOs (a_{1u}, a_{2u}) are almost degenerate (Fig. 4.26A, B). The π-π* transition results in a single Q_o band plus the vibronic overtone Q_v (the Q_o and Q_v bands are also known as α and β bands in some literature). The differences in symmetry, orbital energy, and absorption spectrum between the free-base porphyrin and metalloprophyrin are summarized in Table 4.2.

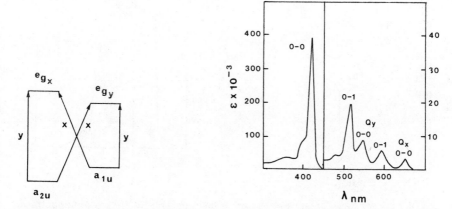

Fig. 4.24. (A) Energy splitting of the π-π* transition, and (B) absorption spectrum of a free-base porphyrin. (Reprinted with permission from Gouterman (20), courtesy of Academic Press, Inc.)

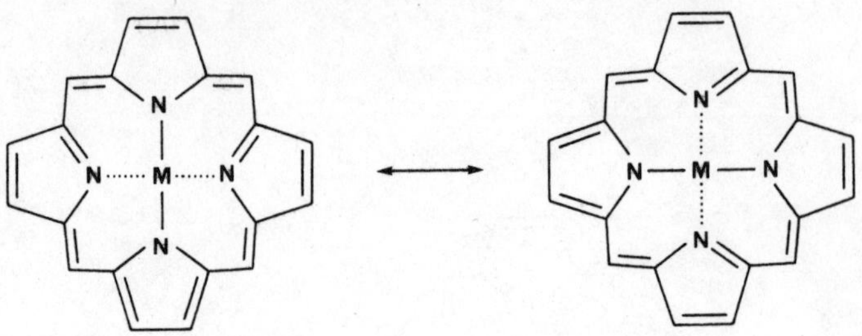

Fig. 4.25. Resonance of metal-porphyrin complex.

In a metal-porphyrin complex, the main effect of the metal on the transitions is the conjugation of the metal $p\pi$ orbital with the porphyrin π orbital having a_{2u} symmetry (Fig. 4.27). The interactions either raise or lower the orbital energy and consequently shift the absorption spectrum.

The porphyrin π-metal transfer occurs in the transitions from the HOMO of the porphyrin (a_{2u}, a_{1u}, a'_{2u}, b_{2u} symmetry) to the metal e_g orbitals (d_{z^2}, $d_{x^2-y^2}$). The d-d transitions are affected since the splitting of the d-orbital symmetry is altered. The splitting of the d orbitals of the metal ion (due to the porphyrin) exhibits additional loss of degeneracy from the theoretical predicted octahedral symmetry. In myoglobin, the octahedral symmetry is distorted to assume tetragonal or square planar symmetry (Fig. 4.28).

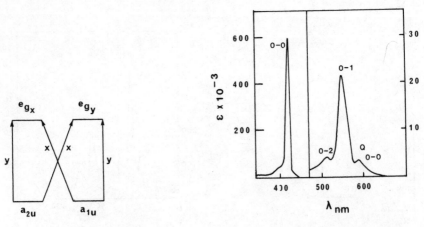

Fig. 4.26. (A) Energy splitting of π-π^* transition, and (B) absorption spectrum of a metal-porphyrin. (Reprinted with permission from Gouterman (20), courtesy of Academic Press, Inc.)

**Table 4.2.   Comparison of Orbital Energy and Spectrum Between
Free-base Porphyrin and Metalloprophyrin**

	ORBITAL ENERGY		VISIBLE BANDS	
	HOMO	LUMO	Q	OVERTONE
Free-base prophyrin $x \neq y$ (polarized) D_{2h} (rectangular)	$e_{gx} > e_{gy}$	$a_{1u} > a_{2u}$	$2(Q_x, Q_y)$	$2Q_v$
Metalloporphyrin $x = y$ (equivalent) D_{4h} (full square)	$e_{gx} = e_{gy}$	$a_{1u} \approx a_{2u}$	$1(Q_o)$	$1Q_v$

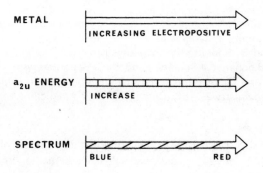

Fig. 4.27.   Effect of metal ion transition energy of the porphyrin π orbital.

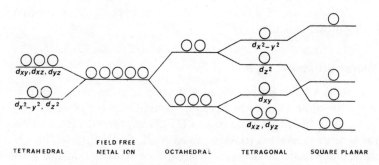

Fig. 4.28.   Energy-level diagram of some common coordinations.

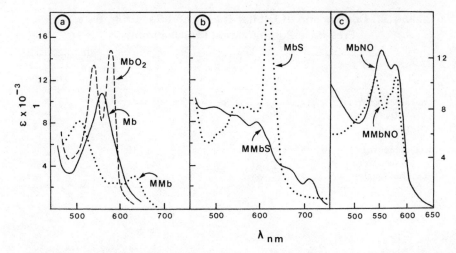

Fig. 4.29. Spectra of (A) deoxymyoglobin, oxymyoglobin, metmyoglobin; (B) sulfmyoglobin, metsulfmyoglobin; and (C) nitrosylmyoglobin, nitrosylmetmyoglobin. (A and B: Reprinted with permission from Nicholls (*23*), copyright 1961 by The Biochemical Society, London. C: Reprinted with permission from Fox, J. B., Jr., and Thomson, J. S. Formation of bovine nitrosylmyoglobin I. pH 4.5–6.5. *Biochemistry* **2**, 466. Copyright 1963 American Chemical Society.)

The visible spectra of myoglobin and their derivatives are also shown in Fig. 4.29 (*22, 23, 24*), which we will discuss further in the following sections.

MYOGLOBIN

Myoglobin is the main pigment in meat, and the structural form, including the prosthetic group, of this protein in meat and processed meat determines the color of the product.

Molecular Structure

Myoglobin is a single polypeptide, globular protein comprised of a protein moiety (globin) and a prosthetic group (heme), with a molecular weight of ~ 18,000 and 153 amino acid residues (*25*). The protein is folded with eight major α-helical segments (referred to as A, B, C, D, E, F, G, and H) and nonhelical segments in between. The folding is such that the prosthetic heme is buried in a hydrophobic cleft. The heme is an iron-porphyrin complex with the iron covalently linked to histidine residue F8 (proximal histidine), and is closely associated with a second histidine E7, distal histidine) (Figs. 4.30, 4.31).

Porphyrins are derivatives of the parent structure porphin, which consists essentially of four pyrrole rings joined by methine ($-CH=$) bridges. The porphyrin of heme consists of methyl, ethenyl, and propionyl side chains; this particular porphin derivative is classified as protoporphyrin IX (Fig. 4.32).

The Heme Iron

Deoxymyoglobin (Mb) contains heme iron in the $+2$ state with only five ligands (without bound oxygen). When oxygen is bound reversibly, it is

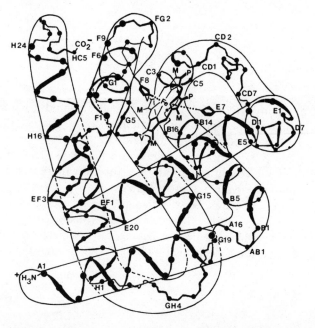

Fig. 4.30. α-Carbon diagram of myoglobin molecule obtained from 2.A analysis. Stretches of α-helix are represented by smooth helix with exaggerated perspective and are given letter-number labels. Nonhelical regions are designated by letter-letter-number symbols and represented by three-segment zigzag lines between α-carbons. Fainter parallel lines outline a high-density region as revealed by 6-A analysis. Heme group framework is sketched in forced perspective, with side groups identified by: M = methyl, V = vinyl, P = propionic acid. Five-membered rings at F8 and E7 represent histidines associated with heme group. (Reprinted with permission from Dickerson, (25), courtesy of Academic Press, Inc.)

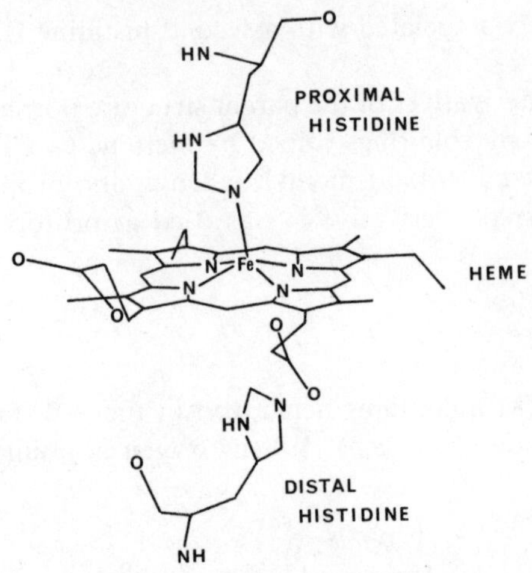

Fig. 4.31. Heme in myoglobin.

referred to as oxymyoglobin (MbO$_2$). Other neutral ligands such as CO and NO can also bind to the sixth coordination position. The Fe(II) is readily oxidized to Fe(III) state to give ferrimyoglobin.

The electronic configurations are shown in Fig. 4.33. The Fe atom can be oxidized to the ferrous ion (Fe^{2+}) or electronically more stable ferric ion (Fe^{3+}) by losing two or three electrons (26).

For an iron atom coordinated with ligands, the 3d orbitals split theoretically into two energy levels. (The octahedral is distorted in metalloporphyrin, but for simplicity, a perfect octahedral symmetry is assumed for now.) The way that the d electrons fill the t_{2g} and e_g orbitals depends on the

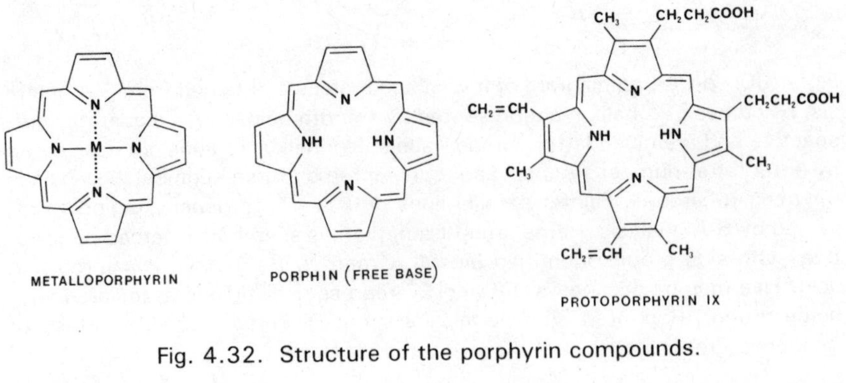

METALLOPORPHYRIN   PORPHIN (FREE BASE)

PROTOPORPHYRIN IX

Fig. 4.32. Structure of the porphyrin compounds.

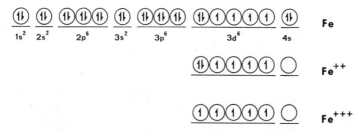

Fig. 4.33. Electronic configuration of iron.

nature of the ligands. Three electronic configurations are of concern to the present discussion (Table 4.3).

In the heme complex, the coordination positions are directed to the four porphyrin nitrogens, and in the fifth and the sixth positions, to histidine F8 and O_2 (or H_2O), respectively. Deoxymyoglobin has only five ligands (four porphyrin nitrogens and axial coordinated histidine F8 nitrogen), and binds O_2 readily to form the low-spin, diamagnetic oxymyoglobin. The molecular orbital scheme is presented in Fig. 4.34.

A σ bond is formed by the overlap of the $3d_z$ orbital of the Fe with π^* of O_2, and a π bond is formed from the overlap of the $3d_{yz}$ orbital of the Fe with the other π^* of the O_2 (Fig. 4.35).

The $Fe^{2+}O_2$ complex in oxymyoglobin is at a lower energy level than the dioxygen or the Fe^{2+}. The $Fe^{2+}O_2$ complex is generally regarded to have the characteristic at least as if it were a ferric-superoxide adduct $Fe^{3+}\text{-}O_2^-$. The oxygen binds in the end-on orientation to give a 135° bend.

The dioxygen is hydrogen bonded to the distal histidine imidazole. The proximal histidine imidazole nitrogen is hydrogen bonded to adjacent

Table 4.3. Electronic Configurations of the Heme Iron

(1) Oxymyoglobin	◯ ◯		e_g
Low spin, ferrous			
$S = 0$	⇅ ⇅ ⇅		t_{2g}
diamagnetic			
(2) Deoxymyoglobin	↑ ↑		e_g
High spin, ferrous			
$S = 2$	⇅ ↑ ↑		t_{eg}
Paramagnetic			
(3) Ferrimyoglobin	↑ ↑		e_g
High Spin, ferric			
$S = 5/2$	↑ ↑ ↑		t_{2g}

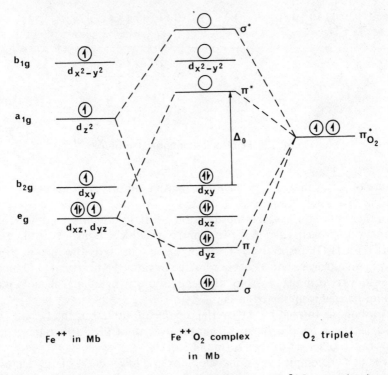

Fig. 4.34. Molecular orbital energy-level diagram of $Fe^{2+}O_2$ complex in oxy-myoglobin.

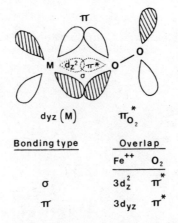

Fig. 4.35. Overlap of dπ orbital of Fe and $\pi^{\cdot}$ orbital of O_2.

amino acid side groups in the globin, thereby increasing its basicity. This increase in electron density strengthens the back-bonding in which the iron donates the $d\pi$ electron back to the π^* orbital of O_2. Since O_2 is a weak σ electron donor, the synergic σ and π bondings tend to provide stability of the complex (Fig. 4.36).

The bonding of O_2 to heme Fe is reversible. In physiological conditions, equilibrium exists between free oxygen and oxygen bound to the heme Fe (Fig. 4.37) (27).

The Role of Globin

The protein globin functions to stabilize the steric and electronic configuration of the iron heme through more than 60 hydrophobic interactions and hydrogen bondings. Changes in the protein conformation inevitably influence the ligand binding property of the heme. Similarly, binding of ligand is expected to induce conformational changes of the globin. Transition from deoxy- to oxymyoglobin is associated with movement of the iron and proximal histidine into the porphyrin phane, as to enhance π and σ overlap with the iron and dioxygen. The proximal histidine nitrogen plays a primary role in the synergistic σ and π bonding within the Fe^{2+}-O_2 complex by feeding electron density that enables the back-bonding of electrons from the iron to the π^* orbital of the oxygen as already noted.

The hydrophobic pocket creates a low dielectric environment that is unfavorable for ionic ligands (e.g., CN^-, OH^-). The hydrophobic environment tends to slow down the autoxidation of $Fe^{2+}O_2$ complex to metmyoglobin. The closely packed amino acid side chains in the pocket also restrict the accessibility of large ligands, and restrict the size and orientation of the ligand bound to the myoglobin. Only small ligands (O_2, CO, NO) can bind to the heme iron. The heme iron, being embedded inside the protein molecules, is spatially isolated from close contact and hence from rapidly oxidizing to the dioxygen-bridged diiron complex (porphyin-Fe^{2+}-O_2Fe^{2+}-porphyrin).

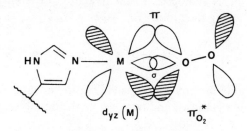

Fig. 4.36. Back-bonding assisted by histidine.

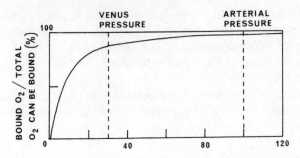

Fig. 4.37. Partial pressure of oxygen (mm mercury) over solution in equilibrium.

The Autoxidation of Oxymyoglobin

Oxymoglobin is known to undergo slow oxidation to ferrimyoglobin. The autoxidation (1) proceeds maximally at the half-saturating oxygen ($pO_2 \sim$ 1–2 torr) pressure where Mb and MbO$_2$ are equal in concentration, and (2) can be accelerated by metal ions, low pH, or high temperature.

The simple one-electron transfer process in which Fe^{2+} is oxidized to Fe^{3+} and O_2 is reduced to the superoxide (Eq. 4-19) does not occur, since ΔG of the reaction is $+30.8$ kcal/mol. Instead, autoxidation involves a two-electron reduction of the oxygen to the peroxide as in Eqs. 4-20 and 4-21 (*28*).

$$Fe^{++} + O_2 \longrightarrow Fe^{+++} + O_2^- \qquad \Delta G = +30.8 \text{ kcal/mol} \quad \text{Eq. 4-19}$$

$$Fe^{++} + O_2 \longrightarrow FeO_2^{++} \left(\text{iron–dioxygen complex}\right) \qquad \text{Eq. 4-20}$$

$$FeO_2^{++} + Fe^{++} \longrightarrow 2Fe^{+++} + O_2^= \left(\text{peroxide}\right) \qquad \text{Eq. 4-21}$$

The two electrons are from the oxymyoglobin FeO_2^{2+} ($Fe^{2+}O_2$) and deoxymyoglobin Fe^{2+}. The one-electron donation from the ferrous iron (Fe^{2+}) of deoxymyoglobin is analogous to the oxidation of oxymyoglobin in the presence of a reducing metal ion (Eq. 4-22).

$$MO_2^{++} + M^{++} \longrightarrow \left(M\text{-O-O-M}\right)^{+4} \rightleftharpoons 2M^{+3} + HOOH \quad \text{Eq. 4-22}$$

The sixth ligand in ferrimyoglobin is H$_2$O (also known as metmyoglobin). The ferrimyoglobin is brown and responsible for discoloration of meat under normal storage conditions.

Reduction of ferrimyoglobin to deoxymyoglobin has been shown to occur enzymatically. Metmyoglobin reductase has been purified from beef heart and dolphin muscle. Nonspecific reducing systems notably include ascorbate, and NADH and flavins. In fresh meat, reducing substances endogenous to the tissue reduce ferrimyoglobin to deoxymyoglobin, which is then oxygenated to oxymyoglobin if oxygen is present as in the case of the surface layer of fresh meat. In the interior, the deoxymyoglobin is the predominant species and the meat color is thus purple. Schematically, the interconversion of Mb, MbO₂, and FMb can be represented as in Fig. 4.38.

The Absorption Spectrum

A comparison of the visible spectra among Mb, MbO₂, and FMb shows that the spectrum of MbO₂ is characterized by a typical porphyrin π-π^* transition with two bands, while Mb exhibits a single broad band and FMb a diffuse band with four peaks from 500 to 650 nm.

Strong ligands such as O_2 increase Δ_0, and the d-d transition is close to the energies of the porphorin π-π^* transition. Consequently, the d-d transition bands are masked by the porphyrin bands.

In the high-spin ferric complex, the weak-field ligands, such as OH or H_2O in FMb, tend to increase the e_g orbital energy, which increases the mixing of porphyrin π and iron d orbitals. There is ~ 35–45% mixing of porphyrin π orbitals with the iron e_g orbitals. Charge transfer transition from the porphyrin π orbitals (a_{2u}, a_{1u}, a'_{2u}, b_{2u}) to iron e_g orbitals accounts for the multiple peaks that appear in the spectrum.

The Hemochromes

In cooked meat, both myoglobin and ferrimyoglobin are converted to ferro- and ferrihemochromes which can exchange ligands within the globin (Fig. 4.39). Amino acid side chains, including histidine, and carboxyl groups may coordinate to the heme iron in hemochromes. Ferrohemochrome has a visi-

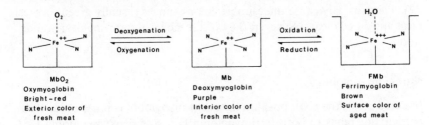

Fig. 4.38. Schematic representation of interconversion of Mb, MBO₂, and FMb.

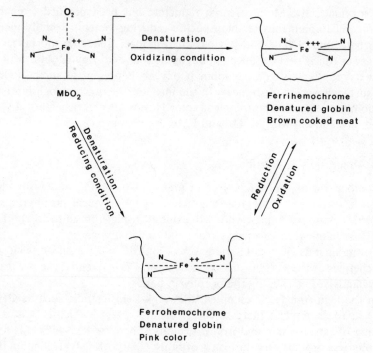

Fig. 4.39. Formation of hemochromes.

ble spectrum similar to oxymyoglobin, but it can be oxidized to ferrihemo-chrome, resulting in browning and color fading.

Processing and storage conditions that cause changes in the structure of the native porphyrin affect the electronic configuration of the heme. Hydrogen peroxide oxidizes the porphyrin to form a green polypyrrolic compound, cholemyoglobin. Hydrogen sulfide reduces the porphyrin of Mb and MbO_2, resulting in the formation of sulfmyoglobin and sulfmetmyoglobin with absorption maxima at 617, and 595 and 715, respectively (23, 29). These usually cause undesirable color changes (usually green) in fresh meat and cured meat. However, the exact mechanisms are not clearly understood.

Nitrosylmyoglobin

In cured meat, the sixth-position ligand of myoglobin is nitric oxide (NO). Nitric oxide binds to both ferrimyoglobin Fe^{3+} and deoxymyoglobin Fe^{2+}. The product in both cases is nitrosylmyoglobin (MbNO), which is low-spin ($s = \frac{1}{2}$) and paramagnetic.

The MO of NO has a single electron in the π^* $2p$ orbital. In the nitrosyl-ferromyoglobin, the complex contains σ bonding between NO π^* and the iron d_z^2 orbitals. The binding in the complex is stabilized via d_{yz} to $p\pi$ back bonding. The MbNO is red, with the visible spectrum similar to MbO_2. The Fe-NO complex is bent at a bond angle of 108–110°. Nitrosylmyoglobin is unstable due to the slow photodissociation of the nitric oxide from the Fe^{2+}NO complex, involving excitation of electrons from the porphyrin π electron cloud and hence the withdrawal of π-electron density from iron to porphyrin (30). This tends to weaken the bond between Fe^{2+} and NO, leaving the Fe susceptible to oxidation. However, cured meat color is stabilized when the protein is denatured to ferrohemochrome. Meat denaturation (or acid pH) of the protein labilizes the histidine F8 ligand to substitution with a second NO ligand, resulting in the formation of a stable diamagnetic complex, $Fe^{2+}(NO)_2$, known as dinitroferrohemochrome.

What is the source of nitric oxide in cured meat? In order to understand the chemistry behind this, we have to look at the reduction-oxidation of nitrogen compounds, which can be represented by the Latimer reduction potential presented below.

Oxidation state	+5	+3	+2	+1	0	−3
	NO_3^-	NO_2^-	NO	N_2O	N_2	NH_4^+
$E°$ (in acidic solution)		+0.94	+0.99	+1.59	+1.77	+0.27
$E°$ (in basic solution)		+0.01	−0.46	+0.76	+0.94	−0.74

Disproportionation occurs if $E°$ to the right of the intermediate is greater than the $E°$ to the left. Thus, in acid solution, HNO_2^- is unstable with respect to NO and NO_3^-. For the reaction $3HNO_2 \rightarrow NO_3^- + H^+ + 2NO + H_2O$, $\Delta E°$ is positive and ΔG $(= -nFE°)$ is negative. In basic solution, NO_2^- is stable with respect to disproportionation to NO and NO_3^-.

In cured meat, the reduction of nitrite to NO requires the presence of a reductant. Reductants such as ascorbate, sulfhydryl compounds, and NADH-flavins can reduce (1) NO_2^- to NO, or (2) Fe^{3+} (FMB) to Fe^{2+} (Mb), which can then reduce NO_2^- to NO. The scheme in Fig. 4.40 summarizes the transformation of myoglobin derivatives during the curing of meat.

CHLOROPHYLL

The chlorophylls are porphyrins complexed to magnesium. A few features of the chlorophylls distinguish them as a unique class of compounds from the nonphotosynthetic porphyrins.

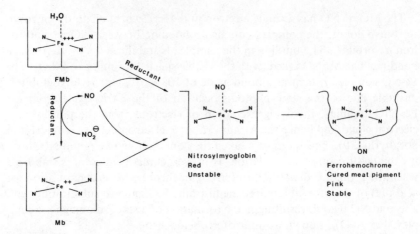

Fig. 4.40. Transformation of myoglobin derivatives during curing of meat.

1. The major chlorophylls of concern (Chl *a* and Chl *b*) contain a reduced pyrrole ring. The porphyrin form with reduction on the pheriphery of one of the pyrrole rings is dihydroporphin (or chlorin) (Fig. 4.41A).

2. Chlorophylls contain an alicyclic ring with a keto carbonyl and a carbomethyoxy group at C-9 and C-10, respectively. (This structural feature is the basis for classification of compounds as chlorophyll.) The porphin form with a cyclopentanone ring is designated as pyroporphin (Fig. 4.41B), and the corresponding chlorin form is pyrrochlorin (Fig. 4.41C).

3. Both Chl *a* and Chl *b* have a long-chain alcohol (phytol) esterified to the propionic acid side chains at C-7 (Fig. 4.41D).

The Magnesium-Ligand Coordination

Since the chlorophyll contains an alicyclic ring with carbonyl substituents, the orbitals do not have square symmetry as in metalloporphyrins. The Q bands are split into x- and y-polarized transitions (Q_x, Q_y) (*31*) (Fig. 4.42).

Unlike myoglobin, hemoglobin, catalase, and peroxidase, which are all complexes of transition metal ions, the chlorophylls are complexed with the alkali earth metal Mg^{2+} (electronic configuration $1s^22s^22p^63s^2$). Magnesium with coordination number of 4 as presented in chlorophyll structure is coordinately unsaturated. At least one of the axial positions must be coordinated with an electron donor ligand.

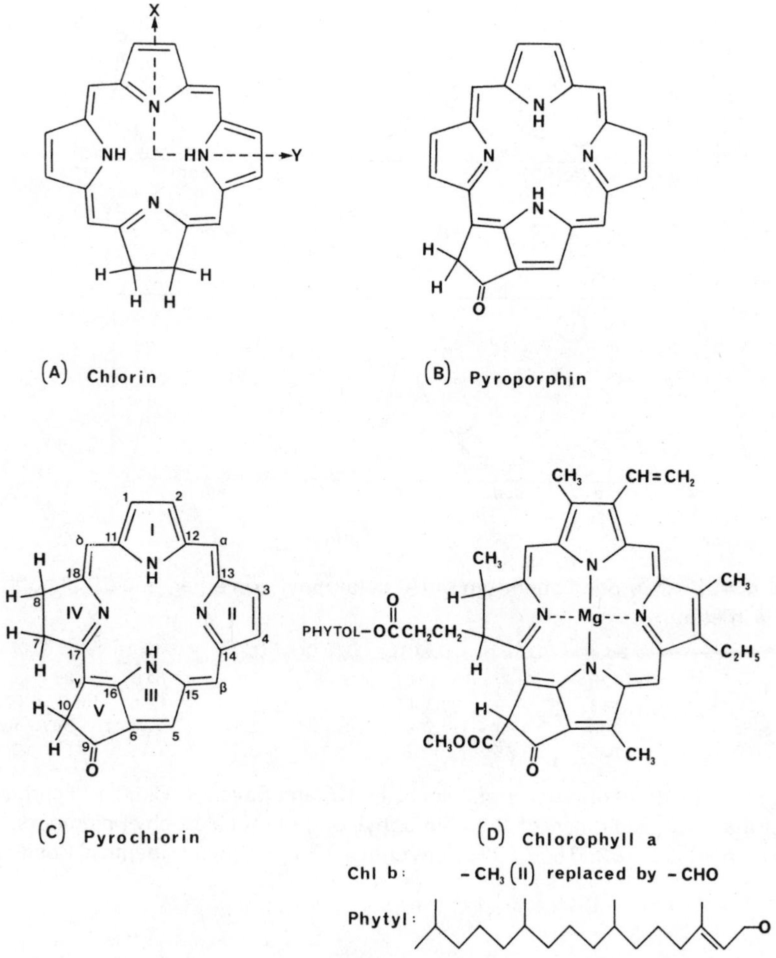

(A) Chlorin

(B) Pyroporphin

(C) Pyrochlorin

(D) Chlorophyll a

Chl b: – CH₃ (II) replaced by – CHO

Phytyl:

Fig. 4.41. Basic structures of chlorophyll.

Dimers and Oligomers

Solvent molecules usually act as electron donor ligands. For example, diethyl ether, acetone, and tetrahydrofuran are nucleophiles with one donor site. The interaction between these molecules and Mg results in monosolvated chlorophyll (with pentacoordinated Mg) (32) (Fig. 4.43A).

Bifunctional nucleophiles such as H_2O and Ch_3OH act as an electron donor and also can hydrogen bond. These $R' - O - H$ ligands can bridge Chl

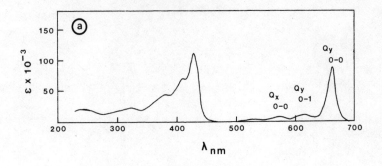

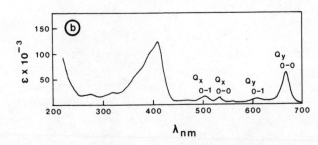

Fig. 4.42. Absorption spectrum of (A) chlorophyll *a* in ether, and (B) pheophytin *a* in ether.

	Soret	$Q_x(0,2)$	$Q_x(0,1)$	$Q_x(0,0)$	$Q_y(0,2)$	$Q_y(0,1)$	$Q_y(0,0)$
Chl *a*	428		530	575		615	661 nm
	113		3.4	6.8		12.6	86.3×10^{-3}
Pheophytin	408	470	506	533	561	609	667 nm
	123	5.85	13.8	12	3.75	9.15	61×10^{-3}

(Reprinted with permission from Houssier, C., and Saver, K. Circular dichroism and magnetic dichroism of the chlorophyll and photochlorophyll pigments. *J. Am. Chem. Soc.* **92**, 781, 782. Copyright 1970, American Chemical Society.)

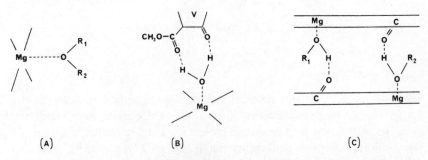

(A) (B) (C)

Fig. 4.43. Cross-linking of chlorophyll via (A) monofunctional nucleophile, and (B) and (C) difunctional nucleophile. (Reprinted with permission from Katz et al. (*33*), copyright 1977, courtesy of National Technical Information Service.)

190

molecules and form oligomeric adducts. The chlorophyll molecule is bridged by one H_2O molecule coordinated to Mg and hydrogen-bonded to the two carboxy oxygens in ring V of another chlorophyll molecule (Fig. 4.43B). Or the chlorophyll molecule (Chl a) is held by two $R' - O - H$ ligands coordinated through its oxygen to the Mg of one Chl a and hydrogen-bonded to the keto $C = O$ of the other Chl a molecule (Fig. 4.43C) (33).

Chlorophyll, being an electron acceptor, also can act as an electron donor. The keto $C = O$ at ring V in the chlorophyll molecule can interact with an Mg atom of another chlorophyll, resulting in the formation of dimers and oligomers. The formation of these aggregates occurs mostly in nonpolar solvents, in which there is little competition for the donor role from other nucleophilic ligands.

The visible spectrum of the Chl a dimer has many overlaps in the Q_x, Q_y transitions, with the latter red-shifted to 675 nm (Fig. 4.44). In fact, all 5- and 6-coordinated Mg complexes have the visible spectrum red-shifted.

Chlorophyll Derivatives

The Mg atom of chlorophyll can be readily replaced by weak acids or Cu^{2+}, Ni^{2+}, Co^{2+}, Fe^{2+}, and Zn^{2+}. The free-base chlorophyll obtained in weak acid is pheophytin (gray-brown color). The visible spectrum of pheophytin shows four bands, Q_x, Q_y, and two overtones. The high-intensity Q_y transition is shifted to a lower wavelength (Fig. 4.42B).

Removal of the phytyl chain from chlorophylls yields the chlorophyllides

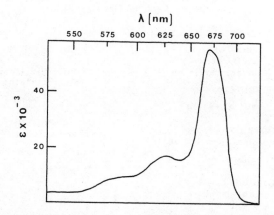

Fig. 4.44. Visible absorption spectrum of chlorophyll a dimer in carbon tetrachloride solution. (Reprinted with permission from Katz et al. (33), copyright 1977, courtesy of National Technical Information Service.)

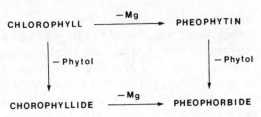

Fig. 4.45. Formation of chlorophyll derivatives.

(34). The hydrolytic reaction can be catalyzed by dilute alkali or by the enzyme chlorophyllase normally present in green plants. Chlorophyllides are water soluble and green-colored, and have the same spectral properties as the chlorophylls. Chlorophyllide with the Mg removed gives the corresponding pheophorbides, which have the same color and spectral properties as those of the pheophytins (Fig. 4.45). The conversion of chlorophyll to pheophytin and pheophorbide has been the most common cause for the loss of the green color in heat-processed green vegetables.

Copper complexes of pheophytin and chlorophyllide are very stable, and the copper ion cannot be removed by acid. Chl *a* and Chl *b*, copper-pheophytin, and copper-chlorophyllide are permitted food colors in many European countries, while in the United States, only Chl *a* and Chl *b* are permit-

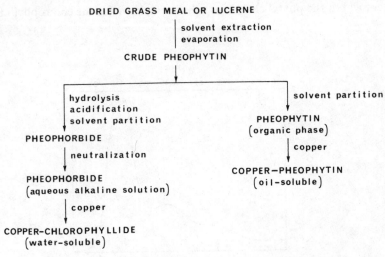

Fig. 4.46. Flowchart of copper-pheophytin and chlorophyllide production. (Reprinted with permission, Humphrey (34), courtesy of Elsevier Applied Science Publishers, Ltd.)

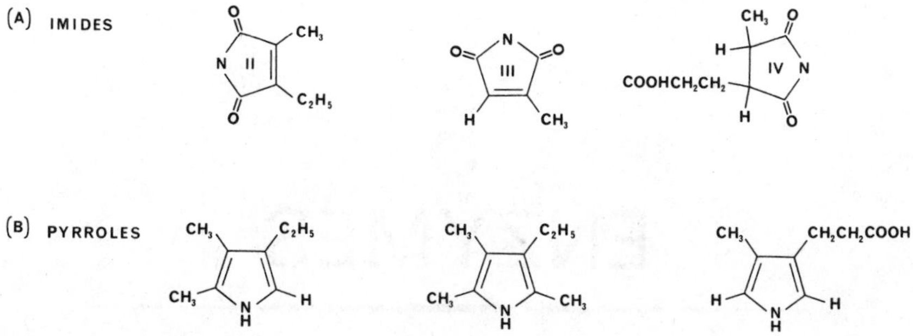

(A) IMIDES

(B) PYRROLES

Fig. 4.47. Products of (A) oxidation, and (B) reduction of chlorophyll.

ted in restricted applications. The commercial production of these color pigments is shown in the scheme in Fig. 4.46. The product should not contain more than 200 ppm free ionizable copper.

Oxidation and Reduction

The vinyl group of chlorophylls a and b can be oxidized by $KMnO_4$ to glycol, formyl, and carboxylic acid substituents. In Chl b, the 3-CHO is also oxidized (Eq. 4-23).

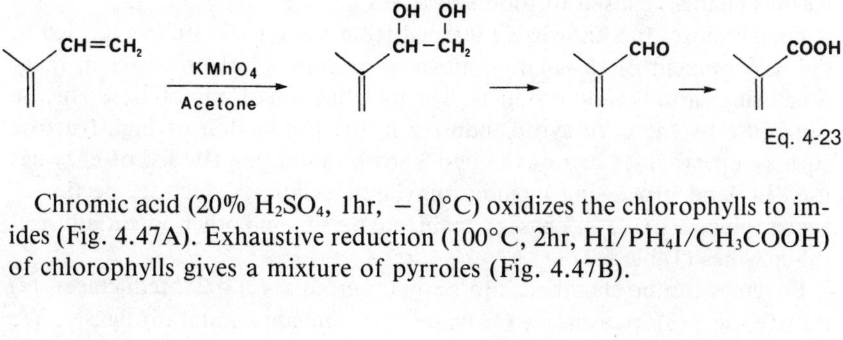

Eq. 4-23

Chromic acid (20% H_2SO_4, 1hr, $-10°C$) oxidizes the chlorophylls to imides (Fig. 4.47A). Exhaustive reduction (100°C, 2hr, $HI/PH_4I/CH_3COOH$) of chlorophylls gives a mixture of pyrroles (Fig. 4.47B).

5
ENZYMES

The study of enzymes is a subject of special importance to food science in two aspects. Enzymes are known to cause numerous changes, desirable or undesirable, to the chemical and physical attributes in a food system. Some notable examples include the many flavor compounds generated by the action of lipoxygenases on unsaturated lipids, the change of color caused by polyphenol oxidase, and the softening of texture in ripening fruits by pectic enzymes. A thorough understanding of the detailed mechanism of catalysis and regulation of these enzymes is essential to the effective control of the various changes caused to food substrates.

Furthermore, the knowledge learned from these basic studies has led to the development of the many innovative technological processes in manufacturing various food products. The introduction of immobilized glucose isomerase by the corn syrup industry in the production of high-fructose corn syrup provides one of the best known examples. The list of enzymes used in food processing includes amylases, cellulases, lactase, pectic enzymes, glucose oxidase, lipase, papain, chymosin, and other potentially useful enzymes (Table 5.1) (1).

Enzymes can be classified into six main groups: (1) oxidoreductases, (2) transferases, (3) hydrolases, (4) lyases, (5) isomerases, and (6) ligases. We will discuss several selected enzymes of importance in food science, with emphasis on the mechanism of catalysis, to provide an overview of the underlying chemical principles of enzymatic reactions.

PAPAIN

The proteolytic enzyme, papain (EC 3.4.22.2), is found in the latex and in the fruit of *Carcia papaya*. The crude latex (which also contains peptidase

A and chymopapain) is used to prevent "chill hazes" in beer and is commonly used as a meat tenderizer.

The Active-Site Region

Papain is a sulfhydryl proteinase, consisting of a single polypeptide chain with 212 amino acid residues. The amino acid sequence of papain is known and its tertiary structure has been determined (2). The protein molecule consists of two domains with a deep cleft, where the active site is situated. The essential cysteine-25 in the left domain is in close proximity to the imidazole of histidine-159 in the opposite wall of the cleft. The histidine is embedded in a hydrophobic region and hydrogen-bonded to asparagine-175 (Fig. 5.1).

The enzyme catalyzes the hydrolysis of amide, ester, and thioester substrates. The reaction is a two-step process, (1) acylation with the formation of an acyl-enzyme intermediate, and (2) deacylation, which is the hydrolysis of the intermediate (Eq. 5-1).

$$E-SH \ + \ R-\overset{O}{\overset{\|}{C}}-X \ \xrightarrow[\text{ACYLATION}]{\text{HX}} \ E-S-\overset{O}{\overset{\|}{C}}-R \ \xrightarrow[\text{DEACYLATION}]{H_2O} \ E-SH \ + \ RCOOH \quad \text{Eq. 5-1}$$

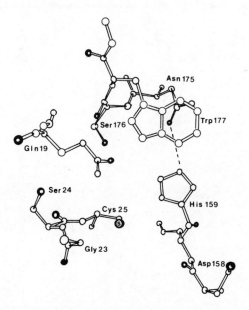

Fig. 5.1. The active-site region of papain near the essential SH group. (Reprinted with permission from Drenth et al. (2), courtesy of Academic Press, Inc.)

Table 5.1. Enzymes Used in Food Processing[a]

ENZYME	SYSTEMATIC NAME[a]	EC	SOURCE
α-Amylase	1,4-α-D-glucan glucanohydrolase	3.2.1.1	*Aspergillus niger*, var. *Aspergillus oryzae*, var. *Rhizopus oryzae*, var. *Bacillus subtilis*, var. barley malt *Bacillus licheniformis*, var.
β-Amylase	1,4-α-D-glucan maltohydrolase	3.2.1.2	barley malt
Bromelain	None	3.4.22.4	pineapples: *Ananas comosus*, *Ananas bracteatus*
Catalase	Hydrogen-peroxide: hydrogen-peroxide oxidoreductase	1.11.1.6	*Aspergillus niger*, var. bovine liver *Micrococcus lysodeikiticus*
Cellulase (endo-1,3(4)-β-D-glucanase)	1,4-(1,3:1,4)-β-D-glucan 3(4)-glucanohydrolase	3.2.1.4	*Aspergillus niger*, var. *Trichoderma reesei*
Ficin	None	3.4.22.3	figs: *Ficus SP.*
β-Glucanase	1,3-(1,3:1,4)-β-D-glucan 3(4)-glucanohydrolase	3.2.1.6	*Aspergillus niger*, var. *Bacillus subtilis*, var.
Glucoamylase (exo-1,4-α-D-glucosidase)	1,4-α-D-glucan glucohydrolase	3.2.1.3	*Aspergillus niger*, var. *Aspergillus oryzae*, var. *Rhizopus oryzae*, var.
Glucose isomerase	D-xylose ketol-isomerase	5.3.1.5	*Actinoplanes missouriensis* *Bacillus coagulans* *Streptomyces olivaceus* *Streptomyces rubiginosus*
Glucose oxidase	β-D-glucose:oxygen 1-oxidoreductase	1.1.3.4	*Aspergillus niger*, var.

Common name	Systematic name	EC number[a]	Source
Hemicellulase	None	None	Aspergillus niger, var.
Invertase (β-D-fructo-furanosidase)	β-D-fructofuranoside fructohydrolase	3.2.1.26	Saccharomyces sp.
Lactase (β-D-galactosidase)	β-D-galactoside galactohydrolase	3.2.1.23	Aspergillus niger, var. Aspergillus oryzae, var. Saccharomyces sp.
Lipase (triacylglycerol lipase)	Triacylglycerol acylhydrolase	3.1.1.3	animal pancreatic tissues Aspergillus oryzae, var. Aspergillus niger, var.
Papain	None	3.4.22.2	papaya: Carica papaya (L)
Pectic enzymes Polygalacturonase	poly (1,4-α-D-galacturonide) glycanohydrolase	3.2.1.15	Aspergillus niger, var. Rhizopus oryzae, var.
Pectinesterase Pectate lyase	Pectin pectylhydrolase Poly(1,4-α-D-galacturonide) lyase	3.1.1.11 4.2.2.2	
Pepsin	None	3.4.23.1	porcine or other animal stomachs
Protease	None		
Microbial serine protease	None	3.4.21.14	Aspergillus niger, var.
Microbial neutral protease	None	3.4.24.4	Aspergillus oryzae, var. Bacillus subtilis, var. Bacillus licheniformis, var.
Chymosin	None	3.4.23.4	fourth stomach of ruminant animals Endothia parasitica Mucor miehei, M. pusillus
Trypsin	None	3.4.21.4	Animal pancrease

[a]IUPAC-IUB Enzyme nomenclature 1978.

The Ionization of the Essential Groups

The kinetics of hydrolysis by papain of several substrates suggest that acylation has a bell-shaped pH rate profile with pK_a values of 3.0–4.0 and 8.0–8.5. Deacylation of the acyl-enzyme intermediate, however, shows a sigmoidal pH rate profile with a pK_a of approximately 4.0. The ionization groups responsible have been identified to be cysteine-25 and histidine-159.

The exceptionally low pK_a (3.0–4.0) of Cys-25 and the shifting of the pK_a from 8.5 to 4.0 for His-159 is attributed to the interactive ionization of these two groups, as evidenced by NMR and fluorescence titration studies (*3*). The pK_a of the thiol group changes from 7.6 to 3.3 on protonation of the histidine. Similarly, the pK_a of the histidine shifts from 8.5 to 4.3 when the thiol group is protonated (Eq. 5-2).

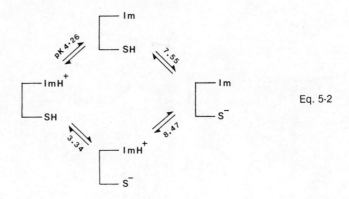

Eq. 5-2

Reaction Mechanism

Therefore, at physiological pH, the imidazole with a pK_a of 8.5 will remain protonated, while the thiol group with its pK_a shifted from 7.6 to 3.3 is deprotonated. The values of these pK's, together with other studies, suggest that Cys-25 and His-159 may exist as an imidazolium-thiolate ion pair (Fig. 5.2) (*4*).

The active form of the enzyme thus has the thiolate anion acting as a strong nucleophile to form a tetrahedral intermediate with the carbonyl car-

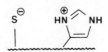

Fig. 5.2. The imidazolium-thiolate ion pair.

bon of the substrate (Eq. 5-3). Formation of the tetrahedral intermediate blocks the thiol group, causing the pK_a of the imidazole of His-159 to decrease from 8.5 to 4.0. This increase in the acidity of the imidazolium cation facilitates the proton transfer from the imidazole to the leaving group and aids the formation of the acyl-enzyme intermediate (5). It is also conceivable that a concerted attack of the ion pair would also result in the formation of the tetrahedral intermediate (3).

Eq. 5-3

Deacylation occurs through general base catalysis assisted by the imidazole (Eq. 5-3), although intramolecular nucleophilic catalysis by the imidazole has also been suggested (6).

Action on Meat Fractions

Besides papain, other sulfhydryl proteinases, such as ficin (from figs) and bromelain (pineapple), are also used to increase the tenderness of meat. All these enzymes are shown to solubilize the various fractions of meat muscle to various degrees (Table 5.2) (7).

The degree of hydrolysis of these fractions varies with the individual enzyme. Papain and ficin show the strongest activity on the salt-soluble fraction, while the other enzymes hydrolyze the insoluble fractions most efficiently. Antemortem injection of papain indicates that approximately 12–19% of the injected enzyme is found in the muscles.

LIPOXYGENASE

Lipoxygenase (linoleate:oxygen oxidoreductase, EC 1.13.11.12) is a dioxygenase that catalyzes the oxygenation of polyunsaturated fatty acids con-

Table 5.2.   Percent of Meat Fractions Solubilized by Enzyme Treatment

| | MEAT FRACTIONS | | |
ENZYME	WATER SOLUBLE	SALT SOLUBLE	INSOLUBLE
Papain	25	60	15
Ficin	9	54	37
Bromelain	16	33	51
Collagenase	13	26	61
Trypsin	19	38	43

Reprinted with permission from Kang and Rice (7), copyright 1970 by Institute of Food Technologists.

taining a *cis,cis*-1,4-pentadiene system to hydroperoxides. The enzyme exists in multiple forms, 4 in soybean, wheat, and peas, and 2 in corn. Lipoxygenases from different sources, as well as their isozymes, may differ in substrate specificity, pH optimum, and oxidation activity, as indicated in Table 5.3 (*8, 9*).

Soybean Lipoxygenase I

Most of the present information regarding lipoxygenase comes from studies on lipoxygenase I, the isozyme first isolated and crystallized from soybeans. The lipoxygenase consists of a single polypeptide chain with molecular weight $\sim$ 100,000, containing one atom of nonheme iron per molecule of protein. Soybean lipoxygenase I has a pH optimum of 9.0, although for most lipoxygenases, the pH range is 5.5–7.0.

Table 5.3.   Characteristics of Lipoxygenases from Soybean and Pea

| | SOYBEAN | | PEA | |
	LOX 1	LOX 2	LOX 1	LOX 2
Mw	98500	98500	95000	95000
pH optimum	9.0	6.5	7.0	6.0
pI	5.5–5.7	5.8–6.2	6.25	5.82
9/13 OOH at pH optimum	10:90	50:50	50:50	—
Carotenoid cooxidation activity	low	high	high	low

From Galliard and Chan (*8*) and Yoon and Klein (*9*).

Regiospecificity and Stereospecificity

Lipoxygenases oxygenate polyunsaturated fatty acids with a *cis,cis*-1,4-pentadiene system. The ω-6 carbon must be a double bond, and the carboxyl group must be unhindered for the enzyme to act. Substrates containing double bonds at the ω-6 and ω-9 positions (such as linoleic acid) are usually oxidized at high rates. Chain length and number of double bonds seem to have little effect.

The orientation of the pentadiene system on the enzyme is considered to be planar, and the hydrogen abstraction at ω-8 (Fig. 5.3) can occur either above or below the plane, H(*Ls*) or H(*Dr*) (*10*). Oxygen then enters from the opposite side of the plane from which the hydrogen is abstracted, at position ω-6 or ω-10. Corn germ lipoxygenase at an optimum pH of 6.6 oxidizes linoleic acid to primarily 9-*Ds*-hydroperoxy-10-*trans,*12-*cis*-octadecadienoic acid—abstraction of H(*Dr*), followed by O_2 attack above the plane at the ω-10 position. Lipoxygenase I from soybeans, at an optimum pH of 9.0, converts linoleic acid predominantly to 13-*Ls*-hydroperoxy-9-*cis,* 11-*trans*-octadecadienoic acid—abstraction of H(*Ls*) followed by oxygen addition to the ω-6 position from below the plane.

In the oxidation of arachidonic acid (all-*cis*-5,8,11,14-eicosatetraenoic acid) by soybean lipoxygenase I, the product of the first oxygenation is 15-*Ls*-hydroperoxy-5-*cis,*8-*cis,*11-*cis,*13-*trans*-eicostetraenoic acid. The second

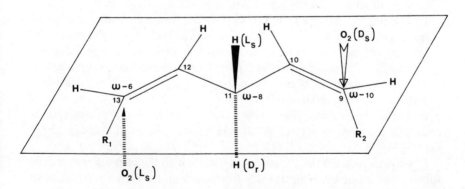

Fig. 5.3. Oxygenation of linoleic acid by lipoxygenase. The pentadiene is planar. The hydrogen at ω-8, H(*Ls*) and H(*Dr*), are above and below the plane, respectively. The oxygen approaches from above or below. $R_x = CH_3(CH_2)_4-$, $R_2 = COOH-(CH_2)_7-$. The symbols *L* and *D* refer to Fischer convention; *R* and *S* refer to Cahn, Ingold, and Prelog convention. (Reprinted with permission from Egmond et al. (*10*), courtesy of Academic Press, Inc.)

oxygenation produces the 8-Ds,15-Ls-dihydroperoxy-5-cis,9-$trans$,11-cis, 13-$trans$, and 5-Ds,15-Ls-dihydroperoxy-6-$trans$,8-cis,11-cis,13-$trans$-eicosatetraenoic acids (*11*).

The Iron in Lipoxygenase

Lipoxygenase in its resting form is colorless and EPR silent (*12*). The iron is high-spin Fe(II) ($S = 4/2$, paramagnetic) with one of the ligands being dioxygen. The iron atom is believed to be coordinated to amino acid residues of the polypeptide chain.

The ferrous form is converted, by the addition of an equal molar concentration of 13-hydroperoxy-octadecadienoic acid, to a yellow ferric enzyme with an EPR signal indicating high-spin Fe(III) ($S = 5/2$). Excess hydroperoxide results in an unstable purple complex between the ferric enzyme and 13-LOOH, which reverts to the activated yellow form with concomitant conversion of the hydroperoxy to a hydroxyepoxy product (*13*). In the case of 13-hydroperoxy-linoleate, the product of conversion is 11-hydroxy-12:13-epoxy-9-cis-octadecenoic acid. The yellow and purple forms differ in symmetry, the former having largely axial ligands and the latter rhombic symmetry.

The Aerobic Reaction Mechanism

The reaction scheme for soybean lipoxygenase I is outlined in Fig. 5.4 (*14*).

1. The native resting enzyme (E-Fe^{2+}-O^2) is first activated to the active form (E-Fe^{3+}). The initial activation reaction is not clear, but the period for the resting enzyme to react with the substrate linoleic acid can be abolished by the addition of its own product, the 13-Ls-hydroperoxide. Presumably, a one-electron transfer occurs from iron to oxygen, which is released as superoxide anion or excited oxygen.

2. The active enzyme (E-Fe^{3+}) catalyzes the stereospecific abstraction of hydrogen from the substrate, and it is reduced to the enzyme radical complex (E-Fe^{2+}-L·).

3. Oxygen then combines stereospecifically with the enzyme-free radical followed by a one-electron transfer from the iron to the peroxy radical to form the enzyme peroxy-anion (E-Fe^{3+}·LOO$^-$).

4. Protonation of the peroxy anion releases the hydroperoxide and regenerates the active enzyme (E-Fe^{3+}).

The mechanism described is referred to as the aerobic cycle in which the product is hydroperoxide. However, under anaerobic conditions, reaction

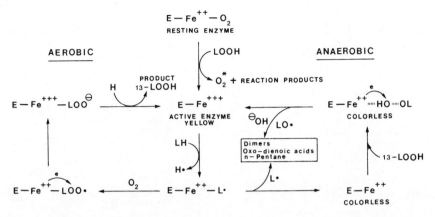

Fig. 5.4. The aerobic and anaerobic reaction mechanisms of soybean lipoxy-genase I. (Reprinted with permission from deGroot et al. (*14*), courtesy of Elsevier Science Publishers, Biomedical Division, Amsterdam, The Netherlands.)

products other than hydroperoxides are formed. These include oxo acids, pentane, and dimers.

The Anaerobic Reaction Mechanism

Two systems can lead to the formation of products other than hydroperoxides: (1) with linoleic acid and the preformed substrate (13-*Ls*-hydroperoxy-9-*cis*, 11-*trans*-octadecadienoic acid) in the absence of oxygen, and (2) oxygen in limiting concentration and the substrate linoleic acid in excess. In the latter system, there is a lag period during which hydroperoxide is formed, and the oxygen is depleted before the conversion reaction starts (*15, 16*).

The anaerobic reaction is believed to proceed according to the following scheme (Fig. 5.4):

1. The enzyme radical complex (E-Fe^{2+}-$L\cdot$) dissociates to yield the free enzyme in the ferrous state (E-Fe^{2+}) and the fatty acid radical ($L\cdot$).

2. The enzyme in the ferrous state (E-Fe^{2+}) is oxidized to the active form (E-Fe^{3+}) by the product hydroperoxide (13-L-LOOH), with the formation of alkoxy radical ($LO\cdot$) and hydroxide ion (OH$^-$).

3. Reaction of fatty acid radicals ($L\cdot$ and $LO\cdot$) and fatty acids (LH) result in the formation of oxodienoic acids and dimeric fatty acids. Oxodienoic acids are formed from the rearrangement and decomposition of hydroper-

oxide via these fatty acid radicals (Eq. 5-4). Nonoxygenated dimeric fatty acids stem from the combination of $2L\cdot$ radicals. Dimers containing oxygen originate from reaction between $LO\cdot$ and $L\cdot$ (Eq. 5-5).

The Fate of Hydroperoxide

The immediate products of lipoxygenase, the hydroperoxides, do not accumulate in the cell tissues but undergo further enzymatic rearrangement or breakdown to secondary products.

Hydroperoxide isomerase catalyzes the conversion of hydroperoxide to

α-keto fatty acids. One of the oxygen atoms from the hydroperoxide is transferred to the vicinal olefinic carbon, incorporated into the oxo group. The mechanism for the action of linoleic acid hydroperoxide isomerase on 9-hydroperoxy-linoleic acid may involve an epoxy-cation intermediate. Nucleophilic substitution at C-9 or C-13 carbon and hydride shift from C-10 to C-11 or C-9 yield the 10-oxo-9-hydroxy-12-*cis*-octadecenoic acid (α-ketol) and 10-oxo-13-hydroxy-11-*trans*-octadecenoic acid (γ-ketol), respectively (Eq. 5-6). The configuration at C-9 is inverted from *Ds* to *Lr* (*17, 18*).

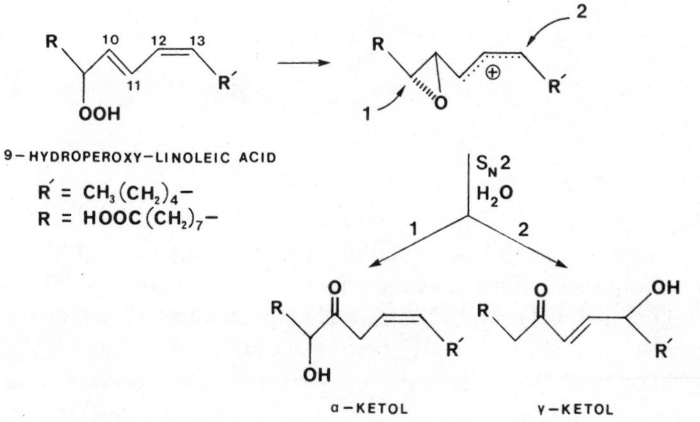

9−HYDROPEROXY−LINOLEIC ACID

$R' = CH_3(CH_2)_4-$
$R = HOOC(CH_2)_7-$

α−KETOL γ−KETOL

Eq. 5-6

However, the enzyme that catalyzes the reactions leading to the formation of volatile compounds with characteristic odors in plants, hydroperoxide lyase, has been identified and partially purified from cucumber and tomato fruits and watermelon seedlings. The enzyme cleaves hydroperoxide to oxoacids and carbonyls. The enzymatic sequences for the formation of carbonyl fragments from linoleic acid in cucumber fruits is proposed in Eq. 5-7 (*19*). *Trans*-2-nonenal is a major flavor compound of cucumber. The other two aldehydes, hexenals and nonadienals, originate from linolenic acid via similar pathways.

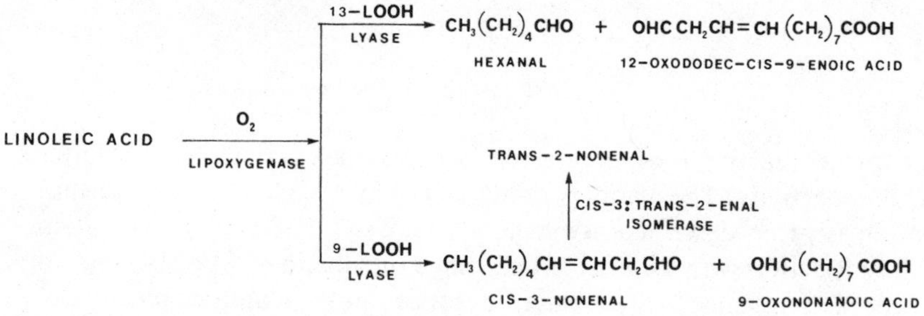

Eq. 5-7

Cooxidation

The ability of lipoxygenase to bleach pigments has been applied to bread making. Commercially, soybean flour is added to wheat flour. The oxidative destruction of pigments (e.g., carotenoids, chlorophyll) during lipid oxidation is believed to be a free-radical mediated reaction (20). Both $LOO\cdot$ and $L\cdot$ radicals have been suggested as the reactive species (Eq. 5-8).

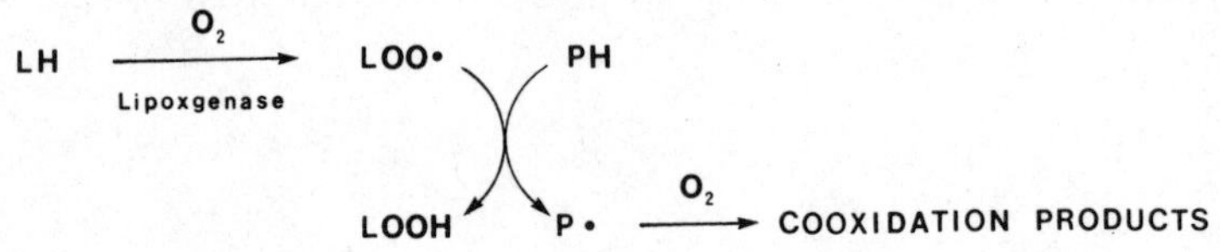

Eq. 5-8

Polyphenol Oxidases

Polyphenol oxidase is a copper-containing enzyme that functions as: (1) a monooxygenase in the o-hydroxylation of monophenols to o-dihydroxyphenols (monophenol, dihydroxyphenylalanine:oxygen oxidoreductase, EC 1.14.18.1) (Eq. 5-9.1), and (2) a two-electron oxidase in the oxidation of o-dipehnols to o-quinones (1,2-benzenediol:oxygen oxidoreductase, EC 1.10.3.1) (Eq. 5-9.2).

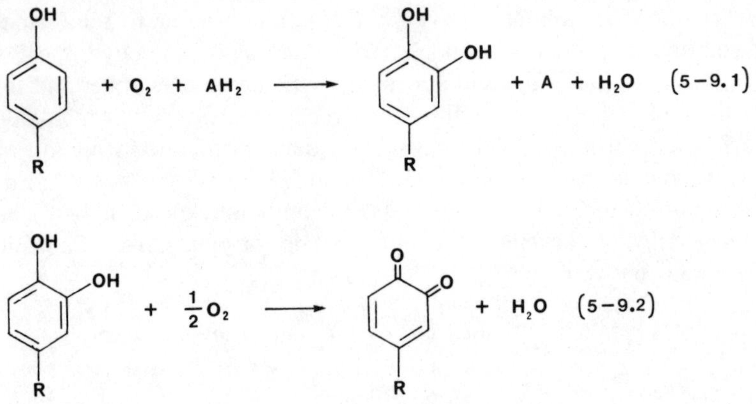

Eq. 5-9

The former reaction is often referred to as cresolase activity and the latter catecholase activity. The complex monophenol hydroxylation and diphenol oxidation is often described as phenolase activity. These enzymes, which generally function with a rather broad range of substrates, are also known variously as tyrosinase, catecholase, cresolase, polyphenolase, phenolase, and catechol oxidase in the literature.

Enzyme Characteristics

Polyphenol oxidases have broad substrate specificity. However, enzymes from different sources tend to differ from each other in their relative activity toward specific substrates. All polyphenol oxidases oxidize o-diphenol, but some may not have cresolase-type activity. Some of the commonly used diphenol and monophenol substrates include catechol, 4-methylcatechol, dopamine (3,4-dihydroxyphenylethylamine), pyrogallol, catechin, caffeic acid, chlorogenic acid, p-cresol, tyrosine, and p-hydroxycinnamic acid.

Most studies have been done on the polyphenol oxidases isolated from *Agaricus bisporus* (common mushroom) and *Neurospora crassa*. Polyphenol oxidase from the common mushroom has a molecular weight of 120,000 and contains four copper atoms per molecule. The enzyme contains two types of polypeptide chains, the heavy chains (H) with MW 43,000 ± 1,000 and the light chains (L) with MW 13,400 ± 600, and has a quaternary structure of L_2H_2 (*21*). Each active site contains a pair of antiferromagnetically coupled cupric ions (binuclear) in the resting enzyme.

The *Neurospora* enzyme, however, is a single-chain protein of 407 amino acids with MW 46,000, and contains a single binuclear copper site. The complete amino acid sequence is known, and a tripeptide containing a thioether bridge between Cys-94 and His-96 has been described (*22*).

Reaction Mechanism

In the oxygenated form, the binuclear cupric site is tetragonal with protein ligands (histidine residues) binding in the equatorial plane and one or two axial ligands such as H_2O. The two coppers are also bridged by an endogenous protein ligand (such as phenolate or carboxylate groups) and an exogenous oxygen molecule bound as peroxide (Eq. 5-10) (*23*). When the copper ions are bridged by an exogenous ligand other than peroxide, the enzyme is in the met form. The deoxy form contains a bicuprous structure without exo- or endogenous bridging.

The met form is converted, by a 2e- reduction, to the deoxy form which binds molecular oxygen reversibly to form oxytyrosinase. Peroxide addition to the met form also yields the oxy enzyme.

Hydroxylation and Oxidation of Monophenol. The initial reaction involves the coordination of a monophenol substrate to the axial position of one of the coppers of oxytyrosinase ($E°O_2$), followed by a distortion of the tetragonal site to a trigonal bipyramidal intermediate (Eq. 5-11). This rearrangement of the copper coordination geometry labilizes the peroxide, introducing polarization into the $O-O$ bond to yield an activated peroxide that can then hydroxylate the phenol substrate. The product thus formed

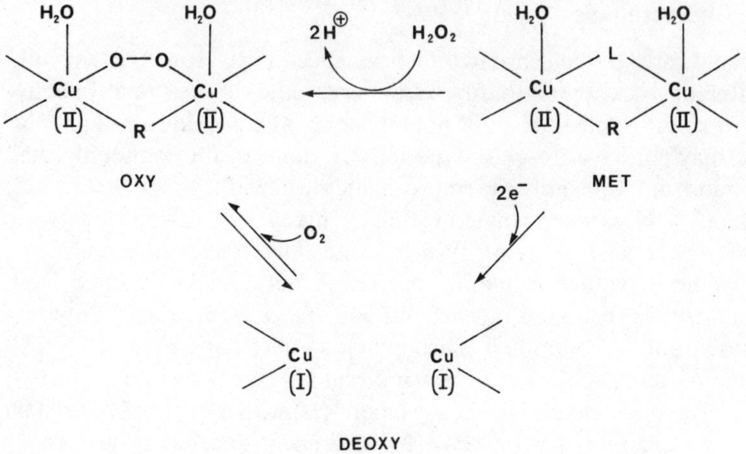

Eq. 5-10

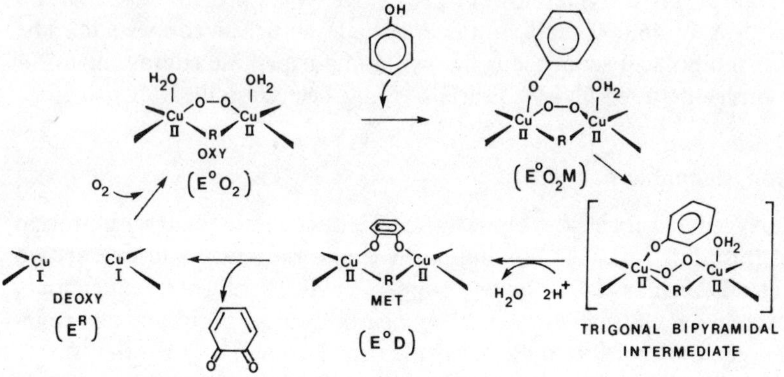

Eq. 5-11

is the met form ($E°D$), where the exogenic ligand, in this case, is the ortho-diphenol substrate (Eq. 5-11) (24).

The protein pocket surrounding the copper site contributes to the stability of the rearranged geometry, and also tends to position the substrate optimally for associative ligand displacement and the electrophilic attack of the peroxide oxygen at the ortho position of the substrate.

Finally, oxidation of the o-diphenol occurs with the product, o-quinone, dissociated from the enzyme, while Cu(II) is reduced to Cu(I). The deoxy form (E^R) can then bind O_2 to give the oxygenated enzyme ($E°O_2$), allowing further turnover.

Oxidation of Diphenol.   In the oxidation of diphenol, the tetragonal site has the correct geometry for both phenolic oxygens to coordinate in the deoxy (E^R), oxy ($E^\circ O_2$), or the met (E_o) forms (25). Rearrangement of the coordination geometry of the copper site is not required for the two-electron oxidation. The catalytic cycle for oxidation of diphenol substrate to o-quinone is presented in Eq. 5-12.

Eq. 5-12

E^R	Reduced enzyme $\left(Cu\ I\right)$
E°	Oxidized enzyme $\left(Cu\ II\right)$
$E^R D$	Reduced, diphenol
$E^\circ D$	Oxidized, diphenol
$E^\circ O_2$	Oxidized, oxygenated
$E^\circ O_2 D$	Oxidized, oxygenated, diphenol
$E^\circ O_2 M$	Oxidized, oxygenated, monophenol

Binding of an o-diphenol to mettyrosinase, followed by the reduction of the bicupric ions, leads to deoxytyrosinase (E^R), with the substrate oxidized to o-quinone. The deoxy form binds an oxygen molecule to form the oxy-tyrosinase $(E°O_2)$ which coordinates a second molecule of o-diphenol to yield $E°O_2D$. Subsequent oxidation of the substrate yields the product o-quinone and regenerates the met form $(E°)$.

Alternately, the deoxytyrosinase (E^R) can coordinate a second molecule of o-diphenol before oxygenation. The coordinated complex (E^RD) binds an oxygen molecule to form $E°O_2D$. Oxidation of the substrate gives the mettyrosinase and the o-quinone.

Secondary Reaction Products

The significance of these enzymes in food arises from the formation of the quinones, which may further take part in secondary reactions: (1) coupled oxidation with other substrates, and (2) condensation and polymerization to cause darkening of the plant tissue.

Compounds such as ascorbic acid, anthocyanins, and many other phenolic compounds reduce o-quinone to diphenol (Eq. 5-13). Therefore, these compounds, which would not have been oxidized directly by the enzyme, are oxidized indirectly via coupling reactions with the o-quinone.

$$\text{O-QUINONE} + \text{RH}_2 \longrightarrow \text{O-DIPHENOL} + \text{R} \qquad \text{Eq. 5-13}$$

Quinones can condense with phenolic compounds, a typical example of which is the formation of theaflavin in leaves, as discussed in Chapter 6. Another example is the oxidation of dopamine (26), the primary substrate (1–2 mg/g dry weight) in banana, to dopamine quinone, which then undergoes nonenzymatic rearrangement to indole-5,6-quinone that in turn is polymerized to form melanin (Eq. 5-14).

Eq. 5-14

GLUCOSE OXIDASE

Glucose oxidase (β-D-glucose:oxygen 1-oxidoreductase, EC 1.1.3.4) has been highly purified from fungi *Penicillium notatum, Penicillium amagasakiense,* and *Aspergillus niger.*

Enzyme Characteristics

The enzyme is a dimer, has a molecular weight of 180,000, and contains 16% (w/w) carbohydrates and two moles of firmly bound FAD per molecule of protein. The enzyme contains no sulfhydryl or disulfide groups. However, a covalently bound phosphorus residue has recently been shown to link two amino acids in the polypeptide chain of the enzyme, similar to disulfide links stabilizing protein secondary structures (27).

Glucose oxidase catalyzes the irreversible oxidation of a number of aldoses to the corresponding lactones (Eq. 5-15). Glucose, deoxyglucose, mannose, galactose are some of the substrates studied (Table 5.4). It is evident that glucose is a far better substrate than the other sugars (28).

$$(5\text{-}15.1)$$

REDUCTIVE HALF-REACTION: $\qquad\qquad\qquad\qquad\qquad (5\text{-}15.2)$

$\text{H}^+\text{E.FAD}$

$\Big\updownarrow \text{K}_1$

$\text{E.FAD} + \text{S}_1 \text{ (GLUCOSE)} \rightleftharpoons \text{E.FAD-S}_1 \longrightarrow \text{E.FADH}_2\text{-P}_1 \longrightarrow \text{E.FADH}_2 + \text{P}_1 \text{(LACTONE)}$

OXIDATIVE HALF-REACTION: $\qquad\qquad\qquad\qquad\qquad (5\text{-}15.3)$

$\text{H}^+\text{E.FADH}_2$

$\Big\updownarrow \text{K}_2$

$\text{E.FADH}_2 + \text{S}_2 \text{ (O}_2\text{)} \longrightarrow \text{E.FAD} + \text{P}_2 \text{ (H}_2\text{O}_2\text{)}$

Eq. 5-15

Table 5.4. Relative Activity of Glucose Oxidase on Various Substrates

	SOURCE OF ENZYME		
SUBSTRATE	A. NIGER	P. NOTATUM	P. AMAGASAKIENSE
Glucose	100	100	100
2-Deoxy-D-glucose	3.3	—	—
Mannose	0.9	1.0	0
Galactose	0.55	0.14	0

From Gibson et al. (28).

Reaction Mechanism

We will limit the present discussion to glucose as the substrate. The product in this case will be δ-D-gluconolactone (5-15.1). The reaction is pH dependent and can be expressed in two redox half reactions (5-15.2, 5-15.3) (29).

Both the reductive and oxidative steps depend on a prototropic group in the enzyme. The substrate (glucose) can only bind with the unprotonated enzyme (EFAD) and the pK_1 of the prototropic group is 5.00. In the oxidative reaction, the binding of O_2 to the reduced enzyme is an acid-catalyzed process, requiring the protonation of a group with pK_2 of 6.90. The ionization groups are postulated to be carboxyl (pK_1) and histidine imidazole (pK_2). The dissociation constant K_2 has been found to be extremely sensitive to ionic strength. Increasing the ionic strength from 0.025 to 0.225 causes a three-fold increase. The intrinsic pK_a of the acidic group is calculated to be 6.7, consistent with the group being a histidine residue.

The Two-Electron Transfer Mechanism

The reduction of flavin by the substrate and the oxidation of the latter to the δ-lactone (E·FAD-S_1 → E·FADH$_2$-P_1) involves hydride transfer from carbon to the flavin. The hydride transfer within the E·FAD-S_1 is assisted by the result of proton transfer from the glucose-1-hydroxyl to a carboxylate anion of the enzyme (Eq. 5-16) (30).

Eq. 5-16

However, in a study of (1) the reduction of glucose oxidase (EFAD) using furoin (Eq. 5-17.1) and (2) the oxidation of reduced glucose oxidase (EFADH$_2$) using nitroxide (5-17.2) as a model, semiquinone intermediates have been detected (*31*).

Failure to detect radical intermediates in the reduction of $^3O_2$ to H_2O_2 by reduced glucose oxidase has been attributed to the high ΔG for the $-C(OH)CO-$ radical formation. The free energy for the second electron transfer is small, and the radical is less likely to be detected. Whereas in the case of electron transfer from furoin to glucose oxidase, the formation of furil has a comparatively small ΔG, the free-energy barrier for the second electron transfer is high and the radical is stable enough to be detected.

A one-electron transfer mechanism based on this study can then be represented by Eq. 5-18. The enzyme EFAD is reduced to the semiquinone by the substrate (glucose in Eq. 5-18), forming a radical-pair intermediate. A second electron transfer results in the formation of the reduced enzyme EFADH$_2$ and the product δ-lactone. Likewise, regeneration of the oxidizing enzyme proceeds via a radical-pair intermediate between EFADH· and O_2^-. However, since this study is carried out with a one-electron reductant and acceptor, the results need to be confirmed as a valid model for the natural substrates, glucose and O_2.

The argument against a semiquinone intermediate is based on the study

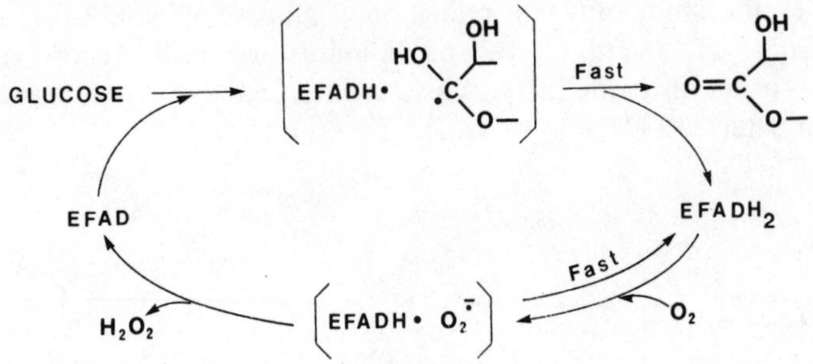

Eq. 5-18

indicating that it is kinetically unfavorable to transfer a second electron in the reduction reaction in which one-electron reductants were used (*32*). The presence of a kinetic barrier is indicated by the fact that in spite of the small difference in the potential (2–40 mV) of the two separate electron transfers (Eq. 5-19), quantitative production (50–70%) of radicals can be obtained using dithionite as the reductant. In addition, since the potential difference is small, it is thermodynamically unfavorable to the formation of a stable glucose oxidase radical, and favors a simultaneous 2*e* transfer mechanism.

	E_m pH 5.3	E_m pH 9.3
$\text{EFAD (Oxidized form)} + e^{\ominus} + H^{\oplus} \rightleftharpoons \text{EFADH}\cdot \text{ (Semiquinone)}$	-0.063	-0.200
$\text{EFADH}\cdot + e^{\ominus} + H^{\oplus} \rightleftharpoons \text{EFADH}_2 \text{ (Reduced form)}$	-0.061	-0.240

Eq. 5-19

Industrial Uses

Application of the enzyme glucose oxidase in food, usually with catalase to decompose the H_2O_2, can be summarized in two categories according to its function (Table 5.5).

Table 5.5. Uses of Glucose Oxidase

FUNCTION	SPECIFIC APPLICATION
1. Removal of glucose	—to prevent Maillard reaction in egg solids, dried meats, potatoes
2. Removal of oxygen	—to prevent oxidation in beer
	—to prevent browning in white wine
	—to stabilize oil in water emulsion against rancidity

AMYLASES

The amylases are glycosidases that catalyze the hydrolysis of α-1,4-glucose polymers by the transfer of a glucosyl residue (donor) to H_2O (acceptor). Alpha-amylase (1,4-α-D-glucan glucanohydrolase, EC 3.2.1.1) occurs widely in all living organisms, while β-amylase (1,4-α-D-glucan maltohydrolase, EC 3.2.1.2) is found in seeds of higher plants and in sweet potatoes, oats, corn, rice, and sorghum. The commercially used α-amylase is mostly obtained from *Bacillus licheniformis* or *Aspergillus oryzae* var. (*33*).

Enzyme Characteristics

Both amylases have molecular weights in the range of 50,000, except sweet potato β-amylase, which is a tetramer of 197,000 daltons. Alpha-amylase contains one Ca^{2+} per mole of protein, which is essential for stabilization of the enzyme. In contrast, β-amylase requires no metals. The optimum pH of α-amylase varies depending on the source (6.0–7.0 for mammalian, 4.8–5.8 for *A. oryzae*, 5.85–6.0 for *B. subtilus*). For β-amylase, the pH optimum for activity ranges from 5.0 for wheat, malt, and sweet potato to 6.0 for soybean.

Alpha-amylase cleaves α-1,4-glucose polymers at internal positions (endo-attack) to yield oligosaccharide fragments with the C-1 hydroxyl group in the α-configuration. Beta-amylase is an exoglycosidase that successively cleaves maltosyl units from the nonreducing end of the polymer to yield maltose in the β-configuration (Table 5.6) (*34*).

Inherent in its sequential attack, the action of β-amylase stops at the α-1,6 branch point in the starch molecule. Alpha-amylase, with its random attack, can bypass the branch points in the polymer. However, the presence of an α-1,6 branching point is known to render the neighboring α-1,4 linkage resistant to attack by amylases (*35*).

Table 5.6. Similarities and Differences of Amylases

CHARACTERISTICS	α-AMYLASE	β-AMYLASE
Source	widespread in nature	plant seeds
Molecular weight	~ 50,000	~ 50,000
Cleavage point	α-1,4 glucosidic bond	
Configuration of new unit	α	β
Mechanism	endo	exo
Product	oligosaccharides	maltose
Action at branch point	bypass	cannot bypass
Decrease in viscosity	rapid	slow

Reprinted with permission from Thomas et al. (*34*), courtesy of Academic Press, Inc.

The α-amylase from *Bacillus amyloliquefaciens* has been partially sequenced (*36*) and the three-dimensional structure of α-amylase from porcine pancreas established by x-ray crystallography (*37*). The enzyme molecule has a dimension of 50 Å, with the active site situated in a deep cleft that runs 30 Å, separating the protein into two different-sized domains. Two specific binding sites are identified, one deeply in the active-site region and the other located on the surface of the molecule.

Reaction Mechanism

The enzymatic hydrolysis is known to proceed by the cleavage of the glycosyl $C-O$ bond. An imidazole group acts as a general acid to protonate the glycosidic oxygen, and a carboxylic anion group serves as a general base, stabilizing group, or nucleophile. Two types of reaction mechanisms have been postulated:

1. Formation of an oxycarbonium ion intermediate followed by hydration.
2. Displacement reaction with the formation of a covalent enzyme-substrate intermediate.

The Oxycarbonium Ion Mechanism. To account for the specific configuration catalyzed by the amylases, an enzyme-induced ring distortion of the substrate to a half-chair intermediate has been suggested (*38*). A distortion to a half-chair conformation relieves the steric strain at C1, enhances the accessibility of the anomeric carbon from the front side (Fig. 5.5), and favors the formation of the oxycarbonium ion intermediate.

Therefore, in the oxycarbonium mechanism, retention occurs if the enzyme directs the H_2O molecule to a front side attack. Likewise, a back side attack directed by the enzyme results in inversion.

Displacement Reaction. The retention of configuration in α-amylase-catalyzed hydrolysis suggests possibly a double displacement reaction. The concerted protonation of the glycosidic oxygen by the imidazole group and the nucleophilic attack of the carboxylate anion at C1 cleaves the glycosidic bond, resulting in the formation of a covalent glycosyl-enzyme intermediate with an inverted configuration (*39*). A second displacement by general base catalysis assisted by the imidazole dissociates the glycosyl-enzyme covalent bonding to give an α-configuration product (Fig. 5.6).

Inversion of configuration by β-amylase-catalyzed hydrolysis can also be explained by double displacement, analogous to that of α-amylase. But in this case, the H_2O molecule is directed towards the carboxyl carbon and assisted by a second imidazole group in the active site (Fig. 5.7) (*40*).

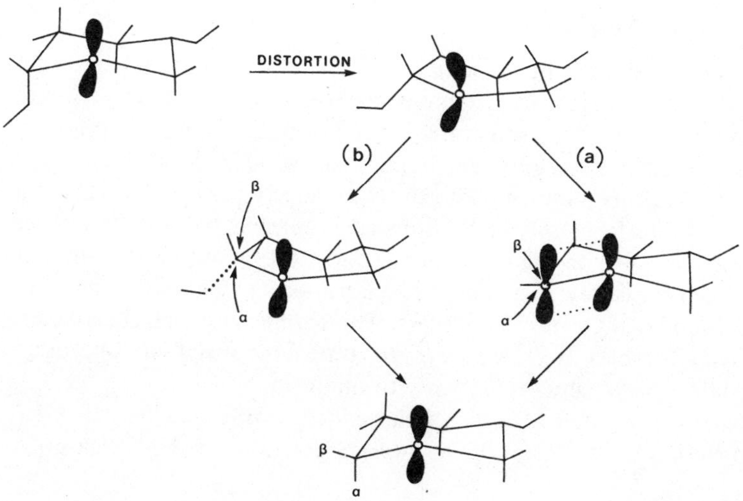

Fig. 5.5. Schematic representation of possible conformations and intermediates for amylase catalyzed hydrolysis of glucopyranoside. (A) Bond formation synchronous to bond rupture. (B) Bond formation before bond rupture. (Reprinted with permission from Thoma (38), courtesy of Academic Press, Inc.)

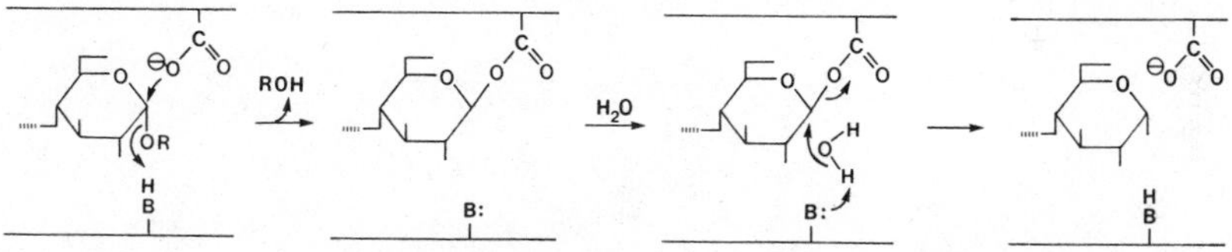

Fig. 5.6. Double displacement mechanism—α-amylase. (Reprinted with permission from Robyt (40), courtesy of Academic Press, Inc.)

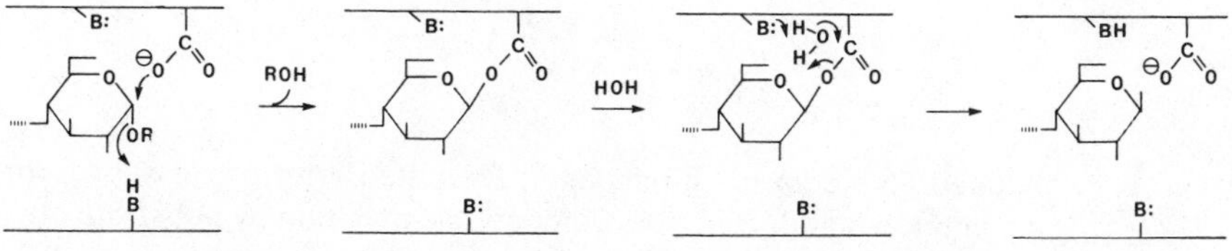

Fig. 5.7. Double displacement mechanism—β-amylase. (Reprinted with permission from Robyt (40), courtesy of Academic Press, Inc.)

The Induced-Fit Model

How do we explain the specificity of β-amylase in producing exclusively maltose? The exogenous enzyme is believed to have the active-site conformation induced by the substrate-enzyme complexation, resulting in a close fit of the catalytic groups for bond cleavage (Fig. 5.8) (41). A substrate with an improper attachment forms an inactive substrate. The catalytic groups fail to fit closely to the substrate molecule to cause bond cleavage. An attachment with the nonreducing end sticking out more than a maltose unit from the catalytic groups, or an α-1,6-branching unit at the active site, would inhibit the catalytic process. The formation of an inactive enzyme-substrate complex therefore serves to control the specificity of β-amylase in the hydrolysis of glucose polymer to maltose.

However, α-amylase, which is an endo enzyme, has its active site in a deep cleft and gives cleavage at internal positions of the starch polymer.

Action Pattern

Another intriguing question involving amylase-catalyzed reactions concerns the relative susceptibility to cleavage of a given substrate molecule in a population.

In a single-chain action pattern the amylase, once complexed with a substrate molecule, continuously acts on it to complete degradation. Another possibility is a multichain action pattern, in which amylase, having cleaved

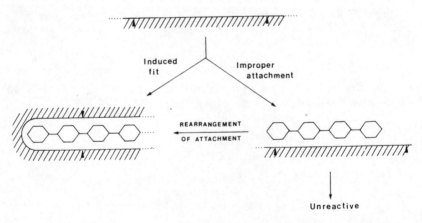

Fig. 5.8. Induced-fit model for β-amylase. (Reprinted with permission from Thoma, J. A., and Koshland, D. E., Jr. Competitive inhibition by substrate during enzyme action. Evidence for the induced-fit theory. *J. Am. Chem. Soc.* **82**, 3332. Copyright 1960 American Chemical Society.)

once a given substrate molecule, dissociates and attacks another substrate molecule. This type of action pattern leads to a simultaneous shortening of all substrate molecules present. It is also conceivable that the enzyme, in a multiple attack pattern, cleaves a given substrate more than once before it dissociates and acts on another substrate molecule. Note that the dissociated products in all these cases can, in themselves, serve as a substrate for the enzyme (34).

Both β- and α-amylases have been shown to have a multiple attack pattern. Porcine pancreatic α-amylase at the optimum pH of 6.9 shows three times the degree of multiple attack of *Aspergillus oryzae* at its optimum pH value. However, at the unfavorable pH of 10.5, the action pattern is predominantly multichain. The direction of multiple attack is towards the nonreducing end of the substrate (42).

The mechanism of multiple attack can be visualized in the following schemes (Fig. 5.9) to consist of three routes (42):

1. The substrate, after initial cleavage, dissociates completely and reassociates to a new position (Fig. 5.9A).

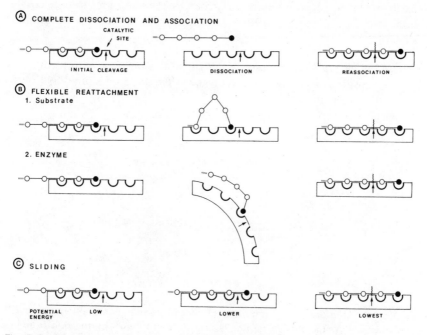

Fig. 5.9. Mechanism of multiple attack. (A) complete dissociation and association. (B) flexible reattachment, and (C) sliding. (Reprinted with permission from Robyt and French (42), courtesy of Academic Press, Inc.)

2. A segment of the product substrate remains attached to the enzyme, while the rest dissociates and realigns to a new position. Such a mechanism requires a certain degree of flexibility either in the substrate or the enzyme or both (Fig. 5.9B).
3. Another possibility is for the product substrate to slide on the enzyme surface to a position of minimum potential energy having all the subsites completely filled (Fig. 5.9C).

If we consider that the molecular motion in each of the individual steps of dissociation, reassociation, and realignment in routes (1) and (2) are infinitesimally small, the product substrate can be visualized as "sliding" on the enzyme to a favorable position where the potential energy is at its minimum as in route 3.

The Multimolecular Process

Amylases do not exclusively cause simple unimolecular hydrolysis. Very often the degradation of oligosaccharide substrates proceeds via a multimolecular process in which more than one substrate molecule is involved (43). Realignment of substrate molecules, such as condensation, transglycosylation, and shifting among substrates, occurs at increasing concentration of substrates.

1. *Condensation:* Two molecules of substrates polymerize to form a new glycosidic bond. The product formed then undergoes rapid hydrolysis to smaller units (Fig. 5.10).

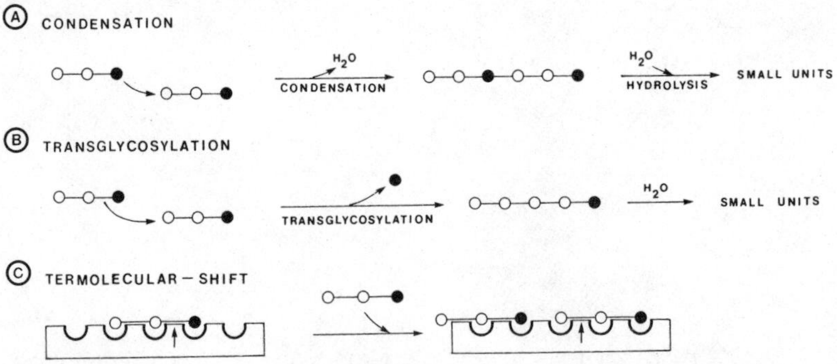

Fig. 5.10. Hydrolysis via multimolecular processes: (A) condensation, (B) transglycosylation, and (C) termolecular. (Reprinted with permission from Allen, J. D., and Thoma, J. A. Multimolecular substrate reactions catalyzed by carbohydrases. *Aspergillus oryzae* α-amylase degradation of maltooligosaccharides. *Biochemistry* **17**, 2341. Copyright 1979 American Chemical Society.)

2. *Transglycosylation:* A glycosyl group of a substrate donor is transferred to another substrate acceptor, similar to hydrolysis, except that water is not the co-substrate.

3. *Termolecular-shift binding:* A second substrate molecule binds to the enzyme in a way that pushes the first bound substrate molecule into a new position, promoting the hydrolysis at a different bond.

As implicated by the process, the binding site consists of subsites geometrically complementary to the monomer units in the substrate. Pancreatic α-amylase has been shown to contain five subsites, with the catalytic site located asymmetrically (*44*). Alpha-amylases from *Bacillus amyloliquefaciens* and *Aspergillus oryzae* are known to have ten and eight subsites, respectively (*45*).

Industrial Uses

Amylases are used in both the baking industry and the corn syrup industry (Chapter 7). Yeast fermentation ceases when the sugar in the dough is depleted. There are three sources of sugar: the sugar originally present in the flour, added sugar, and breakdown from starch by amylases. Flour from grains grown in North America has a very low α-amylase content. The addition of enzymes provides a gradual and constant supply of maltose to allow continuous fermentation.

Alpha-amylase is also used in the clarification of starch haze in beer and wine processing. In combination with glycoamylases and isomerase, α-amylase is essential in the production of various corn syrups.

PECTIC ENZYMES

Pectic enzymes constitute a unique group of enzymes which catalyze the breakdown of pectic substances in plant cell walls. Pectic enzymes can be classified according to the type of cleavage (*46*).

1. Pectinesterase (PE) (pectic pectylydrolase, EC 3.1.1.11)—catalyzes the deesterification of pectin.

2. Polygalacturonases (PG)—catalyze the hydrolytic cleavage of the α-1,4-glycosidic linkage next to a free carboxyl group. The exo-PG (poly (1,4-α-D-galacturonide)galacturonohydrolase, EC 3.2.1.67) cleaves from the nonreducing end, while the endo-PG (poly(1,4-α-D-galacturonide) glycanohydrolase, EC 3.2.1.15) cleaves the pectic chin randomly.

3. Pectate lyases—catalyze the cleavage of the glycosidic linkage next to a free carboxyl group via β-elimination. Both exo (poly(1,4-α-D-galacturo-

nide)exo-lyase, EC 4.2.2.9) and endo (poly(1,4-α-D-galacturonide)lyase, EC 4.2.2.2) enzymes exist, and pectate is the best substrate.

Pectinesterase

All pectinesterases are highly specific for methyl esters of polygalacturonate (Eq. 5-20). Methyl esters of other uronides or polymers of less than ten galacturonic acids are not de-esterified. Deesterification starts (1) from the reducing end or (2) at some secondary locus, next to free carboxyl groups, and proceeds along the chain, creating blocks of free carboxyl groups (47).

Eq. 5-20

Most plant pectinesterases have pH optima between 7 and 9. Divalent cations increase the activity of PE in higher plants several times, and the product, polygalacturonic acid, has been shown to act as a competitive inhibitor in tomato PE and orange PE. Pectinesterases are also found in molds and bacteria.

Polygalacturonases

Polygalacturonases catalyze the hydrolysis of α-D-1,4-glycosidic bonds of nonesterified residues (5-20), and low methoxy pectins are the preferred substrates.

Both exo- and endo-PGs have weakly acidic pH optima—pH 4.6 for both papaya exo- and endo-PGs, pH 5.5 and 4.0 for peach exo- and endo-PGs, respectively, pH 5.0 for the endo-enzyme from *rhizopus arrhizus,* and pH 5.5 for *Aspergillus niger* endo-PG (48, 49, 50, 51).

Endo-polygalacturonases. Endo-PGs usually exist in multiple forms, and the molecular weights vary widely depending on the source. For example, the molecular weight of PG from papaya has been determined to be 164,000 daltons, while for *Rhizopus arrhizus* PG, the MW is 30,000. All endo-PGs depolymerize pectic acid randomly, accompanied by a rapid decrease in viscosity of the substrate solution.

The action pattern of the endo-PG from *Aspergillus niger* has been studied using oligogalacturonic acids as substrates (Fig. 5.11). The binding site of the enzyme is composed of four subsites, with the catalytic site situated between subsites 1 and 2 (*51*).

Trimer substrate does not occupy the complete set of subsites. In cases where the trimer complexes with the subsites without the catalytic groups, the resulting complex is inactive (i.e., a nonproductive complex is formed). For tetramers, the most probable complex is to have all the subsites occupied, and indeed the products of hydrolysis are a trimer and a monomer. Pentamers can complex with the active site in two ways, both satisfying the complete occupancy of all four subsites. With hexamers, three productive complexes can be formed.

In high-molecular-weight substrates, the number of productive binding modes in which all the subsites (m) are occupied by the substrate with the degree of polymerization (n), equals ($n - m + 1$) (*52*), the mode of hydrolysis becomes random.

Exo-polygalacturonases.   Exo-PGs hydrolyze the substrate starting from the nonreducing end to yield galacturonic acid. However, the degradation

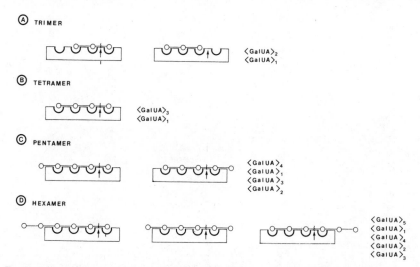

Fig. 5.11. Action pattern of endo-polygalacturonase on (A) trimer, (B) tetramer, (C) pentamer, and (D) hexamer substrates. (Reprinted with permission from Lubomira (*51*), courtesy of European Journal of Biochemistry.)

of pectate is usually not complete, because the catalyzed hydrolysis is restricted only to the α-D-1,4-linkage and is interrupted by branching. The rate of hydrolysis increases with substrate size and reaches a maximum for polygalacturonate with a degree of polymerization of 20 in carrot and peach exo-PGs (53). The terminal action pattern is shown by a large increase in reducing groups formation and a slow increase in viscosity.

Pectate Lyases

Pectate lyases catalyze the degradation of glycosidic bonds next to a free carboxyl group by the mechanism of β-elimination (54). Pectate or low-methoxy pectin are the preferred substrates for the enzyme.

In the β-elimination mechanism, the proton at C5 is first removed to form a carbanion which is stabilized by the C6 carboxyl group. Cleavage of the glycosidic linkage in the β-position to the carboxyl group results in the formation of a double bond between C4 and C5 (Eq. 5-21).

Eq. 5-21

· Both the exo- and the endo-enzymes have an alkaline pH optimum ranging from 8.0 to 9.5 and a requirement for Ca^{2+} for activity. The exo-enzyme starts the cleavage from the reducing end of the substrate. And in contrast to the endo-enzyme, the rate of cleavage does not depend on the size of the substrate. The pectate lyases are not found in higher plants but in bacteria and molds, some of which are involved in the softening of fruits and vegetables. Pectin of a high degree of esterification is the best substrate. Pectin lyases have a pH optimum ranging from 5.0 to 6.6 and most require Ca^{2+} activation.

Industrial Importance

The study of pectic enzymes is important to food science in three aspects. Pectic enzymes are known to play an important role in textural changes during the ripening of various fruits. The softening of fruits is correlated with the increasing enzyme activity on the degradation of pectic substances.

Second, post-harvest loss due to rotting of fruits and vegetables has been attributed, in part, to the action of microbial pectic enzymes. The best known examples include the softening of cucumber and olives in brine due to yeast PG and bacterial PAL, and the softening of canned apricots caused by the fungus *Rhizopus arrhizus* PGs (50).

The pectic enzymes are studied also for reasons of technological applications, the largest use being in the wine and fruit industry. During the production of apple juice, for example, the enzyme mix is added to the pressed juice for clarification. In the manufacture of red wine, enzyme treatment of grape pulp results in better color extraction and higher juice yield.

LIPOLYTIC ENZYMES

Lipolytic enzymes consist of two major groups, the lipases which are triacylglycerol acyl hydrolases, and the phospholipases A_1 and A_2 which are phosphoglyceride acyl hydrolases (55). Although phospholipases C and D are not acyl hydrolases, they are nonetheless commonly included as lipolytic enzymes (Table 5.7). The present discussion will mostly concentrate on the extensively studied porcine pancreatic lipase.

Pancreatic Lipase

Porcine pancreatic lipase is a single polypeptide containing 449 amino acids with a molecular weight of 49,900. A single carbohydrate chain of fucose, galactose, mannose, and *N*-acetylglucosamine is linked to Asn-66 (56).

Pancreatic lipase requires a protein cofactor, colipase, to exert its catalytic function. Colipase in its native form (procolipase) contains 101 amino acid residues. It is activated by low concentrations of trypsin, which hydro-

Table 5.7. Lipolytic Enzymes, Substrates, and Products

ENZYME	SYSTEMATIC NAME[a]	SUBSTRATE	PRODUCTS
Triacylglycerol lipase (3.1.1.3)	triacylglycerol hydrolase	triacylglycerols	diglyceride, fatty acid anion
Phospholipase A_2 (3.1.1.4)	phosphatide 2-acylhydrolase	lecithin	1-acylglycerol-phosphocholine, fatty acid anion
Phospholipase A_1 (3.1.1.32)	phosphatidate 1-acylhydrolase	lecithin	2-acylglycerol-phosphocholine, fatty acid anion
Lysophosphopholipase (3.1.1.5)	lysolecithin acylhydrolase	lysolecithin	glycerophosphocholine, fatty acid anion
Phospholipase C (3.1.4.3)	phosphatidylcholine cholinephosphohydrase	phosphatidylcholine	1,2-diacylglycerol, choline phosphate
Phospholipase D (3.1.4.4)	phosphatidylcholine phosphatidohydrolase	phosphatidylcholine	choline, phosphatidate
Sphingomyelin phosphodiesterase (3.1.4.12)	sphingomyelin cholinephosphohydrolase	sphingomyelin	N-acylsphingosine, choline phosphate

[a]IUPAC-IUB Enzyme Nomenclature 1978.

lyzes the N-terminal pentapeptide to form colipase$_{96}$. Higher concentrations of trypsin result in the formation of colipase$_{85}$, with the loss of 11 amino acid residues at the C-terminal. The specific activities of both are about five times higher than that of the procolipase (57).

The Role of Colipase

Contrary to conventional enzyme reactions that normally take place in aqueous solution, lipolytic enzymes catalyze the hydrolysis by heterogeneous catalysis at the interface between two immiscible phases, the aqueous phase and the apolar phase containing the lipid substrate.

Lipolytic enzymes, like other proteins, however, are irreversibly denatured at these interfaces. In the physiological environment, the adsorption and hence the denaturation of pancreatic lipase is prevented by the presence of bile salts at these interfaces. The bile salts inhibit the adsorption by physically excluding the enzyme from the interface or by a general detergent effect. Inhibition may also result from the increasing surface pressure due to bile salts. Lipase has been shown to be unable to penetrate the lipid film above a critical surface pressure.

For the lipase to act on the triacylglycerol substrate at the interface in the presence of bile salts, colipase is required. Colipase and lipase bind in a 1:1 complex with a dissociation constant of $10^{-6}M$. Both hydrophotic and ionic interactions are involved. The presence of colipase has been shown to increase the critical surface pressure to allow the lipase penetration. It has also been shown that colipase has a high affinity for bile salt micelles and monolayers. The binding of colipase to the interface leads to a conformational change, unmasking a specific binding site for lipase to complex (58), and the resulting complex is capable of adsorbing to bile-salt-covered interface. The formation of the bile salt–colipase-lipase ternary complex stabilizes the lipase against surface denaturation and thus anchors the lipase to the interface where hydrolysis can occur (Fig. 5.12). (It is also conceivable that the colipase and lipase complex before anchoring to the interface.)

Mechanism of Catalysis

The hydrolysis reaction catalyzed by pancreatic lipase consists of a mechanism resembling that of the serine proteases such as chymotrypsin, in which the hydroxyl group of serine is activated by a histidine-carboxylate anion protein-relay system (Fig. 5.13). The serine hydroxyl group thus reacts with the substrate to form an acyl-enzyme, which is then deacylated to give the product (59).

The role of serine and the carboxylate anion have been questioned (60). Recently, however, a similar reaction mechanism is proposed for lipoprotein lipase catalysis, which is shown in Eq. 5-22, for p-nitrophenyl ester substrate (61). The acylenzyme is formed via the formation of a tetrahedral intermediate. The transition state is stabilized by a general base proton

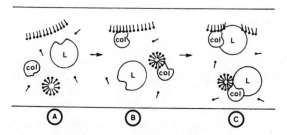

Fig. 5.12. The role of colipase for the functioning of lipase in the presence of bile salts. (A) When alone, lipase recognizes neither the micelles nor the interface. (B) Colipase can recognize both, creating a binding site for lipase. (C) Binding of lipase mediated with colipase. (Reprinted with permission from Chapus et al. (58), courtesy of Elsevier Science Publishers, Biomedical Division, Amsterdam, The Netherlands.)

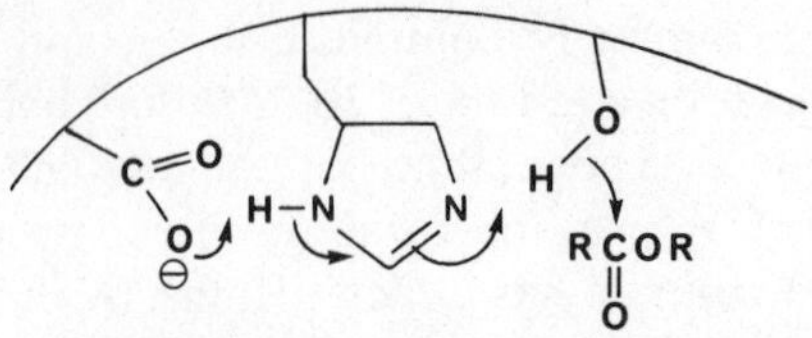

Fig. 5.13. Proposed mechanism of catalysis for pancreatic lipase. (Reprinted with permission from Brockerhoff (*59*), courtesy of Elsevier Scientific Publishers Ireland Ltd., Limerick, Ireland.)

bridge between the serine and the histidine. Expulsion of the *p*-nitrophenoxide breaks down the tetrahedral intermediate to form the acyl-enzyme. Hydrolysis of the acyl-enzyme occurs by general base catalysis.

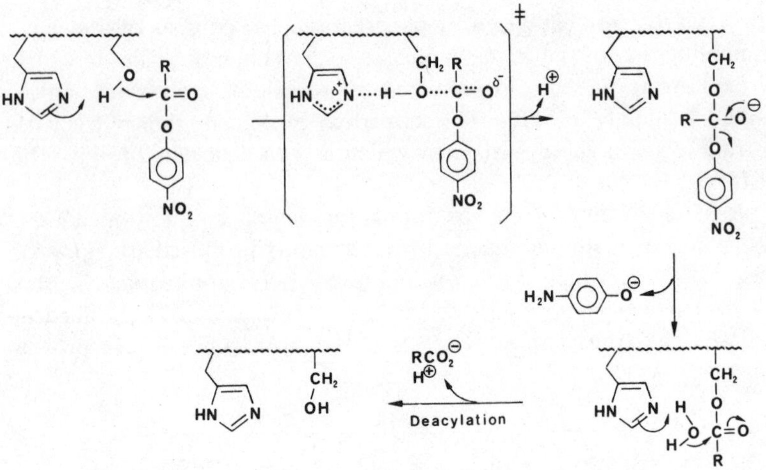

Eq. 5-22

Specificity

Lipases exhibit various types of specificity (*62*), including the following:

1. *Substrate specificity:* The enzyme hydrolyzes the acylglycerols (TG, DG, MG) or types of fatty acids at different rates. For example, pancreatic lipase from *Geotrichum candidum* hydrolyzes oleic acid and palmitic acid in preference to stearic acid.

2. *Positional specificity:* The enzyme catalyzes the release of fatty acids at preferential positions on the acylglycerol molecule. Pancreatic lipase is specific for the primary esters of acylglycerols, and the sequence of hydrolysis follows TG > 1,2(2,3)DG > 2-MG. Lipases from *G. candidum* and *Penicillum cyclopium* show no specificity with regard to the position,

whereas lipases from *Aspergillus niger* and *Rhizopus delemar* do not hydrolyze the ester bond in position 2. In general, 1,3-specificity is common among microbial lipases.

3. *Stereospecificity:* Pancreatic lipase and most microbial lipases show no distinction in catalytic activity at positions 1 and 3. However, lipoprotein lipase and heparin-releasable hepatic lipase preferentially attack the *sn*-1 position, and lingual lipase prefers the *sn*-3 position of the acylglycerol molecule. (The stereochemistry of glycerol derivatives is expressed by *sn*, the stereospecific numbering nomenclature, which recognizes the fact that the two primary carbinol groups of the parent glycerol are not identical in their reaction with dissymmetric structures.)

The Acyl Transfer Reaction

Under specific, controlled conditions, lipase catalyzes acyl transfer between acylglycerols and fatty acids. In these types of reactions, the acyl group is transferred to the glycerol instead of water. Obviously, the presence of water in the reaction system favors the hydrolysis pathway. Increasing the proportion of organic solvent facilitates the acyl transfer reaction.

The acyl transfer reaction offers a novel method of synthesizing acylglycerols starting with glycerol and fatty acids (*63*). Using lipase with the desirable specificity, it is possible to manipulate the synthesis to obtain a designated esterification product. Lipases from *A. niger* and *R. delemar* synthesize only 1(3)-MG and 1,3-DG (Eq. 5-23.1) whereas lipases from *G. candidum* and *P. cyclopium* make ester bonds at all three positions (5-23.2). Esterification of glycerol with oleic acid using *Candida rugosa* lipase gives 2-MG, 1-MG, 1,2-DG, 1,3-DG and TG (*64*).

Eq. 5-23

The acyl transfer reaction, however, is most useful to the food industry in its application in the interesterification process (*65*). Hydrolysis and resynthesis cause acyl exchange (1) between acylglycerol molecules to form interestified products (Eq. 5-24), or (2) between acylglycerol and free fatty acids to produce new acylglycerols enriched with the added fatty acid (Eq. 5-25).

$$
\begin{array}{c}
\begin{array}{l} \text{A} \\ \text{B} \\ \text{A} \end{array}
+
\begin{array}{l} \text{C} \\ \text{B} \\ \text{C} \end{array}
\xrightarrow[\text{LIPASE}]{4,3 \text{ SPECIFIC}}
\begin{array}{l} \text{A} \\ \text{B} \\ \text{A} \end{array}
\begin{array}{l} \text{A} \\ \text{B} \\ \text{C} \end{array}
\begin{array}{l} \text{C} \\ \text{B} \\ \text{C} \end{array}
\quad \text{Eq. 5-24}
\end{array}
$$

$$
\begin{array}{c}
\begin{array}{l} \text{A} \\ \text{B} \\ \text{A} \end{array}
+ \text{C}
\xrightarrow[\text{LIPASE}]{1,3 \text{ SPECIFIC}}
\begin{array}{l} \text{A} \\ \text{B} \\ \text{A} \end{array}
\begin{array}{l} \text{A} \\ \text{B} \\ \text{C} \end{array}
\begin{array}{l} \text{C} \\ \text{B} \\ \text{C} \end{array}
\quad \text{Eq. 5-25}
\end{array}
$$

It is quite clear that, using lipases with the appropriate specificity, one can direct the process of interesterification to produce the desired product. While chemical interesterification gives products in which the fatty acyl residues are randomly distributed, the enzymatic process provides interesterification at specific positions with specific acyl groups. Thus, the products unobtainable by the chemical process can be produced. Considering that the physical characteristics of fats and oils depend on the position of a particular fatty acid in the glyceride, the advantage of enzymatic interesterification and its practical importance are significant.

6
FLAVORS

Flavor is the sensation produced by a material perceived principally by the senses of taste and smell. In certain cases, flavor also denotes the sum of the characteristics of the material which produces that sensation. The primary role of flavor in food processing is to make the food palatable. Many food products may become unattractive for consumption without the supplementation of flavors, either unintentionally or intentionally. The use of flavors also adds variety to the diet and the functional and economic values of the food products. The application of flavor technology depends largely on the identification of the sensory active compounds responsible for the natural flavors. Most synthetic flavor compounds are the chemical imitates of the key flavor constituents in the natural sources.

The biochemical and chemical reactions that generate the many characteristic flavors in food are extremely complex. There are nearly 4000 volatiles identified in approximately 200 types of fruits, spices, and other foods. Few pathways are known, and many of these mechanisms are postulated from model studies. Furthermore, the structure-activity relationships of the many known flavor compounds largely remain to be elucidated.

THE SENSATION OF TASTE

Taste is a combination of chemical sensations perceived by the papillae of the tongue. There are four basic taste sensations: acidity, salinity, sweetness, and bitterness. Most substances do not have a single taste, but a complex sensation comprising more than one of the four basic sensations.

Sour taste is caused only by hydrogen ions; acids, acid salts, and other substances that give hydrogen ions in contact with water give the sour taste. The threshold is about $0.002N$ acid. Sour taste can also be induced by passing electric current through the tongue; presumably this is due to the hydrolysis of acid or water with the generation of hydrogen ions. Strong currents, however, induce a bitter taste. Some acid substances, for example, potassium acid oxalate and protocatechuic acid, stimulate both sour and bitter tastes.

Saltiness is stimulated by soluble salts. Most salts of high molecular weight are bitter rather than salty. Low-molecular-weight salts—notably chlorides of sodium, potassium, and calcium—are salty. The threshold level is 0.007–0.016% salt solution.

The sweetness of a substance can be related to the functional groups that make up the glycophore units in a tripartite model. The molecular theory of sweet taste is well developed and will be dealt with in Chapter 7.

Bitter compounds require a polar (electrophilic or nucleophilic) group and a hydrophobic group. The structure-activity relationships of bitter-tasting compounds are not firmly established (1). There are large numbers of bitter substances in plants; these are mainly classified into the alkaloids and glycosides.

ODOR—THE SITE-FITTING THEORY

In order to be odorous, a compound must be sufficiently volatile, and there has to be physical interaction between the odorous compound and the receptor site. Although many theories have attempted to explain the mechanism of odor perception at the molecular level, the present discussion is mostly limited to the site-fitting theory.

According to this theory (2), there are seven primary odors—camphoraceous, ethereal, musky, floral, minty, pungent, and putrid. For each class of these odorous compounds, there are receptor sites that are complementary to the size, shape, and electronic status of the molecule. The ethereal, camphoraceous, and musky odors depend primarily on the size of the molecule, and the floral and the minty depend on the shape, whereas pungent and putrid are caused by electrophilic and nucleophilic molecules, respectively. The minty structure also has an additional requirement of possessing, near the point of the wedge-shaped molecule, a group capable of forming a hydrogen bond with the receptor site. The primary odors are perceived when there is a fit between a molecule and the corresponding receptor site, in a way similar to the lock-and-key mechanism (Fig. 6.1).

A compound can develop more than one primary odor if the molecule can fit more than one type of receptor site. For example, acetylenetetrabromide fits either site for camphoraceous and ethereal. It is also possible for molecules to come together to fill a common site, giving complex odors. In short, a given complex odor is a mixture of the appropriate primary

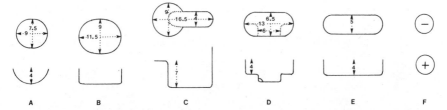

Fig. 6.1. Olfactory receptor sites: (A) Camphoraceous (δ-camphor, 3,3-dini-tropentane, acetylenetetrabromide, (B) musky (androstan-3α-ol, 3-methyl-cyclopentadecanone, undecyl-γ-butyrolactone), (C) floral (diphenylamine, ger-aniol, anisole), (D) minty (ℓ-methone, methyl 2,4-dimethylphenyl ketone, cyclopentanone), (E) Ethereal (diethyl ether, ethylene, acetylenetetrabromide), (F) pungent (electrophilic compounds, such as isocyanate, isothiocyanate, and chloramine), and putrid (nucleophilic compounds such as mercaptan and amines). From Amoore (2).

odors. The following are some examples of complex odors and their pri-mary odor components.

Odor	Components (primary odor)
Almond	camphoraceous, floral, minty
Lemon	camphoraceous, floral, minty, pungent
Garlic	ethereal, pungent, putrid
Rancid	ethereal, minty, pungent

The molecular shape requirements are inadequate to explain the distinct odors exhibited by many small molecules. The use of electrophilic and nucleophilic properties to relate pungent and putrid recognizes, to a certain degree, the importance of a particular functional group in the molecule.

The profile-functional group theory (3) postulates that while the shape and size of an odorous molecule is responsible for the quality of the odor, the functional group determines the orientation of the molecule at the re-ceptor site. It has an important influence on the homogeneity of the orienta-tion pattern and on the affinity of the interaction complex. Removal of the functional group changes the orientation pattern, resulting in randomness and decreased affinity in absorption to the receptor site. For example, the isochromane in Fig. 6.2 has a musk odor, but substitution with a methyl group at positions causing steric hindrance of the functional ether oxygen results in the loss of odor. Replacing the oxygen with nitrogen decreases the intensity.

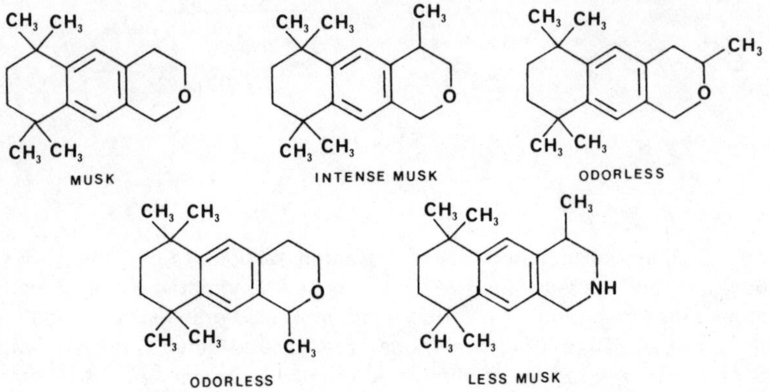

Fig. 6.2. Effect of functional groups on odor.

CHARACTER-IMPACT COMPOUNDS

A characteristic taste in a food can usually be related to a particular compound or a class of compounds, while an odor can be attributed to a combination of numerous volatile compounds each of which individually smells very differently. The difference in characteristics of certain odors is partially due to varying proportions of these many widely distributed volatiles, such as esters, acids, alcohols, aldehydes, and ketones that occur in all foods. These volatiles are called "contributing flavor compounds." However, most substances contain trace amounts of a few unique volatile compounds which possess the characteristic essence of the odor. These are the "character-impact compounds." Unfortunately, there are not too many character-impact compounds that have been identified. Nonetheless, discussion will concentrate on these unique compounds which contribute the most significant characteristic flavor to a particular food.

ORIGIN OF FLAVOR

Biosynthesis

Many flavors, especially those in fruits and vegetables, are the products or by-products of various metabolic pathways. Schematically, the biosynthesis of these flavors can be represented by Fig. 6.3.

The Shikimic Pathway. The initial step is the condensation of phosphoenolpyruvate, an intermediate from glycolysis, and erythrose-4-phosphate

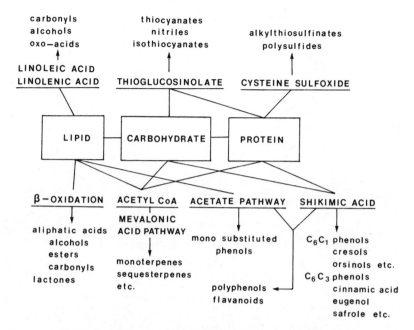

Fig. 6.3. Scheme of biosynthesis of flavor compounds.

(from pentose phosphate pathway). The C7 product cyclizes and undergoes further dehydration and reduction to yield shikimic acid. Condensation with a second molecule of phosphoenolpyruvate gives chorismic acid (Eq. 6-1.1).

Chorismic acid is the precursor of the aromatic amino acids phenylalanine and tyrosine, which can undergo deamination to yield *trans*-cinnamic acid, the parent compound of many C_6C_3 phenolic compounds (Eq. 6-1.2).

The Polyketide Pathway. The main reaction involves the addition of malonyl CoA to acyl CoA. Successive additions of malonyl CoA extend the chain to give a ketide unit, which then cyclizes to form various phenolic compounds (Eq. 6-2). All phenols synthesized via polyketide pathways are meta-substituted.

The Isoprene Pathway. The isoprene pathway involves the condensation between acetyl CoA and acetoacetyl CoA with the formation of mevalonic acid. Decarboxylation and phosphorylation of mevalonic acid gives isopentenyl pyrophosphate, which through self-condensation yields the terpenes

6 − 1.1

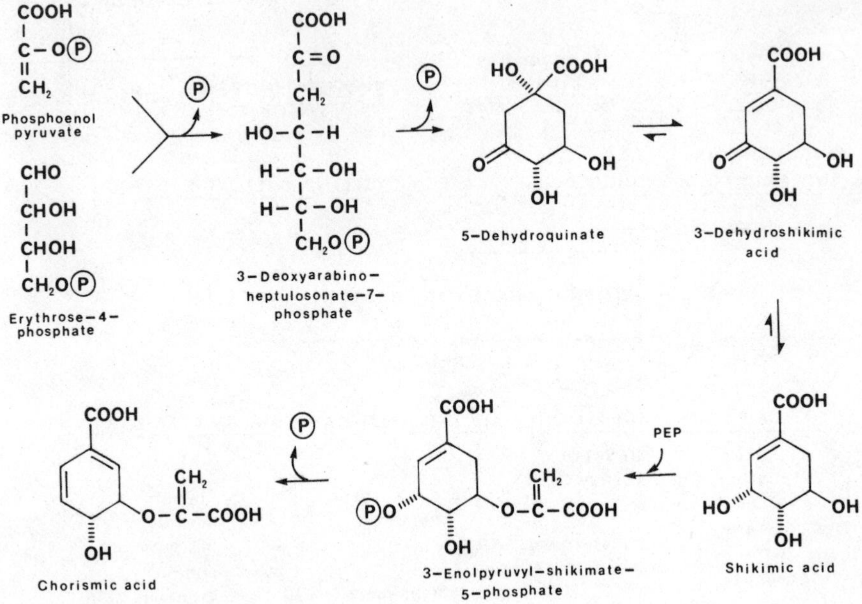
Phosphoenol pyruvate
Erythrose−4−phosphate
3−Deoxyarabino−heptulosonate−7−phosphate
5−Dehydroquinate
3−Dehydroshikimic acid
Shikimic acid
PEP
3−Enolpyruvyl−shikimate−5−phosphate
Chorismic acid

6 − 1.2

Eq. 6-1

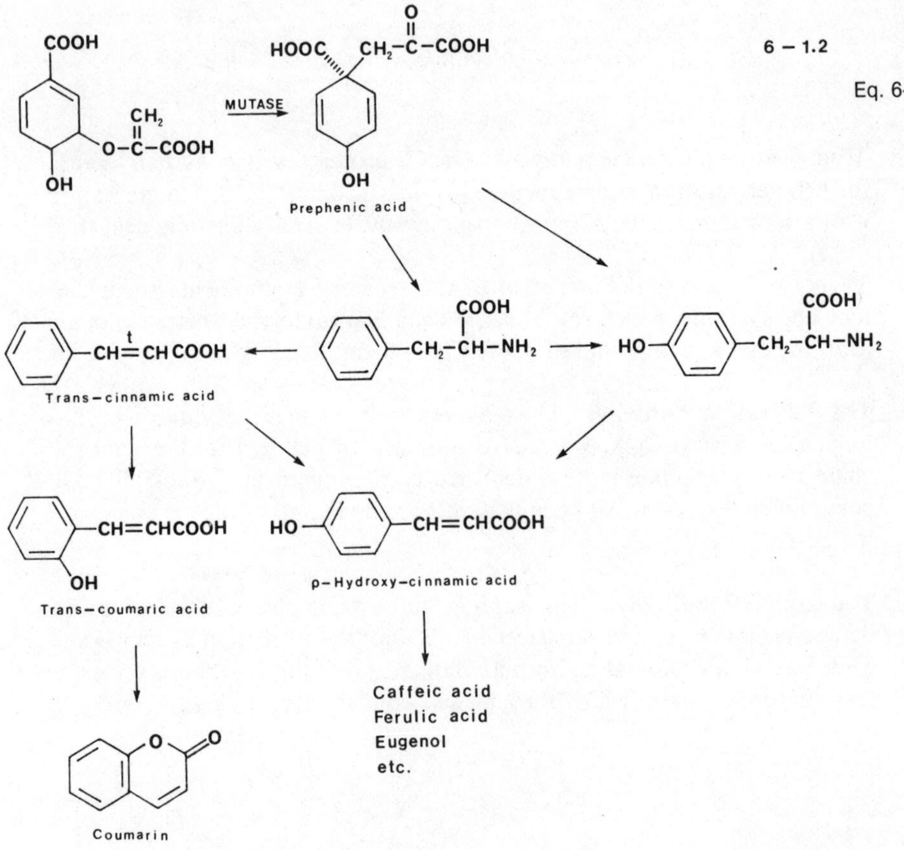
MUTASE
Prephenic acid
Trans−cinnamic acid
Trans−coumaric acid
ρ−Hydroxy−cinnamic acid
Caffeic acid
Ferulic acid
Eugenol
etc.
Coumarin

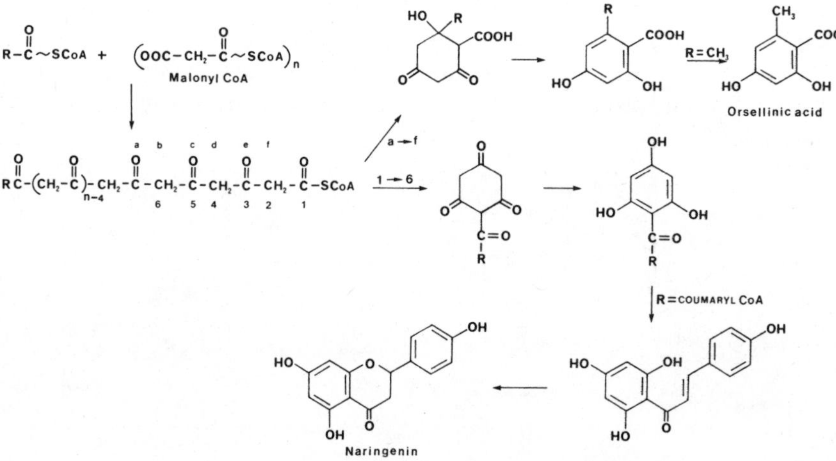

Eq. 6-2

(Eq. 6-3). Monoterpenes are usually volatile and odorous. Mevalonate is also the key intermediate in the synthesis of cholesterol in animals.

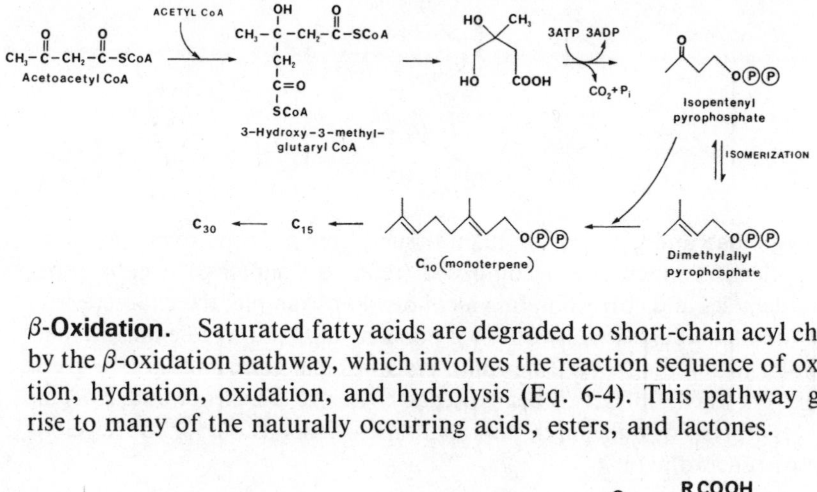

Eq. 6-3

β-**Oxidation.** Saturated fatty acids are degraded to short-chain acyl chains by the β-oxidation pathway, which involves the reaction sequence of oxidation, hydration, oxidation, and hydrolysis (Eq. 6-4). This pathway gives rise to many of the naturally occurring acids, esters, and lactones.

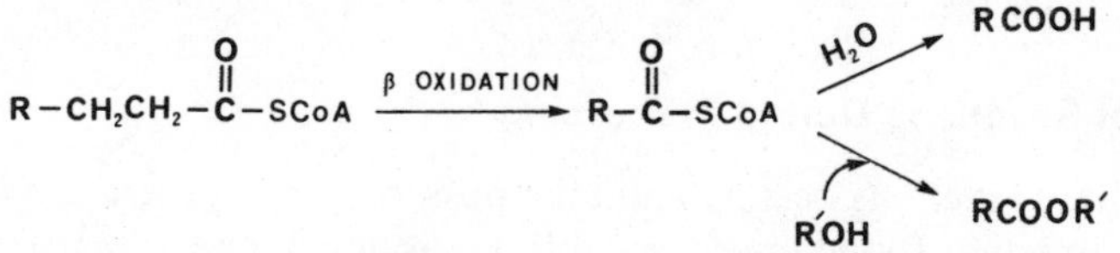

Eq. 6-4

The oxidation of unsaturated fatty acids undergoes many of the reactions similar to those for saturated fatty acids. However, as is now well known, cis-$\Delta^{3,4}$ enonyl CoA is not a substrate for acyl CoA dehydrogenase. The position and configuration of the cis-$\Delta^{3,4}$ double bond must be isomerized in position and configuration. A similar rearrangement also occurs with the cis-$\Delta^{2,3}$ double bond (Eq. 6-5). Unsaturated fatty acids are the precursors of numerous unsaturated esters, which characterize many flavors in various food systems.

Eq. 6-5

Lipoxygenase and Lipase.   In plant tissues, lipid hydroperoxides are both enzymatically formed and decomposed (refer to Chapter 5), yielding specific aldehydes and corresponding alcohols. For example, the characteristic flavor of cucumber is mainly due to 2-nonenal and 2,6-nonadienal and their alcohols. In tomatoes, cis-2-hexanal, cis-3-hexenal contribute to the fresh flavor. The flavor of green beans is partly due to 2-hexenal and 1-octen-3-ol. Hydrolysis of milk fat in milk by lipase is responsible for the development of rancid flavor.

Chemical Reactions During Processing

The Maillard reactions occupy a unique position in the generation of many flavor compounds found in processed food products. The primary reactions have been dealt with previously in connection with the chemistry of monosaccharides (Chapter 3). Only the secondary reactions that are best understood and are significant in their contribution to flavor will be presented. Flavor products of the Maillard reactions include furans, pyrones, carbonyls, and acids from dehydration and/or fragmentation, and pyrroles, pyrazines, oxazoles, thiazoles, and sulfur compounds formed through Strecker degradation, condensation, and further reactions. The reaction intermediates in the Maillard reactions that are most responsible for generating flavor products are the dicarbonyl compounds. For simplicity, glucose is the aldohexose used in the reaction schemes.

Formation of Pyrrole and Pyrazine. The 3-deoxyglycosulose and the unsaturated glycosulos-3-ene formed in the reactions initiated by the 1,2-enolization of the amadori compound produce furfurals, pyrroles, and pyrazine derivatives. Formation of the latter two types of compounds is due to Strecker degradation, whereas the heterocyclic nitrogens are derived from the α-amino acid. Pyrroles are formed by the cyclization of the aminocarbonyl product of Strecker degradation. Condensation of the aminocarbonyl compound, followed by oxidation, gives pyrazines (Eq. 6-6)(4).

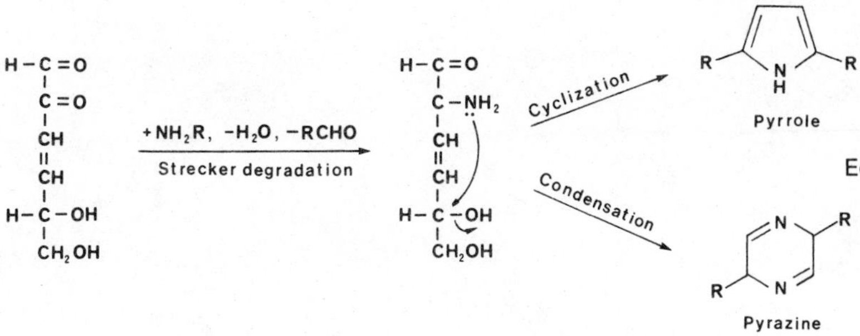

Eq. 6-6

Oxazoles and Derivatives. Oxazoline formation is favored when, in the reaction in Eq. 6-6, cyclization occurs before hydrolysis of the Schiff base, followed by protonation of the resulting oxazolinide ion (Eq. 6-7) (5).

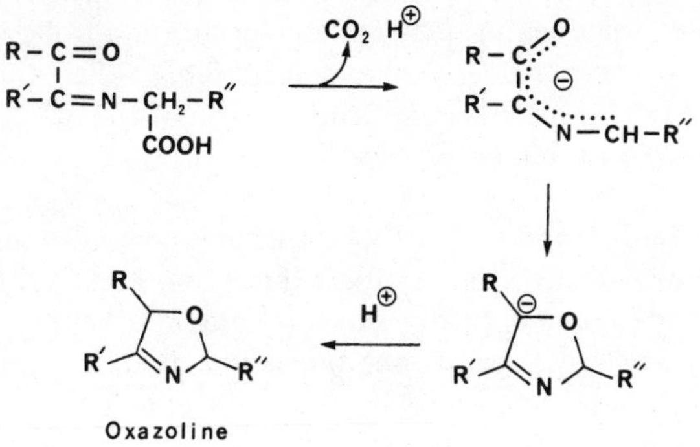

Eq. 6-7

Pyrrolines and Pyrrolidines. Both pyrrolines and pyrrolidines have been proposed to form via Strecker degradation of proline with a dicarbonyl compound. Condensation of the aldehyde and the secondary amine forms an iminium carboxylate intermediate that is transformed by decarboxylation into a reactive ylide or iminium ion (6). Further hydrolysis gives pyrroline and pyrrolidine, while reduction yields *N*-acylpyrrolidine (Eq. 6-8).

Eq. 6-8

Formation of Pyrones. The methyl dicarbonyls formed via 2,3-enolization in the Maillard reactions may undergo cyclization, followed by degradation to give pyrones (7). Or they may form the 4,5-dienol, which readily loses the C6-OH to give triketone. Ring closure of the 2,3-enolic form of triketone gives the furanone (Eq. 6-9).

Formation of Reductones. The 2,4-dicarbonyl may also undergo allylic loss of a hydroxyl group at C3 to form the α,β-unsaturated intermediate, which reacts with amino derivatives to form amino triketone. Scissions between C2, C3 or C4, C5 lead to the formation of amino reductones (Eq. 6-10).

$$\text{Eq. 6-9}$$

$$\text{Eq. 6-10}$$

Thiazole and Thiazoline. Dicarbonyls also react with H_2S and NH_3 to form thiazole or thiazoline. Both H_2S and NH_3 are derived from the degradation of amino acids. Alternatively, the condensation of cysteine and a carbonyl compound, followed by cyclization, also yields thiazole (Eq. 6-11) (8).

$$\text{Eq. 6-11}$$

Polysulfide Heterocyclic Compounds. Cyclic polysulfide compounds are found in many cooked meat products. The formation of these may result from the reaction of H_2S with aldehyde and ammonia via various condensations (Eq. 6-12).

Eq. 6-12

Lipid degradation during food processing contributes a significant part in flavor formation, both desirable and undesirable. Numerous saturated and unsaturated acids, aldehydes, alcohols, ketones, esters, hydrocarbons, lactones, and aromatic compounds originate from lipid constituents in food. These may be due to autoxidation, photosensitized oxidation, irradiation, thermal degradation (Chapter 2), or enzymatic breakdown (Chapter 5). All these processes have been covered and will not be discussed further.

BEVERAGE FLAVOR

Tea

The formation of flavor (and color as well) during fermentation in tea manufacturing is related predominantly to the oxidation of the phenolic compounds. The most important reactions are the oxidation of the flavanols catalyzed by catechol oxidase. The tea flavanols constitute 15–25% of the dry weight and are mostly catechins and their derivatives (Fig. 6.4).

In many oxidations of phenols, the first step is the abstraction of a phenolic hydrogen, forming a semiquinone radical. Removal of a second elec-

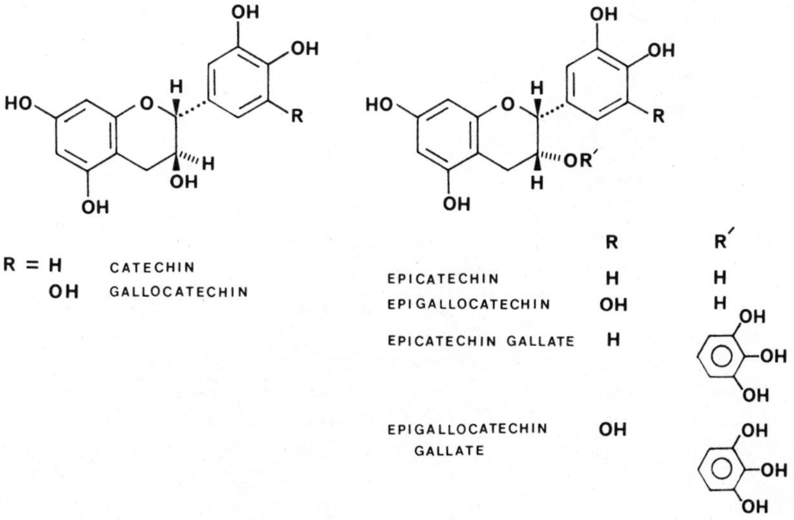

Fig. 6.4. Structure of catechin and derivatives.

tron forms a stable quinone. The semiquinone radical can react with another radical to form a dimer. In tea leaf tissues, the enzymatically oxidized flavanols condense to form the dimer theaflavin and polymeric proanthocyanidin (thearubigins) (Eq. 6-13). The orange-red theaflavins are astringent and contribute to the distinct tea flavor known as "briskness" (9).

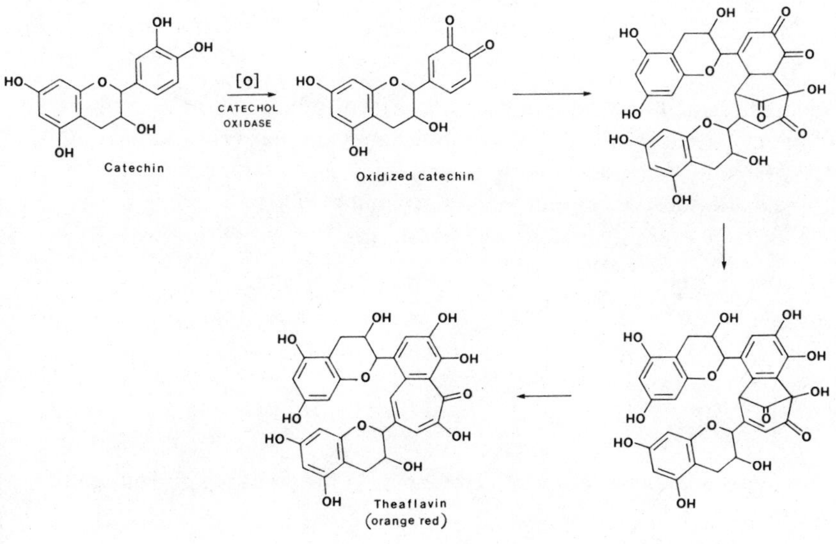

Eq. 6-13

Alternatively, the oxidized flavanols can react with amino acids to yield various aldehydes via Strecker degradation. Isobutanal, 2-methylbutanal, isovaleraldehyde, and phenylacetaldehyde found to be present in fermented tea aroma are proposed to be produced from the amino acids valine, isoleucine, leucine, and phenylalanine (Eq. 6-14).

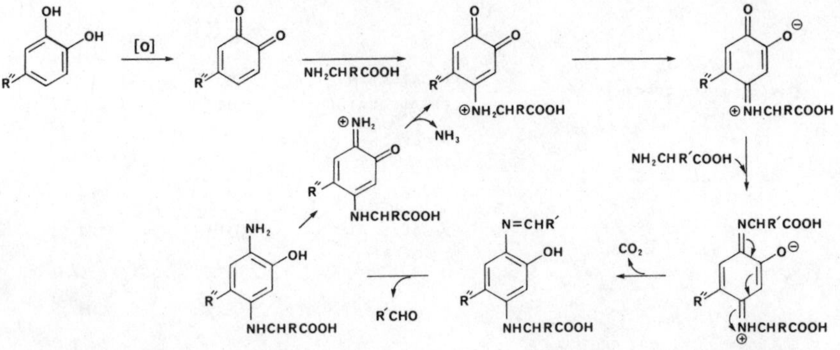

$$\text{Eq. 6-14}$$

A number of volatile compounds are known to be derived from the degradation of carotenoids, which occurs only in the presence of oxidized flavanols. The major volatile products are ionones and their epoxy derivatives (Fig. 6.5). It is suggested that the oxidized tea flavanols formed during fermentation cause the degradation of carotenes present in the tea leaves.

Figure 6.6 shows the relationship between various oxidative changes induced by the oxidation of tea flavanols (*10*).

Beer

The characteristic bitter note in beer is ascribed to hop "resins." Hop "resins" are classified into soft (soluble in hexane) and hard (insoluble in hexane). Soft resins include α-acids (humulone, cohumulone, adhumulone) and β-acids (lupulone, colupulone, adlupulone) (Fig. 6.7).

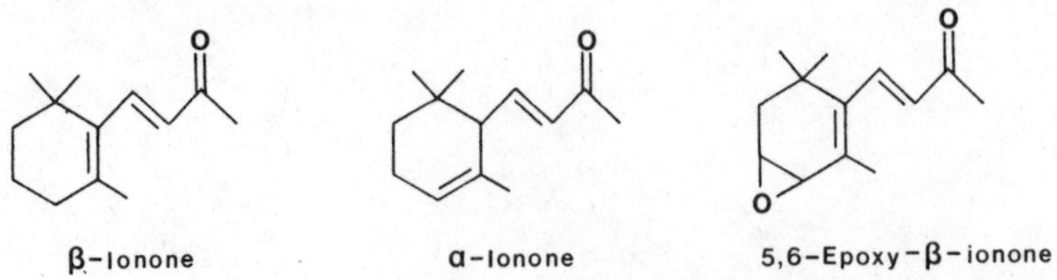

β-Ionone   α-Ionone   5,6-Epoxy-β-ionone

Fig. 6.5. Degradation products of carotenoids in the presence of oxidized flavanols.

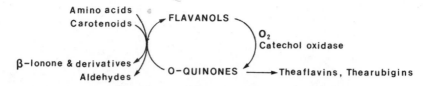

Fig. 6.6. Oxidative changes induced by the oxidation of tea flavanols.

In the brewing process, the α-acids isomerize into iso-α-acids, which possess a very bitter taste. Isomerization of an α-acid produces two pairs of stereoisomers as indicated for humulone in Fig. 6.7.

Beer, when exposed to sunlight, develops an unpleasant sulfur odor, termed "sunstruck" flavor. Its formation has been ascribed to 2-methyl-2-butene-1-thiol, which is derived from the photolysis of iso-α-acids in the presence of sulfur-containing amino acids (11) (Eq. 6-15).

The characteristic hop flavor in beer is due to the transfer of some essential oil-derived compounds to beer. More than 100 aroma constituents have been identified, including esters, ketones, alcohols, ethers, terpenoids, and sesquiterpenoids. The cyclic esters and the oxygenated sesquiterpenoids are

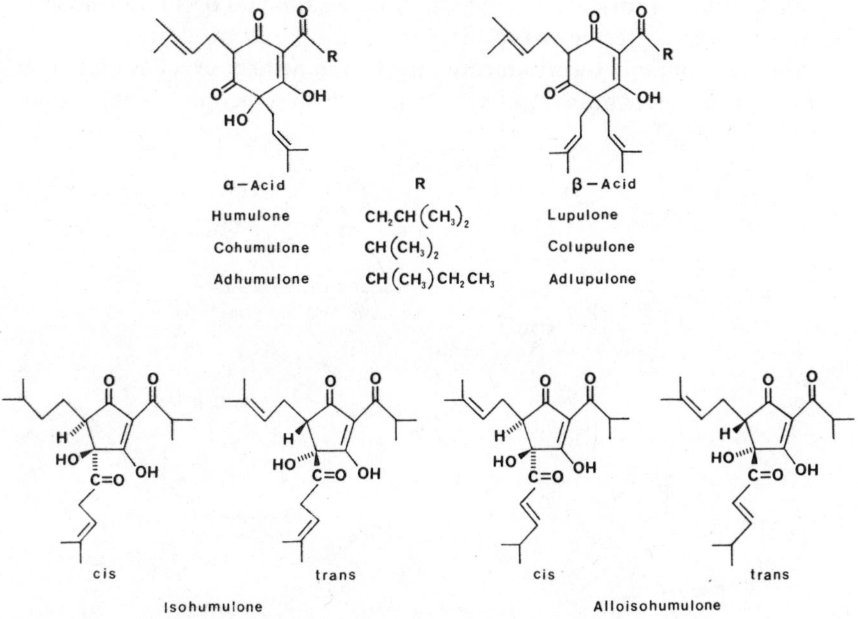

α-Acid	R	β-Acid
Humulone	$CH_2CH(CH_3)_2$	Lupulone
Cohumulone	$CH(CH_3)_2$	Colupulone
Adhumulone	$CH(CH_3)CH_2CH_3$	Adlupulone

cis   trans   cis   trans
Isohumulone   Alloisohumulone

Fig. 6.7. Structure of α-acid and β-acids of hop resins.

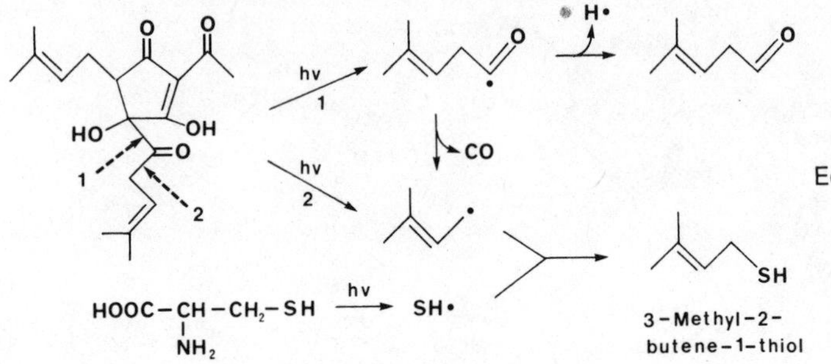

Eq. 6-15

HOOC-CH-CH₂-SH $\xrightarrow{hv}$ SH•
     |
     NH₂

3-Methyl-2-
butene-1-thiol

constituents with a strong hop aroma. The esters have predominantly bi-cyclic structures, and the sesquiterpenoids have either an epoxide or a hy-droxyl constituent (*12*). A few examples are given in Fig. 6.8.

Coffee

A number of sulfur-substituted furans have been characterized in green and roasted coffee. The compound 2-furylmethanethiol is possibly a character-impact compound of roasted coffee (*13*) (Fig. 6.9). The threshold is 0.01 ppb in water, and at concentrations of 0.1-5 ppb, the compound has the aroma of roasted coffee. At increasing concentration (as occurs during stor-age of roasted coffee beans), it develops a strong sulfur odor.

Another compound worth mentioning is trigonelline, which is present at a 1% level in coffee beans. Unlike caffeine, which is thermally stable, trigo-

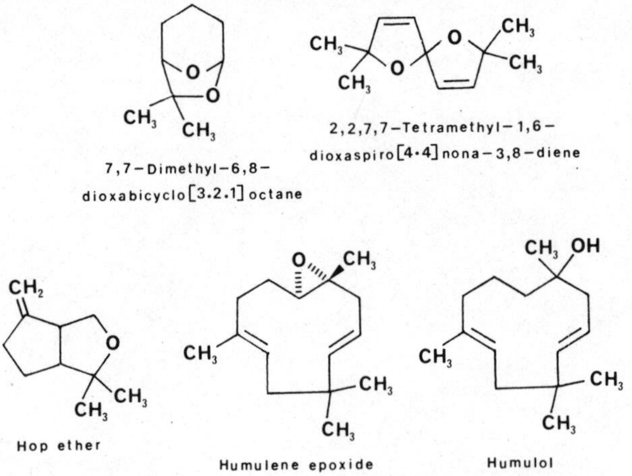

7,7-Dimethyl-6,8-
dioxabicyclo[3.2.1]octane

2,2,7,7-Tetramethyl-1,6-
dioxaspiro[4.4]nona-3,8-diene

Hop ether

Humulene epoxide

Humulol

Fig. 6.8. Cyclic esters found in hop flavor.

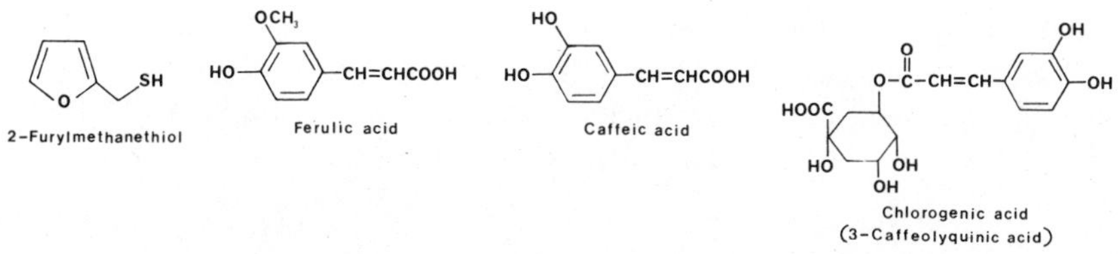

Fig. 6.9. Flavor compounds found in coffee beans.

nelline decomposes readily at roasting temperature to give a series of pyridines and pyrroles (*14*) (Eq. 6-16).

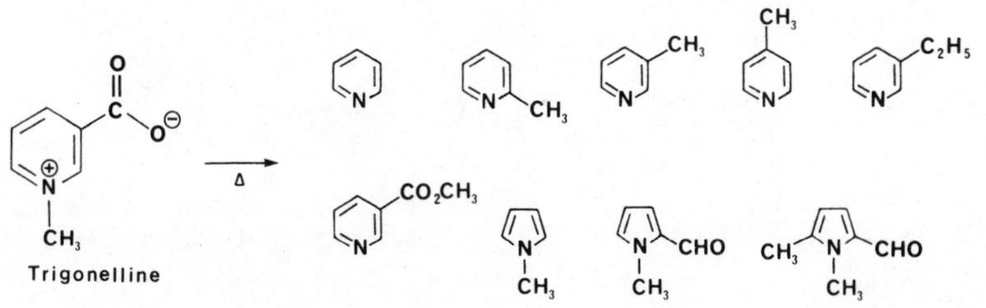

Eq. 6-16

The "acid" flavor in coffee is ascribed to the many organic acids. Phenolic acids, in particular, average about 7.5% total dry weight in coffee beans. These include ferulic, caffeic, and the major component, chlorogenic acid (Fig. 6.9).

SPICE FLAVOR

In food processing, spices are often applied in the form of essential oils or oleoresins. Essential oils are prepared by water and steam distillation of the dried ground spices. The oil contains the volatile flavor compounds. Oleoresins are extracts of freshly ground spices. The ground spice is repeatedly extracted with an organic solvent that is eventually removed to give a product that consists of volatile essential oil, nonvolatile resinous materials, and the active principle characteristics of the spice.

Garlic and Onion

Among the most extensively studied flavors are those of garlic and onion. The odor of garlic is derived from allyl-2-propenethiosulfinate (allicin). The odor develops only when the garlic is cut or crushed, when the enzyme alliinase catalyzes the breakdown of an odorless substrate, (+)-*S*-allyl-L-cysteine sulfoxide (alliin), to 2-propene sulfenic acid (Eq. 6-17.1), which dimerizes to form allicin (Eq. 6-17.2). Allicin readily decomposes to 2-propenesulfenic acid and thioacrolein. The latter self-condenses via a Diels-Alder reaction to form cyclic sulfur compounds (Eq. 6-17.3) (*14*).

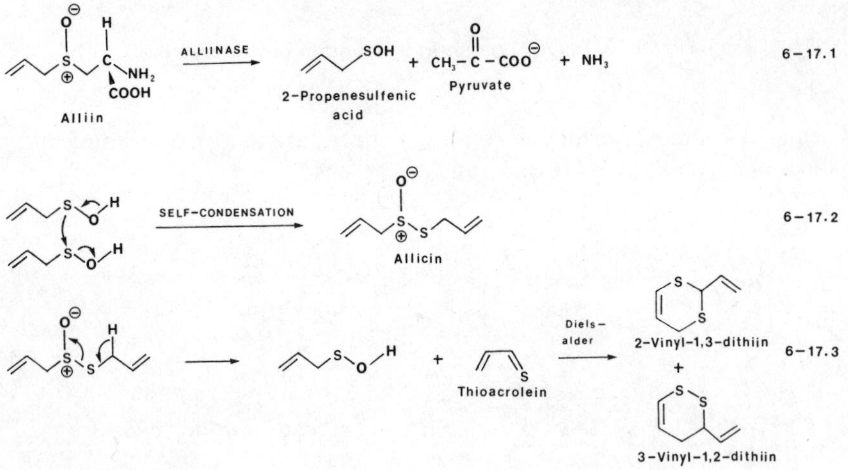

Eq. 6-17

In onion, the precursor is *trans*-(+)-*S*-(1-propenyl)-L-cysteine sulfoxide, a positional isomer of alliin. The alliinase in onion converts it into 1-propene-sulfenic acid, which then rearranges to form *syn*-propanethiol *S*-oxide, the lacrimatory factor that causes the eye to tear. The oxide readily undergoes hydrolysis to yield propionaldehyde and hydrogen sulfide (Eq. 6-18).

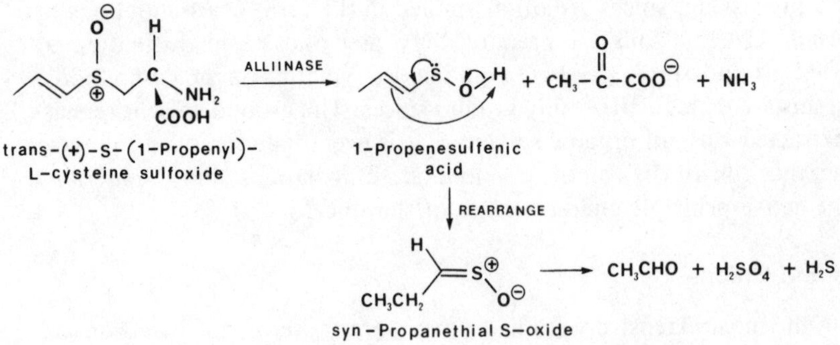

Eq. 6-18

The flavor precursor, *trans*-(+)-(1-propenyl)-L-cysteine sulfoxide, is biosynthetically derived from valine and cysteine (Eq. 6-19). In nature, as much as half of the precursor is bound as a peptide, γ-L-glutamyl-S-allyl-L-cysteine sulfoxide, which is not susceptible to the action of alliinase.

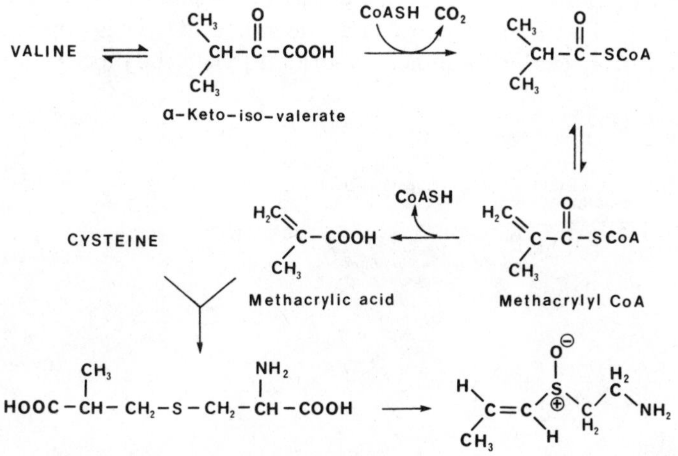

Eq. 6-19

Many other compounds, including sulfides (mono-, di-, tri-, and tetra-), thiophenes, and thiosulfonates, found in the essential oils also contribute to the flavor (*15*). Disulfides and thiosulfonates can be formed by disproportionation of the unstable thiosulfinates (allicin is an alkenyl thiosulfinate) (Eq. 6-20). Thiosulfonates with four or more carbon atoms possess the distinct odor of freshly cut onions. Propyl and propenyl di- and tri-sulfides possess a cooked-onion flavor.

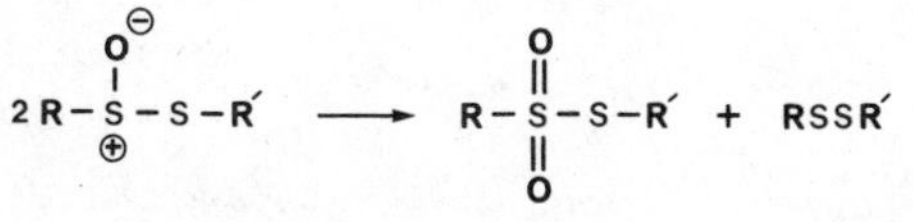

Eq. 6-20

Thiophenes are formed, along with mono- and tri-sulfides, from the decomposition of alkyl and alkenyl disulfides. Dimethylthiophenes display a distinct fried-onion flavor (Eq. 6-21).

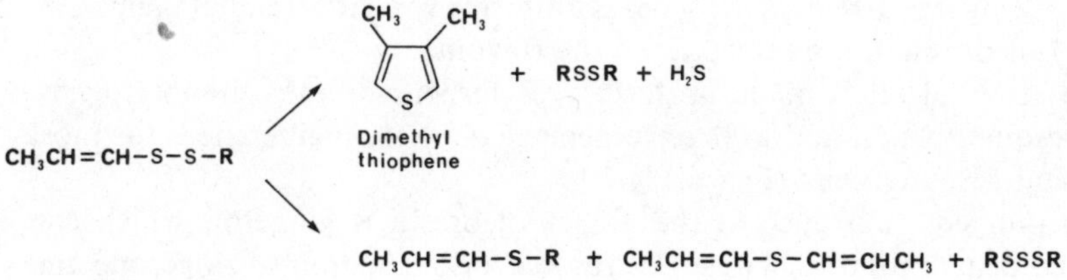

Eq. 6-21

Pepper

Pepper (*Piper nigrum* L.) oleoresin contains the pungent alkaloid piperine. Piperine can be synthesized by the aldol condensation of piperonal with acetaldehyde. Further reaction with acetic anhydride yields piperinic acid, which can then react with an acid chloride to yield the piperine (*16*) (Eq. 6-22).

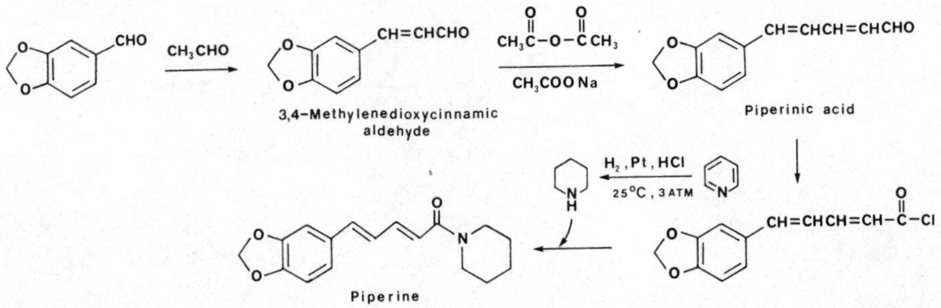

3,4-Methylenedioxycinnamic aldehyde

Piperinic acid

Piperine

Eq. 6-22

Chili

Chili spice is obtained from the fruits of *Capsicum* species. The pungent principle of chili is ascribed mainly to a class of alkaloid compounds collectively known as capsaicinoids (*17*) (Fig. 6.10).

Synthetically, capsaicinoids can be made by reacting vanillyl amide with an acid chloride (Eq. 6-23).

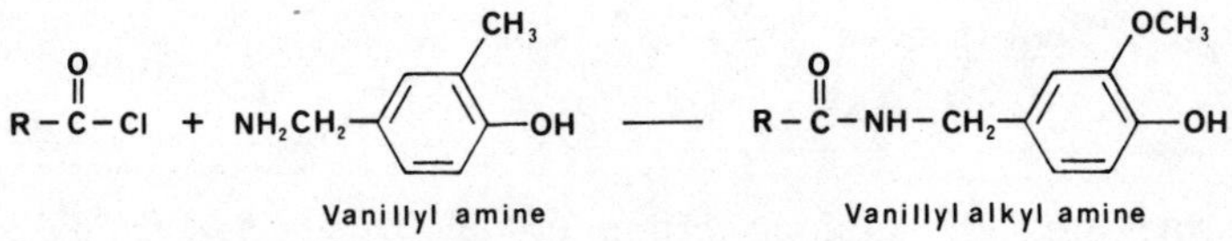

Vanillyl amine

Vanillyl alkyl amine

Eq. 6-23

Ginger

Ginger (*Zingiber officinale* Roscoe) is a fibrous-rooted perennial plant. The oil and oleoresin are obtained from the rhizome.

The essential oil is made up largely of terpenoids (4% monoterpenes, 63% sesquiterpenes, and 17% of terpene alcohols). Zingiberene is the most abundant sesquiterpene (Fig. 6.11).

The pungent substance in the ginger oleoresin is gingerol, which can undergo dehydration to shogaol or reverse aldol reaction to zingerone and

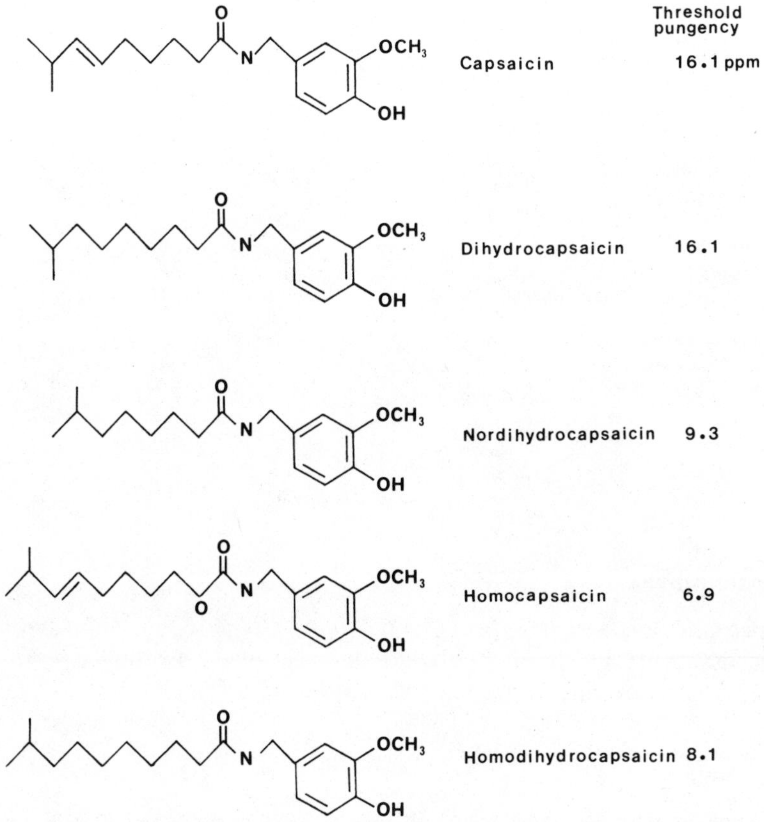

Fig. 6.10. Structure of capsaicinoids and their pungency thresholds.

aldehyde (Eq. 6-24). The former reaction is accelerated by alkaline pH, and the latter reaction occurs at a high temperature (*18*). Both zingerone and shogaol are less pungent than gingerol. The aliphatic aldehyde causes the formation of off-flavors. These undesirable changes occur rarely in fresh ginger but more often in preparing and storing oleoresins.

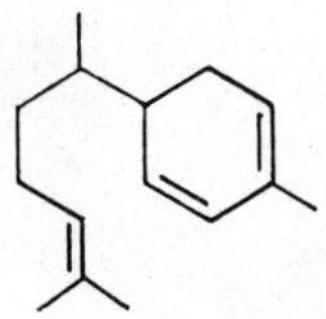

Fig. 6.11. Structure of zingiberene.

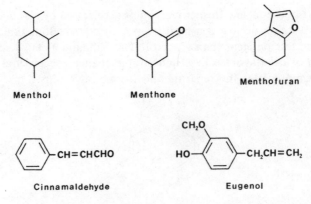

Eq. 6-24

Peppermint

Peppermint oil is obtained from *Mentha piperita* plants. The important flavor constituents are the monocyclic monoterpenoids, menthol (50–60%), and to a lesser extent, menthone and menthofuran (Fig. 6.12).

Cinnamon

The spice cinnamon is obtained from commercial species *Cinnamomum cassia* (China) and *C. zeylanicum* Blume (Ceylon). The traditional commer-

Menthol

Menthone

Menthofuran

Cinnamaldehyde

Eugenol

Fig. 6.12. Flavor compounds in peppermint and cinnamon.

cial cinnamon is the "quills"—rolled pieces of peeled bark. Essential oils are produced by distillation of bark oil or leaf oil. Cinnamaldehyde and eugenol are two main flavor compounds in the oil (Fig. 6.12).

FRUITS AND VEGETABLES

Essential Oil and Essence

The most important category of flavor ingredients obtained from fruits, especially citrus, is the essential oils. An essential oil is an oil substance obtained from a plant material, and it retains the characteristic flavor of that material.

There are two common processes for the production of essential oils of citrus fruits. Most of the oil is found in the rind of the fruit. Citrus fruits that contain high concentrations of essential oil (>3%) include orange, lemon, lime, tangerine, mandarin, and grapefruit. Commercially, the oil is obtained as a by-product of fruit juice production. When juice is pressed from the fruit, the oil is also carried through. It is removed by centrifugation and becomes what is known as the cold-pressed peel oil. Most of the essential oils used in industry are distilled oils obtained as a by-product of various types of essence recovery processes.

The natural essences of citrus fruits are obtained when the fruit juice is concentrated in high-temperature evaporators. Approximately 25% of the juice water is removed. The volatile flavors carried in the vapor are recovered and concentrated in the essence units (fractionation stills with condensers). The flavor components collected are separable into an aqueous (aroma) and oil phase (essence oil). Essence is a common name that includes fractions of both aroma and essence oil. Since essence oil is distilled oil, it lacks the nonvolatile components and, therefore, many natural antioxidants found in cold-pressed oils. For this reason, essence oil is relatively less stable than cold-pressed oil.

Fruit Flavor

Citrus essential oils consist mainly of aldehydes, ketones, esters, alcohols, and acids (Table 6.1). Many of these are isoprenoids. The cyclic terpene δ-limonene is one of the most abundant constituents (~90%) present in oranges and other citrus oils. It possesses the flavor characteristic of lemon, orange, and caraway (19). Another major sesquiterpene present in orange oil is valencene (1–2%). Interestingly, nootkatone, the character-impact compound of grapefruit, is chemically synthesized from valencene (Fig. 6.13). Recently, the importance of nootkatone has been questioned. A new

Table 6.1. Components of Orange (Valencia) Oil

COMPOUND	CONCENTRATION
Ethanol	0.1%
Ethyl acetate	50 ppm
Acetal	20 ppm
Hexanal	200 ppm
Ethyl butyrate	0.1%
Trans-2-hexenal	50 ppm
α-Pinene	0.4%
Sabinene	0.4%
Myrcene	1.8%
Octanal	0.5%
δ-Limonene	93.6%
Linalool	0.5%
Decanal	0.6%
Neral	0.2%
Geranial	0.1%
Valencene	1.7%

Reprinted with permission from Johnson and Vora (*19*), copyright, 1983 by Institute of Food Technologists.

compound, 1-*p*-menthene-8-thiol, has been isolated which displays a character aroma of fresh grapefruit juice. The flavor threshold is in the range of 1×10^{-4} ppb in water, among the lowest detection thresholds for naturally occurring flavor compounds (*20*).

Essential oils of tangerine and mandarin contain thymol, dimethylan-

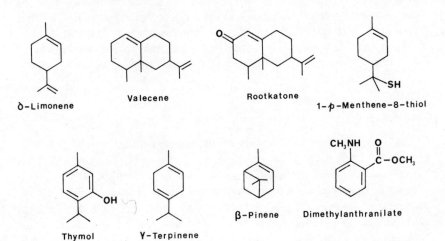

δ-Limonene Valecene Rootkatone 1-*p*-Menthene-8-thiol

Thymol γ-Terpinene β-Pinene Dimethylanthranilate

Fig. 6.13. Isoprenoids identified in fruit flavor.

thranilate, γ-terpinene and β-pinene (Table 6.2). The latter two compounds are shown to contribute to the flavor of mandarin oranges as distinct from other oranges (*21*).

Citral, a terpene aldehyde with a powerful lemon flavor, is present in the essential oils of lemon, lemon grass, and lime. Citral is commercially synthesized from acetylene and acetone (Eq. 6-25).

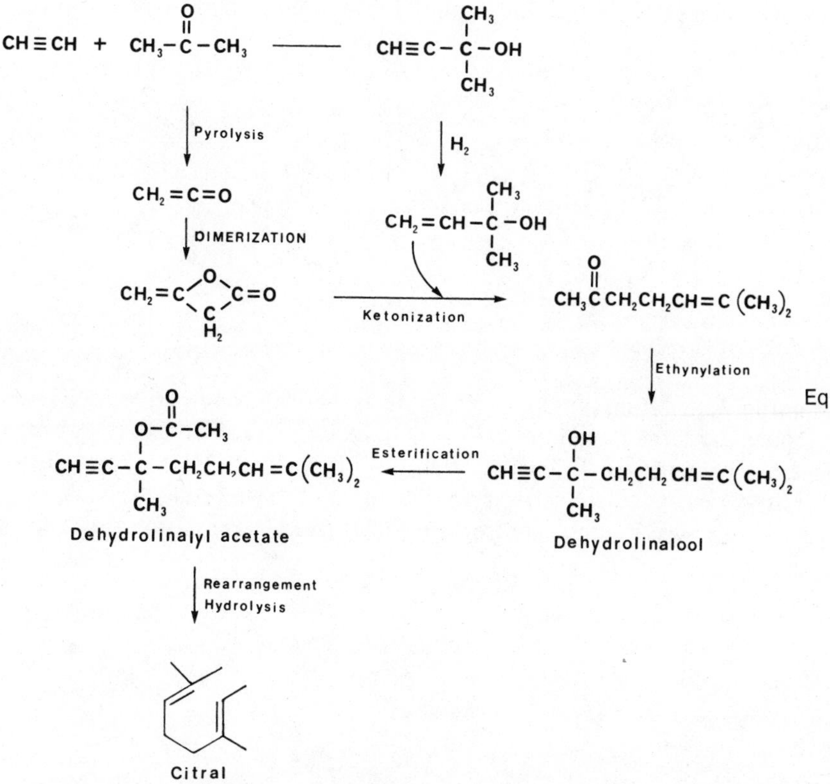

Eq. 6-25

Citral

Table 6.2. Major Components of Tangerine and Mandarin Oil

COMPONENT	TANGERINE OIL (WT%)	MANDARIN OIL (WT%)
Thymol	0.022	0.182
Methyl-N-methylanthranilate	0.072	0.652
γ-Terpinene	1.74	14.0
β-Pinene	0.17	1.8

Reprinted with permission from Wilson, C. W. III, and Shaw, P. E. Importance of thymol, methyl N-methyl-anthranilate, and monoterpene hydrocarbons to the aroma and flavor of mandarin cold-pressed oils. *J. Agric. Food Chem.* **29**, 495. Copyright 1981 American Chemical Society.

The bitterness of citrus fruits is ascribed to flavonoid compounds, especially naringin and limonin. Limonin belongs to a class of triterpene derivatives known as limonoids. It consists of two lactone rings, one of which has a furan substituent and an epoxide group.

Limonin is the cause of the problem known as delayed bitterness, in which citrus fruit juice turns bitter after extraction. This is due to the conversion of the major natural limonoate A-ring lactone in citrus fruit to limonin catalyzed by the enzyme limonin D-ring lactone hydrolase (22) (Eq. 6-26).

Eq. 6-26

Limonoate A-ring lactone

Limonin

Another bitter compound is naringin, a flavanone neohesperidoside. The aglycone part is the flavanone naringenin, and the disaccharide part is neohesperidose, which is 2-O-α-L-rhamnosyl-β-D-glucose (23). Alkaline hydrolysis of naringin converts it to the intensely sweet dihydrochalcone (~300 times sweeter than sucrose) (Eq. 6-27) (see Chapter 7 for a detailed discussion).

Eq. 6-27

Naringin

Naringin dihydrochalcone

The typical flavor compounds of noncitrus fruits such as banana, pears, peaches, and apples are produced during the climacteric rise. Many aliphatic acids and amino acids in the unripe fruits are converted to esters, alcohols, and ethers that are mostly responsible for the characteristic fruit odor (24). Esters, especially acetates and butyrates, are most abundant in bananas, and so are some phenol ethers. The main component that gives a banana aroma is isobutyl acetate (Fig. 6.14). In Bartlett pears, ethyl *trans*-2-*cis*-4-decandienoate is identified as the important flavor compound. Ethyl 2-methylbutyrate in apple, 2,5-dimethyl-4-hydroxy-3(2H)-furanone in pine-

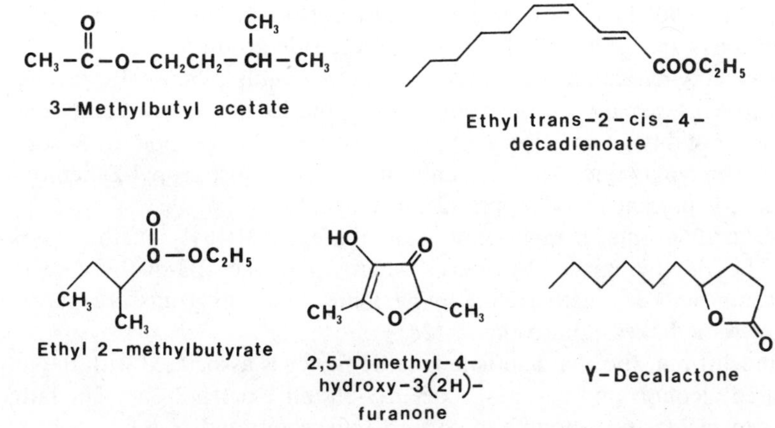

3-Methylbutyl acetate

Ethyl trans-2-cis-4-
decadienoate

Ethyl 2-methylbutyrate

2,5-Dimethyl-4-
hydroxy-3(2H)-
furanone

γ-Decalactone

Fig. 6.14. Character compounds in some noncitrus fruits.

apple and strawberry, and γ-decalactone in peaches are some more examples of character-impact compounds.

Vegetables

Most *Cruciferous* vegetables and all *Brassicas,* such as cabbage, cauliflower, brussels sprouts, turnips, and mustards, contain glucosinolates that can be enzymatically degraded to yield the pungent sulfur compound isothiocyanate.

The formation of isothiocyanate involves a process similar to the Lossen rearrangement in which the alkyl group of a hydroxamic acid salt migrates from the carbon to the nitrogen (Eq. 6-28) (see Chapter 8).

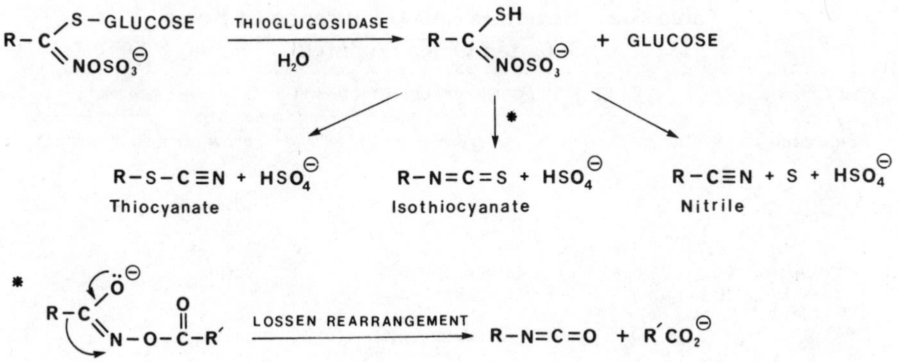

$$R-S-C\equiv N + HSO_4^{\ominus}$$
Thiocyanate

$$R-N=C=S + HSO_4^{\ominus}$$
Isothiocyanate

$$R-C\equiv N + S + HSO_4^{\ominus}$$
Nitrile

LOSSEN REARRANGEMENT $\longrightarrow$ $R-N=C=O + R'CO_2^{\ominus}$

Eq. 6-28

More than 90 glucosinolates have been identified. Some of the more important ones are listed in Table 6.3 (only the side group R is shown).

Alkoxyalkylpyrazines are widespread in vegetables, especially in peas, pepper, beans, asparagus, beetroot, carrot, and lettuce. These compounds possess a notably green odor. Bell pepper contains 20,000 ppm of 3-isobutyl-2-methoxypyrazine. Another common pyrazine, 3-isopropyl-2-methoxypyrazine, is present in 1400 ppm (25) (Fig. 6.15).

The earthy aroma of raw potato is attributed to 3-ethyl-2-methoxypyrazine. The alkylpyrazines, 2-isobutyl-3-methylpyrazine, 2,3-diethyl-5-methylpyrazine, and 3,5-diethyl-2-methylpyrazine, taken as a mixture, have a characteristic baked-potato flavor (26).

In mushroom, the characteristic volatile flavor is associated with the unsaturated alcohols and ketones, 1-octen-3-ol and 1-octen-3-one. The latter occurs in increasing amounts in cooked mushroom and also has been related to metallic or mushroom off-flavors in dairy products. Another unsaturated alcohol, cis-hept-4-en-2-ol is the major component in the volatile flavor in corn (27).

Asparagus consists of sulfur-containing acids and esters as principal flavor compounds. The strongly concentrated (~7 ppm) component, methyl 1,2-dithiolane-4-carboxylate, possesses an aroma characteristic of raw asparagus. During cooking of asparagus, the acid, 1,2-dithiolane-4-carboxylic acid (asparagusic acid), is decomposed to 1,2-dithiacyclopentene and 1,2,3-trithiane-5-carboxylic acid (Fig. 6.16) (28).

Besides the breakdown products of asparagusic acid, methyl disulfide has also been suggested to a major aroma constituent (2–10 ppm) in cooked asparagus, formed by thermal fragmentation of S-methylmethionine.

Phthalides and their derivatives are implicated in the characteristic odor

Table 6.3. Structure and Occurrence of Some Selected Glucosinolates

SIDE CHAIN	GLUCOSINATE (COMMON NAME)	OCCURRENCE
$CH_2=CHCH_2-$	sinigrin	brown or black mustard
$HO-\langle \bigcirc \rangle-CH_2-$	sinalbin	white mustard
$CH_2=CHCH{\overset{OH}{\mid}}CH_2-$	progoitrin	turnip
$\langle \text{indole} \rangle-CH_2-$	glucobrassicin	cabbage

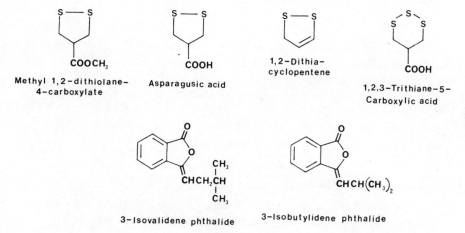

3–Isobutyl–2–
methoxypyrazine

3–Isopropyl–2–
methoxypyrazine

Fig. 6.15. Alkoxyalkylpyrazines in pepper.

of celery. These are the 3-isovalidene phthalide, 3-isobutylidene phthalide, and their dihydro-derivatives (*29*) (Fig. 6.16).

MEAT FLAVOR

The chemistry of meat flavor represents the utmost complexity in flavor research. Meat must be cooked before it develops flavor, and numerous reactions can occur during the process.

There are well over 200 volatile compounds identified in cooked (boiled, roasted) beef. It is generally believed that the heterocyclic compounds are the important components in meat flavor. Table 6.4 list classes of volatile compounds in beef aroma (*30*).

All the compounds have been shown to contribute to the cooked meat flavor, but only the thiazoles and thiazolines possess meaty odor. The compound 4-hydroxy-5-methyl-3(2*H*)-furanone (DMHF) and its derivatives are worth mentioning. These are α-dicarbonyl compounds that can thermally

Methyl 1,2–dithiolane–
4–carboxylate

Asparagusic acid

1,2–Dithia–
cyclopentene

1,2,3–Trithiane–5–
Carboxylic acid

3–Isovalidene phthalide

3–Isobutylidene phthalide

Fig. 6.16. Sulfur-containing compounds and phthalides in asparagus.

Table 6.4. Classes of Volatile Heterocyclic Compounds of Beef Aroma

Thiophene		Furan	
Thiazole	Thiazole	Oxazole	Oxazole
	Thiazoline		Oxazoline
Thiopane		Pyrrole	
Pryazine		Pyridine	
Benzenoid		Lactone	γ
			δ

From MacLeod and Seyyedain-Ardebili (*30*).

react with H_2S to form a whole array of compounds with meaty flavors in model systems (Table 6.5). Those identified include mercapto-substituted furans and thiophenes (*31*).

The reaction between the dihydrofuranone and hydrogen sulfide involves substitution of the ring oxygen by sulfur to give the thio analog, via the intermediate, 2,4-diketone (Eq. 6-29). Study of the reaction using cysteine also results in the formation of roasted meat flavor.

Simulated Meat Flavors

Most simulated meat flavorings have been produced by thermal processing a mixture of "precursor" compounds, some of which are listed below.

Table 6.5. Meaty Flavors Formed in the Reaction of 4-Hydroxy-5-Methyl-3 (2*H*)-Furanone with Hydrogen Sulfide

4-Mercapto-2-methylfuran		green, meaty
3-Mercapto-2-methyl-4,5-dihydrofuran		roasted meat
4-Mercapto-3-oxo-tetrahydrofuran		green, meaty
3-Mercapto-2-methyl-thiophene		roasted meat
4-Mercapto-2-methyl-2,3-dihydrothiophene		rubbery, meaty
3-Mercapto-4-hydroxy-2-methyl-2,3-dihydro-thiophene		meaty, savoury

Reprinted with permission from Van den Ouwedand, G. A. M., and Peer, H. G. Components contributing to beef flavor. Volatile components produced by the reaction of 4-hydroxy-5-methyl-3(2*H*)-furanone and its thio analog with hydrogen sulfide. *J. Agric. Food Chem.* **23**, 502–503. Copyright 1975 American Chemical Society.

Eq. 6-29

Amino acids—cysteine, cystine, methionine, glutamic acid, glycine, valine
Proteins—glycoproteins, hydrolyzed vegetable, yeast, animal proteins
Nucleotides—Adenosine-5'-monophosphate, guanosine-5'-monophosphate
Carbohydrates (reducing sugars)—ribose, glucose, xylose, ribose-5-phosphate
Acids—α-ketobutyrate, succinate, lactate, aliphatic carboxylic acid
Vitamins—thiamin
Sulfur compounds—thiols, sulfides, furanones, sulfur amino acids

In general, the reaction mixture always contains (1) an amino-containing compound (such as amino acid), (2) a sulfur-containing compound (such as cystine, thiamin), and (3) a reducing sugar or carbonyl compound.

In recent years, numerous chemicals have been synthetically produced to be added to the reaction mixture to improve the flavor. These meat flavor chemicals include many of the naturally occurring flavor compounds, structures of which have been mentioned above, and also compounds that are not reported as naturally occurring in meat. Some of these are listed in Fig. 6.17 (*32*).

MICROENCAPSULATION OF FLAVORS

There are increasing numbers of food products containing microencapsulated flavors and ingredients. The flavor ingredients are coated with edible matrix film. The coated particle has a size of less than 5000 μ. The most widely used coating materials are the following:

1. Polysaccharides, such as gum arabic, starch
2. Malto-dextrins
3. Protein—gelatin
4. Hydrolyzed gelatin
5. Modified polymers—succinylated gelatin, alkylated starch

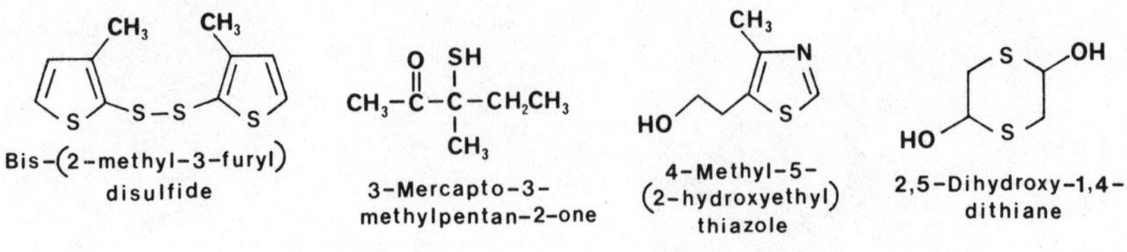

Fig. 6.17. Some synthetic flavoring compounds.

Microencapsulation has the advantages of (1) converting liquid flavor concentrate to solid or powder form, (2) protecting flavor loss during food processing and storage, and (3) controlling the rate of release of volatiles and nonvolatiles into a food system. An example of the last is the use of slow-release capsules of ascorbic acid and calcium peroxide in bread making.

7
SWEETENERS

Sweeteners can be classified into two basic categories: nutritive and nonnutritive. Nutritive sweeteners include sugars, sugar alcohol, corn syrups, and high-fructose corn syrup. Nonnutritive sweeteners include saccharin, cyclamate, and acesulfame K (Table 7.1). Most nutritive sweeteners are obtained from plant sources, and nonnutritive sweeteners are synthetic. However, stevioside, a sweet diglycoside extracted from the wild plant *Stevia rebaudiana*, is noncaloric. Aspartame is a synthetic but low-calorie sweetener.

Each individual sweetener has its specific properties and limitations. Development of new and sweet-tasting compounds with desirable properties becomes increasingly possible only due to the numerous investigations in recent years on the chemistry of the nature of sweetness. Theoretically, sweet compounds could be constructed if the basic chemistry of sweetness was understood. In fact, sweet compounds designed for research studies provide much of present knowledge on the theory of taste.

MOLECULAR THEORY OF SWEETNESS

The chemical nature and sweetness of a substance can be correlated based on the widely accepted tripartite model. Basically, the glycophore (sweetcarry) in a sweet-tasting compound consists of two units—AH and B—together with a third component, γ, arranged in a tripartite structure. The unit AH is a group consisting of oxygen or nitrogen carrying a hydrogen atom such as OH, NH, or NH_2, and B is a group consisting of O, N, or any electronegative center that is capable of attracting a hydrogen atom to form hydrogen bonding. The γ unit is a hydrophobic site and need not be present as a specific functional group. The distance parameters are A, B =

Table 7.1. Commercially Available Sweeteners

NAME	SWEETNESS (SUCROSE = 1)	NUTRITIVE	STATUS IN UNITED STATES
Sorbitol	0.5	+	+
Xylitol	1.0	+	+
Cyclamate	40	−	−(Approved for use in many European and Asian countries)
Saccharin	450	−	+
Aspartame (NutraSweet)[a]	200	+	+
Acesulfame-K (Sunnett)[a]	200	−	−(Approved for use in UK, Ireland, Germany, Belgium, and Australia and recently United States)
Stevioside (Steviosin)[a]	300	−	−(Approved for use in China and Japan)
Glycyrrhizin (MagnaSweet)[a]	50	+	+
Thaumatin (Talin)[a]	2000	+	+

[a]Trade name.

Reprinted with permission from *Food Technol.* **40**(8), 199, copyright 1986 by Institute of Food Technologists.

2.6 Å, B, γ = 5.5 Å, and A, γ = 3.5 Å. The orbital distance between the AH proton and B is about 3.0 Å (Fig. 7.1) (*1*).

In a three-dimensional picture, the glycophore binds to the receptor site. The sweet taste is then initiated by intermolecular hydrogen bonding between this glycophore (AH, B) and a similar AH, B units on the receptor. The γ component functions to direct and align the molecule as the AH, B glycophore approaches the receptor site (Fig. 7.1).

The AH, B glycophore can be found in many carbohydrates with 1,2-glycol structure. In the sugars, the 1,2-glycol has to be in the gauche confor-

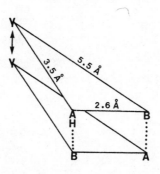

Fig. 7.1. The tripartite model.

mation to satisfy the distance requirement. For a pyranose sugar, both α-D-C1 and α-D-1C acquires a 60° angle. Anti conformation as in β-D-1C, elicits no sweet taste. Eclipsed vicinal OH groups as in α-D-B3 conformer tastes less sweet due to strong intramolecular hydrogen bonding.

Experimentally, it is possible to locate the AH, B glycophore in a particular sweet compound. Some representative compounds and the AH, B glycophore assigned to be responsible for the sweet taste are listed in Fig. 7.2.

Note that for glucose, the C4-OH has the AH function, and C3-O is the B unit. The reason for this assignment is based on the studies that C6-OH is so positioned as to hydrogen bond to C4-O, and consequently the proton of C4-OH becomes more acidic and more capable of intermolecular hydrogen bonding with receptor site B in the initial reaction (Fig. 7.3A). This "activation" of the proton of C4-OH by C6-OH, and thus the enhancement of the AH, B functions is called "Lemienx effect." Analogous bonding in fructose fixes the position of C1-O in space and enhances the proton-donating group C2-OH (Fig. 7.3B) (2).

Sucrose contains eight hydroxyl groups arranged in the crystal in which the fructose and glucose units are bridged by two intramolecular hydrogen bonds from C6'-OH to C5-O, and C1'-OH to C2-O (Fig. 7.3C). In solution, the sucrose exists in a form wherein the C1'-OH to C2-O hydrogen bond is maintained (Fig. 7.3D) (3). In addition to the glycophore in the glucopyranosyl unit, a second glycopore can be established involving C2-OH (AH), C3-O (B) and C1' (γ).

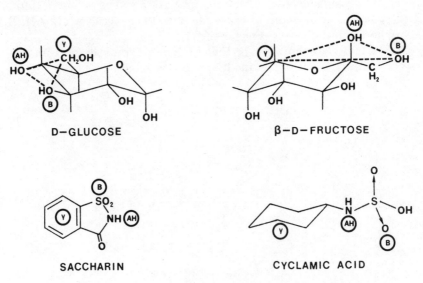

Fig. 7.2. Glycophore of some sweet compounds.

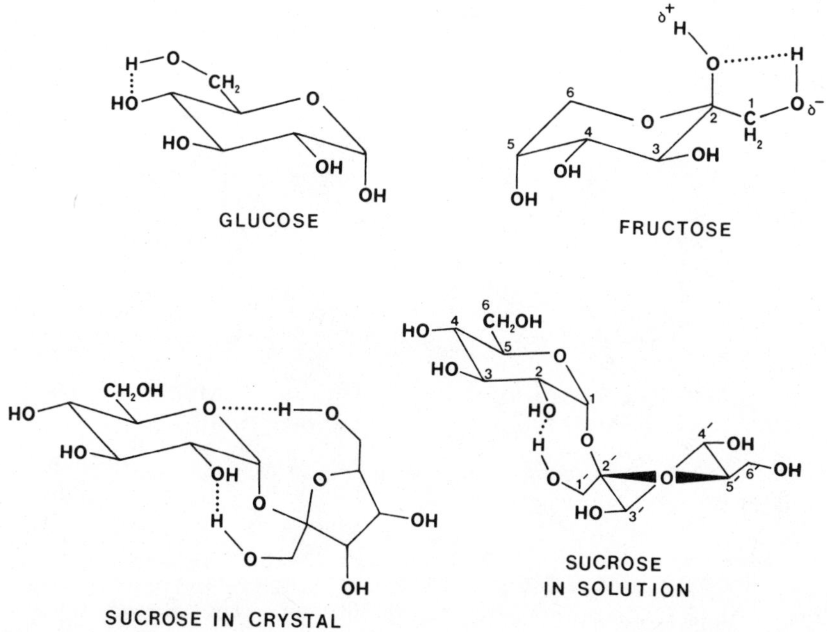

Fig. 7.3. The "Lemienx effect" in (A) glucose, (B) fructose, (C) sucrose crystal, and (D) sucrose in solution.

AMINO ACIDS AND DIPEPTIDES

A number of D-amino acids, including Trp, His, Phe, Tyr, Leu, Gly, and Apn, are sweet, but their enantiomers, the L-amino acids, are known to be tasteless or bitter (Fig. 7.4). The difference in sweetness in the enantiomeric amino acids can be explained by the fact that the D-amino acids are superimposable upon the receptor sites while the L-amino acids are not, as shown in Fig. 7.5.

The low-calorie sweetener aspartame (marketed under the brand name NutraSweet) is a methyl ester of a dipeptide L-aspartyl L-phenylalanine (L-Asp-L-Phe-OCH$_3$). The dipeptide is ~ 180–200 times sweeter than sucrose, and exhibits no bitter aftertaste generally associated with other artificial sweeteners. Furthermore, aspartame acts synergistically with other sweeteners.

Aspartame contains an ester linkage that at high temperature and pH may hydrolyze to aspartylphenylalanine or cyclize to diketopiperazine before being converted to aspartylphenylalanine (Eq. 7-1). The resulting dipeptide can then be hydrolyzed to the individual amino acids. Aspartame is more stable in the weak acid range between 3.0 and 5.0 than in the basic

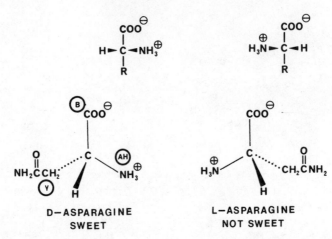

Fig. 7.4. Fischer projection formula and the corresponding configurational drawing of D- and L-asparagine.

range. At 25°C, the optimum pH for stability is 4.3. The pH effect is markedly increased with increasing temperature. The tendency for aspartame to hydrolyze or cyclize limits its use in products that require high-temperature processing or other adverse conditions (4).

In aspartame, the AH, B system is present at the aspartyl residue where the protonated α-amino group and the ionized β-carboxyl group represent the AH and B unit, respectively (Fig. 7.6) (5). The phenylalanine −OCH₃ portion of the molecule could be replaced by various esters of amino acids

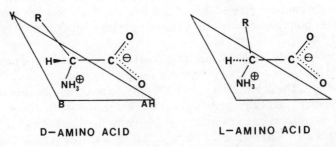

Fig. 7.5. Binding of D-amino acid and L-amino acid with receptor. (Reprinted with permission from Shallenberger (2), courtesy of Elsevier Science Publishers, Biomedical Division, Amsterdam, The Netherlands.)

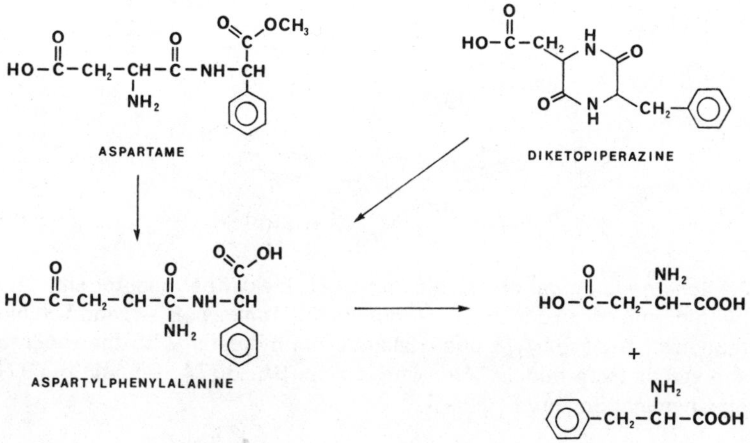

ASPARTAME

DIKETOPIPERAZINE

Eq. 7-1

ASPARTYLPHENYLALANINE

without losing the sweetness, since the taste receptor site recognizes only the shape and size and the hydrophobic portion of the molecule.

Only the L-L isomer is intensively sweet. L-Asp-D-Phe-OCH₃ is not sweet. This structural-taste relationship can readily be explained by considering the receptor site as a "pocket" with the AH, B, γ system inside a spatial barrier. The L-D isomer has the methyl ester group so positioned that the

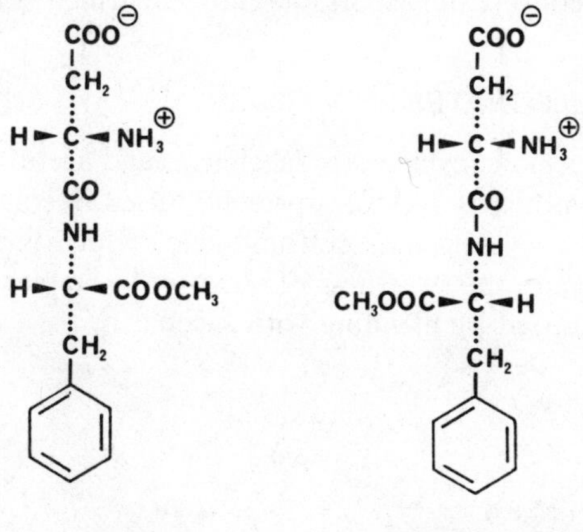

L-Asp-L-Phe-OCH$_3$
(200 X Sucrose)

L-Asp-D-Phe-OCH$_3$
(not sweet)

Fig. 7.6. L-Asp-L-Phe-OCH$_3$ and L-Asp-D-Phe-OCH$_3$ represented by Fischer projection formula.

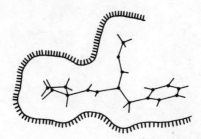

Fig. 7.7. Schematic model of L-Asp-L-Phe-OCH$_3$ inside the receptor site. (Reprinted with permission from Lelj, F., Tancredir, T., Temussi, P. A., and Toniolo, C. Interaction of α-L-aspartyl-L-phenylalanine methyl ester with the receptor site of the sweet taste bud. *J. Am. Chem. Soc.* **98**, 6674. Copyright 1976 American Chemical Society.)

molecule cannot fit into the pocket for the interaction with the receptor site to occur. In the case of the L-L isomer, the conformation of the molecule allows it to interact with the receptor site within the spatial barrier (Fig. 7.7) (6).

Aspartame has a pI of pH 5.2, where minimum solubility occurs. The phenylalanine portion of aspartame is highly hydrophobic, since the carboxyl group is esterified. Consequently, aspartame is only slightly soluble in water (about 1% at 24°C, pH 5.2). In practice, food grade citric or malic acid is used to convert the aspartame into salt which is readily soluble.

THE AMINOSULFONATES

Aside from aspartame, cyclamate, saccharin, and acesulfame K (Fig. 7.8A, B, C) are the synthetic sweeteners presently used in commercial products.

Cyclamate is a cyclohexane sulfonic acid, which is a white crystalline powder 30–60 times the sweetness of sucrose. Its unpleasant aftertaste can usually be minimized by blending with saccharin.

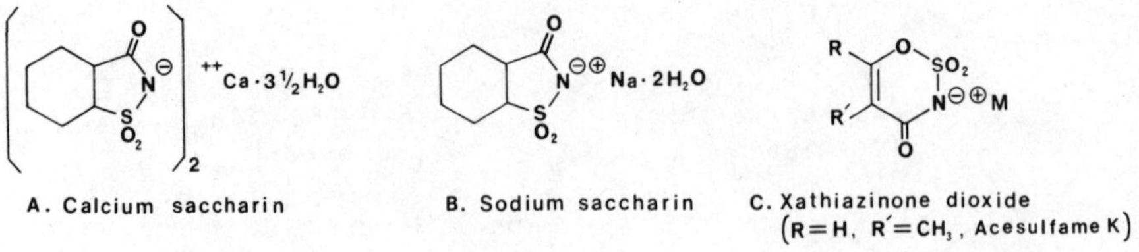

A. Calcium saccharin

B. Sodium saccharin

C. Xathiazinone dioxide
(R = H, R′ = CH$_3$, Acesulfame K)

Fig. 7.8. Structure of some synthetic sweeteners.

Saccharin is usually supplied in the sodium or calcium form, with a solubility of 6.5% in water at 25°C and a sweetness intensity 450 times that of sucrose.

Acesulfame K differs from saccharin by having an oxygen atom in between the SO_2 group and the π ring. In both compounds, altering the imide function destroys the sweet taste. Acesulfame K is 200 times sweeter than sucrose, and more stable. It shows no loss of stability at pH values of 3 and higher, and only decomposes at a temperature higher than 225°C. Its solubility at 20°C in water is about 27%.

These aminosulfonate sweet compounds have the AH group (NH) and B group (oxygen of the SO_3) in the gauche conformation. The great sweetness found in these compounds are likely due to the large hydrophobic surface contact provided by the ring structure. Ring substituents with various lengths indicate a loss of sweetness when the length of the hydrophobic group on the nitrogen exceeds 0.7 Å. This relationship may be explained by the presence of a spatial barrier in the receptor site, located at 0.7 Å from the NH interaction point (7). A bulky substituent would displace the molecule that interaction with the receptor AH, B units becomes impossible.

DIHYDROCHALCONE

The bitter flavanone glycosides in citrus fruits include notably the neohesperidin of oranges and lemons and naringin of grapefruits.

In neohesperidin, the disaccharide component linked to the 7-hydroxy group of the aglycone is neohesperidose (2'-O-α-L-rhamnopyranosyl-D-glucose) (Fig. 7.9A). The aglycone in these structures has a 2(S) configuration. When the rhamnose is linked to the 6-position of the glucose instead, the compounds become tasteless. The disaccharide now becomes 6-O-α-L-rhamnosyl-β-D-glucose with the trivial name rutinose (Fig. 7.9B).

The compounds naringin, neohesperidin, and other neohesperidose-containing flavanones can be hydrolyzed to the chalcones and then dihydrochalcones upon hydrogenation (Eq. 7-2). Both the chalcones and dihydrochalcones are sweet (8). The rutinose-containing flavanones such as hesperidin and naringenin rutinoside yield tasteless dihydrochalcones which become sweet when the rhamnose is removed enzymatically or by acid hydrolysis (Eq. 7-3).

Rhamnose linked to the C-2 hydroxy glucose enhances sweetness but causes a complete loss of sweetness when it is linked to the C-6 hydroxyl group. The C-6 substituents are not necessary for sweet taste. Hesperetin dihydrochalcones with the disaccharide replaced by xylose or galactose are twice as sweet as a glucose substituent (Fig. 7.10) (9).

The C-3 and C-4 hydroxyl groups are the AH, B glycophore unit. Aside

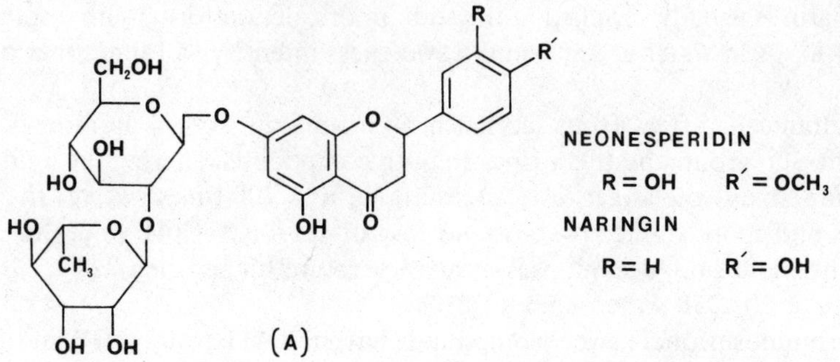

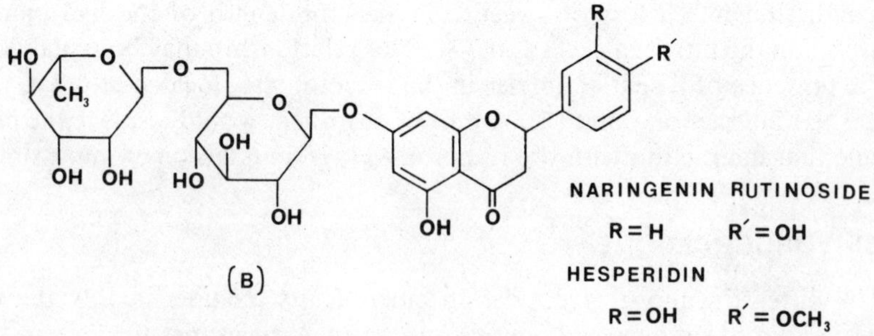

Fig. 7.9. Structure of (A) neohesperidin and naringin, and (B) hesperidin and naringenin rutinoside.

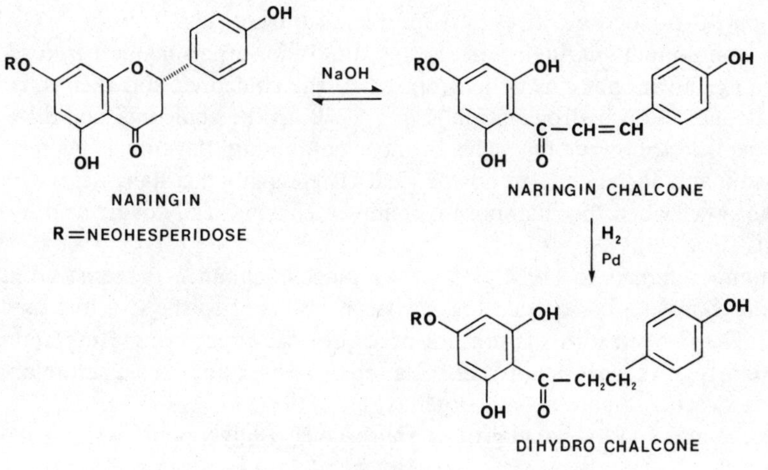

Eq. 7-2

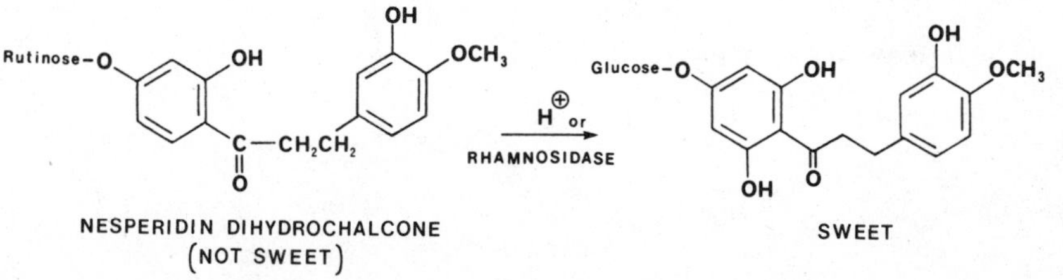

$$\text{Eq. 7-3}$$

from the glucoside position, there is a very specific requirement of the substituents in ring B of the flavanone of the dihydrochalcone, as indicated in Fig. 7.11 (9). A hydroxyl substitution at the B ring and its position are important for the instensity of sweetness.

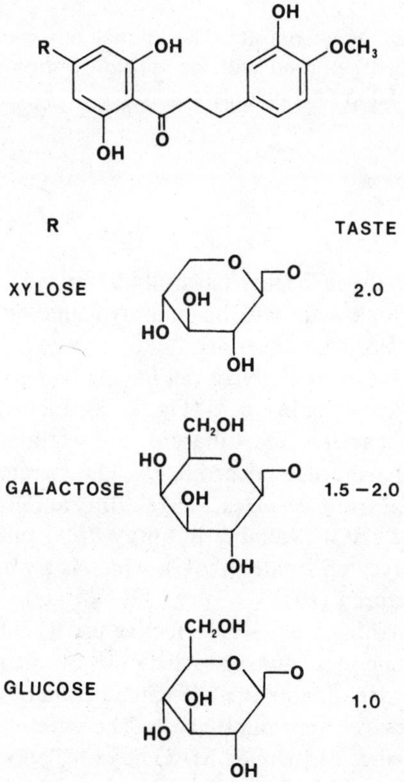

Fig. 7.10. Sweetness of hesperetin dihydrochalcone modified with various monosaccharide substituents.

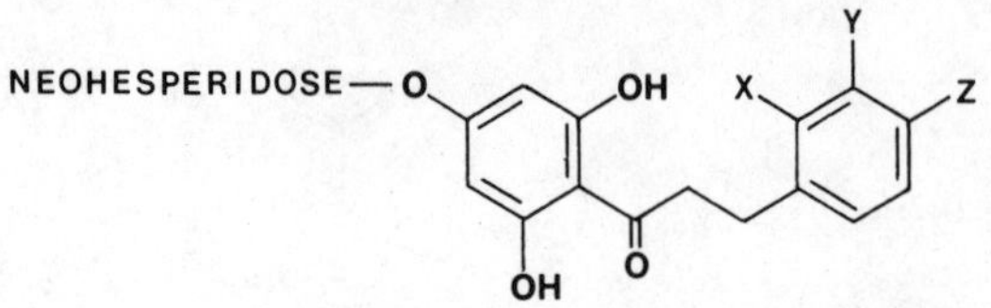

X	Y	Z	SWEETNESS (sucrose=1)
H	H	H	0
OH	H	H	0
H	OH	H	110
H	H	OH	110
H	OH	OH	0
H	OCH$_3$	OH	0
H	OH	OCH$_3$	950
H	OH	OCH$_2$CH$_3$	1100
H	OH	OCH$_2$CH$_2$CH$_3$	2000

Fig. 7.11. Sweetness of hesperetin dihydrochalcone modified with various substituents in ring B. (Reprinted with permission from Inglett (9), copyright 1969 by Institute of Food Technologists.)

GLYCYRRHIZIN

The sweetener glycyrrhizin occurs naturally as mixed calcium and potassium salts of glycyrrhizic acid, which is mainly found 6–10% in the licorice roots (*Glycyrrhiza glabra* L.).

Glycyrrhizin is a triterpenoid glycoside, having two glucuronic acid units. The aglycone is the glycyrrhetinic acid (Fig. 7.12). Licorice extracts are used in the flavoring of cigarettes and tobaccos, confectionary, beverages such as root beer, and pharmaceutical products. The sweetener glycyrrhizin is commercially isolated from the extract as the fully ammoniated salt (ammonium glycyrrhizinate, AG) available in spray-dried brown powder form. Mono-ammonium glycyrrhizinate (MAG), which is a white crystalline powder, is also manufactured (10).

Ammonium glycyrrhizinate has an effective use level of 20–1000 ppm. It is soluble in water, alcohol, and propylene glycol. At pH below 4.5, AG tends to precipitate. Heating above 105°C causes deterioration of the flavor. AG is ~50 times sweeter than sucrose. The sweetener is potentiated to 100 times in the presence of sucrose. MAG has very poor solubility in water and alcohol, but it is stable over a wide pH range. Both AG and MAG are on the FDA list of GRAS food additives and marketed in modified forms under the brand name "MagnaSweet."

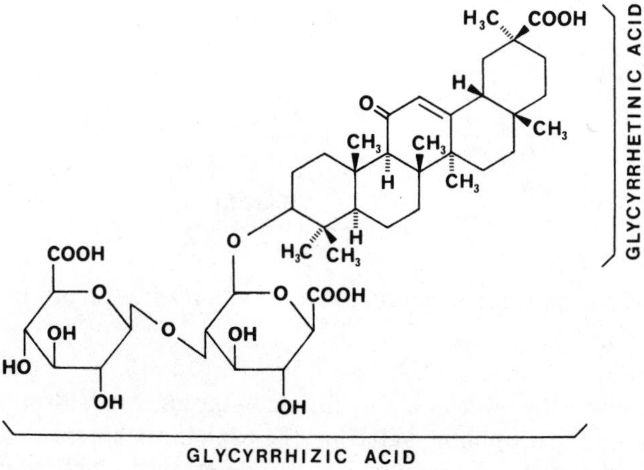

Fig. 7.12. Structure of glycyrrhizic acid.

STEVIOSIDE

Stevioside is the sweet diterpenoid glycoside found in the leaves of *Stevia rebaudiana* (Bert.). It is a white crystalline, hygroscopic powder 250–300 times sweeter than sucrose, but has a bitter and unpleasant aftertaste. The aglycone is a diterpenoid called steviol (Fig. 7.13A). The α-hydroxyl group at C-13 is linked to the disaccharide sophorose (2-O-β-D-glucopyranosyl-β-D-glucopyranose), and the α-carboxyl at C-4 is condensed with β-D-glucopyranose (Fig. 7.13B). The aglycone steviol has been shown to be weakly antiandrogenic.

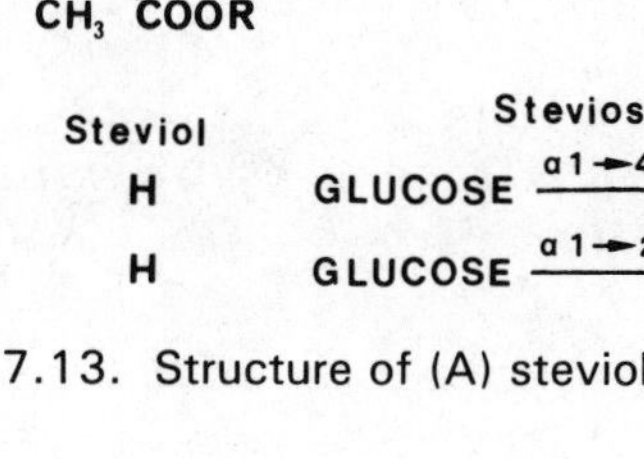

	Steviol	Stevioside
R	H	GLUCOSE $\xrightarrow{\alpha 1 \to 4}$
R′	H	GLUCOSE $\xrightarrow{\alpha 1 \to 2}$ GLUCOSE $\xrightarrow{\alpha 1 \to 13}$

Fig. 7.13. Structure of (A) steviol, and (B) stevioside.

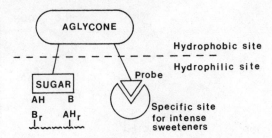

Fig. 7.14. Schematic representation of interaction between glycoside and receptor.

All the aglycoside sweeteners mentioned contain the hydrophilic disaccharide and the hydrophobic aglycone. To explain the intense sweetness of these large molecules, the tripartite theory is extended to include another hydrophilic function, designated as the polar "probe." The probe may be the β-D-glucose in stevioside, the C-20 carboxylate anion of ammonium glycyrrhizinate, the phenolate ion of naringin dihydrochalcone, and neohesperidin dihydrochalcone. It is postulated that the probe is also to bind to a more specific type of receptor site, which by electron transfer or charge displacement illicits a strong response to the taste of sweetness. Schematically, the structural interactions of glycosides and receptor sites are presented in Fig. 7.14 (11). The entire molecule positions so that the hydrophobic and hydrophilic units are attracted to the corresponding sites of the receptor, placing the probe to a specific receptor for inducing the intensely sweet taste.

SUGAR ALCOHOL

Most sugar-derived polyhydric alcohols are slightly less sweet when compared with sucrose as shown in Table 7.2. The last two listed are polyhydric glycosides, although they are usually regarded as alcohols.

Table 7.2. Sweetness of Polyhydric Alcohols

ALCOHOL	SWEETNESS
Sucrose	100
Xylitol	95
Sorbitol	54
Galactitol	46
Mannitol	62
Maltitol	63
Lactitol	34

Xylitol is commercially produced by the hydrogenation of xylose obtained from the hydrolysis of xylan. Raw materials such as corn cobs, hardwood chips, and sugar cane bagasse typically contain 20–35% xylan (Eq. 7-4) (*12*).

$$
\text{XYLAN} \xrightarrow[\substack{H_2O}]{\substack{H^{\oplus}}}
\begin{array}{c}
\text{CHO} \\
| \\
\text{H}-\text{C}-\text{OH} \\
| \\
\text{HO}-\text{C}-\text{H} \\
| \\
\text{H}-\text{C}-\text{OH} \\
| \\
\text{CH}_2\text{OH}
\end{array}
\xrightarrow[\text{CATALYST}]{\substack{H_2}}
\begin{array}{c}
\text{CH}_2\text{OH} \\
| \\
\text{H}-\text{C}-\text{OH} \\
| \\
\text{HO}-\text{C}-\text{H} \\
| \\
\text{H}-\text{C}-\text{OH} \\
| \\
\text{CH}_2\text{OH} \\
\text{XYLITOL}
\end{array}
\qquad \text{Eq. 7-4}
$$

Lactitol can be synthesized by the catalytic hydrogenation of lactose under high pressure and temperature (Eq. 7-5).

LACTITOL   Eq. 7-5

CORN SWEETENERS

Sweeteners prepared by hydrolysis of cornstarch have found large-scale applications in various food industries. The corn sweeteners can be classified into three basic categories: (1) conventional corn syrups, (2) dextrose solids, and (3) high-fructose corn syrups.

The key step in the manufacture of the above sweeteners is hydrolysis, using a combination of a starch-liquefying enzyme (α-amylase) and a starch-saccharifying compound (glucoamylase) (Eq. 7-6). The former enzyme hydrolyzes the α-1,4 bonds randomly along the chain, while the latter breaks down the α-1,4 links and, more slowly, the α-1,6 links from the nonreducing end.

$$
\text{STARCH} \xrightarrow[\substack{\alpha-\text{amylase}}]{\substack{105°C}} \underset{\text{DE} \sim 10}{\text{DEXTRINS}} \xrightarrow{\text{Glucoamylase}} \text{GLUCOSE} \qquad \text{Eq. 7-6}
$$

The process results in a mixture of dextrose, maltose, and a whole range of higher saccharides depending on the degree of hydrolysis. The total reducing sugar in the product is determined and compared to the reducing power of pure dextrose to yield a number known as the dextrose equivalent (D.E.). Thus the higher the D.E. of the product, the more low-molecular-weight saccharides it contains. The product is more sweet and more fermentable, has a greater freezing point depression and lower viscosities, and is more susceptible to nonenzymatic browning reaction. Commercially, the hydrolysate is refined (removal of impurities) and concentrated to yield the conventional corn syrups. There are four types of corn syrups made suitable for various applications (13).

Type	D.E.	Application
I	20–38	Relatively nonsweet, used as bulking agents, provide body and mouth feel
II	38–58	Commonly used as sweeteners in combination with sucrose or HFCS
III	58–73	Used mainly as sweeteners
IV	>73	Used for further processing into HFCS

Alternatively, the dextrose in the refined hydrolysate can be crystallized out to give a commercial prodcut of alpha dextrose monohydrate or alpha dextrose anhydrate. Dextrose is three-fourths as sweet as sucrose.

High-fructose corn syrup is derived from the high-D.E. (92–96 D.E.) corn syrups. The corn syrup is treated with glucose isomerase which catalyzes the isomerization of glucose and fructose. The product contains a liquid mixture of glucose and fructose that is similar to sucrose in sweetness and chemical composition. Since the fructose comprises 42% of the total solids, these syrups are known as 42% HFCS. The 42% FHCS can be pumped past a calcium or other cation affinity carrier, where it is further concentrated to yield the 90% HFCS. Blending the 90% and 42% HFCS gives the 55% HFCS. The high-fructose corn syrups are rapidly replacing sucrose in many food applications.

SWEET PROTEINS

There are a few proteins that are extremely sweet. These include monellin found in red berries (serendipity berries) of the tropical plant *Dioscoreophyllum cumminsii,* thaumatin isolated from the fruit of *Thaumatococcus danielli* in western Africa, and miraculin from miracle fruit, *Synsepalum dulcificum.* These proteins are about 100,000 times sweeter than sucrose on a molar basis and several thousand times on a weight basis.

The most studied sweet protein is thaumatin I. Thaumatin I is a basic protein (pI = 12) consisting of a single polypeptide chain (MW 22,000) of

207 amino acids, with eight disulfide bonds and no histidine. The complete amino acid sequence of the protein and its tertiary structure are known.

The protein consists of a very low α-helix or random coil. The main domain (I) is a long β sheet folded into flattened β barrel (Fig. 7.15). The β strands are arranged antiparallel except the N- and C-terminal strands, which are parallel to each other. The two other domains, II and III, are loops linked and stabilized by disulfide bonds (14).

Monellin is composed of two polypeptide chains held by noncovalent bonds. Subunits I and II contain 50 and 42 amino acid residues, respectively. There is only one single sulfhydryl group (in subunit II), and contrary to thaumatin I, no disulfide bond. Only the undissociated protein is found to be sweet; subunits are not sweet. Chemically blocking the −SH group results in the loss of sweetness.

Five homologous amino acid sequences are found in monellin and thaumatin I. Four of these are located in domain I of thaumatin I, and one is in between domains I and II (Fig. 7.16). In monellin, three homologous sequences are in subunits I and two in subunit II (15). The significance of this homology in sequences is not yet clear.

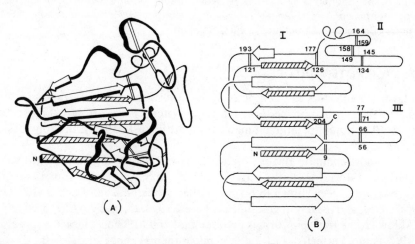

Fig. 7.15. (A) Backbone structure of thaumatin I. The main structure consists of two β sheets forming a flattened β barrel. β strands in the top sheet are clear, and those in the bottom sheet are shaded. Open bars represent disulfide bonds. (B) Topological structure of thaumatin I. There are two β sheets in the structure. The β strands of the top sheet are indicated by wide arrows. There are three domains of the protein and a crystallographic assignment of disulfide bonds shown in open bars. (Reprinted with permission from de Vos et al. (14), courtesy of National Academy of Sciences and S.-H. Kim.)

```
               116
Thaumatin    - Pro - Thr -|Thr - Arg - Gly|- Cys - Arg -
               27
Monellin A   ʼ- Tyr - Lys -|Thr - Arg - Gly|- Arg - Lys -

               126
Thaumatin    - Cys - Ala -|Ala - Asp - Ile|- Val - Gly -
               19
Monellin A   - Phe - Arg -|Ala - Asp - Ile|- Ser - Glu -

               98
Thaumatin    - Asp - Tyr -|Ile - Asp - Ile|- Ser - Asn -
               4
Monellin B   - Glu - Ile -|Ile - Asp - Ile|- Gly - Pro -

               99
Thaumatin    - Tyr - Ile -|Asp - Ile - Ser|- Asn - Ile -
               20
Monellin A   - Arg - Ala -|Asp - Ile - Ser|- Glu - Asp -

               92
Thaumatin    - Leu - Asn -|Gln - Tyr - Gly|- Lys - Asp -
               26
Monellin B   - Ile - Gly -|Gln - Tyr - Gly|- Arg - Leu -
```

Fig. 7.16. Sequence similarities between thaumatin and monellin. (Reprinted with permission from Iyengar et al. (*15*), copyright 1979 by Federation of European Biochemical Societies, Liege, Belgium.)

FLAVOR POTENTIATORS

Flavor potentiators are compounds that function to modify (intensify or mask) the sensory properties of a food system. They are also known to possess a unique, palatable umami taste distinguished from the four basic tastes.

The most common flavor potentiator is monosodium glutamate (MSG). Only the L-form possesses activity, and dicarboxylates with four to seven carbons have been shown to have a similar property. The pK_a of α-COOH, γ-COOH, and α-NH$_2$ in glutamic acid are 2.19, 4.25, and 9.67, respectively. At low H, the free acid form is predominant and MSG becomes less soluble. MSG is most effective in food systems in the pH range of 5.5–8.0.

Another group of flavor potentiators are the nucleotides, inosine 5'-monophosphate (IMP) and guansine 5'-monophosphate (GMP). Whereas the MSG is prepared in monosodium form, the IMP and GMP are disodium salts at neutrality (Fig. 7.17).

Only the 5'-nucleotides are active, although three possible isomers (2', 3', and 5') can exist. The 6-OH and the phosphate group are required for potentiation. The threshold levels for the nucleotides are in the range of

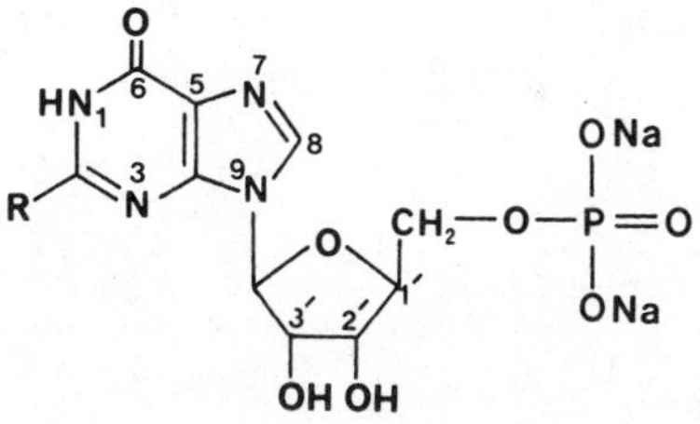

Fig. 7.17. Structure of nucleotide flavor potentiators.

0.01–0.03%. The threshold levels are greatly reduced when the nucleotides are combined syngergistically with MSG (Table 7.3).

Maximum flavor activity is obtained when the nucleotide and MSG are in a 1-to-1 mix. Lower or higher rates of mixing tend to reduce the effectiveness. GMP shows about four times the effectiveness in the synergistic action as IMP (Table 7.4) (16).

GMP and IMP are industrially produced by the enzymatic degradation of yeast RNA, using 5'-phosphordiesterases. The products commercially available are usually a 50:50 mixture of IMP and GMP.

It has been postulated that flavor nucleotide interacts with its receptor at three sites. Sites A and B are electrophilic and interact with the two phosphoryl oxygens and the C6 oxygen, respectively. Site X interacts with the substituent at C2 (17). The distance between sites A and B is about 8 Å. The flavor-enhancing activity depends on the electron density of the C6

Table 7.3. Synergistic Interaction Between MSG and 5'-Nucleotide in the Reduction of Individual Threshold levels.

	THRESHOLD LEVEL (%)		
	IMP	GMP	MSG
Water	0.012	0.0035	0.03
0.1% MSG	0.0001	0.00003	—
0.01% IMP	—	—	0.002

From Kuninaka (16, 17).

Table 7.4. Flavor Activity of Mixtures of MSG and 5'-Nucleotide

MIXTURE		RELATIVE FLAVOR ACTIVITY PER UNIT WEIGHT	
MSG	IMP OR GMP	IMP	GMP
1	0	1.0	1.0
1	1	7.5	30.0
10	1	5.0	19.0
20	1	3.4	12.4
50	1	2.5	6.4
100	1	2.0	5.5

From Kuninaka (*16,17*).

oxygen. Furthermore, site X interaction depends largely on the ability of delocalization of the substituent at C2. Thus GMP, with the C2 substituent being $-NH_2$, has a comparatively higher activity than IMP. Association of the molecule with a receptor initiates an umami sensation, and simultaneously activates other receptors whose sensitivity for their respective flavor compounds are enhanced.

8
NATURAL TOXICANTS

We consume a large number of toxic chemicals daily in our perfectly natural diet. The toxicants may be chemical constituents of the food itself, contaminants from microbial infestation, or degradation products from chemical changes during food processing. Toxicants vary in chemical structures ranging from amino acids to proteins, from simple amines to alkaloids, glycosides, and many phenolic compounds. The biological effects of all these chemicals are diverse and complex, and only a small percentage of these studies has been directed to the mechanisms of action at the molecular level. A thorough understanding, however, of the structural activity and biochemical mechanisms of these naturally occurring toxicants is essential to ensure proper preparation and processing of foods beneficial to the public. Caution must be taken to the fact that toxicity is determined not only by the chemical and biological properties of the compound, but also the level and duration of exposure an individual is subjected to. While it is true that many food plants contain toxicants, the generally low level of these compounds combined with the variety of choices in human diet usually precludes a high risk of intoxification.

CYANOGENIC GLYCOSIDES

Cyanogenic glycosides are known to be present in several plant species used for food. They are found in (1) seeds of bitter almond, apricot, and peach; (2) green leaves of sorghum; and (3) cassava and lima bean. Poisoning due to consumption of a large amount of cyanogenic glycosides is rare, since the edible portion of the plants mentioned above, in general, contains no glycosides. In the case of cassava, the edible tuber contains far less cyanogenic glycosides than the leaves. Proper processing, such as boiling, roast-

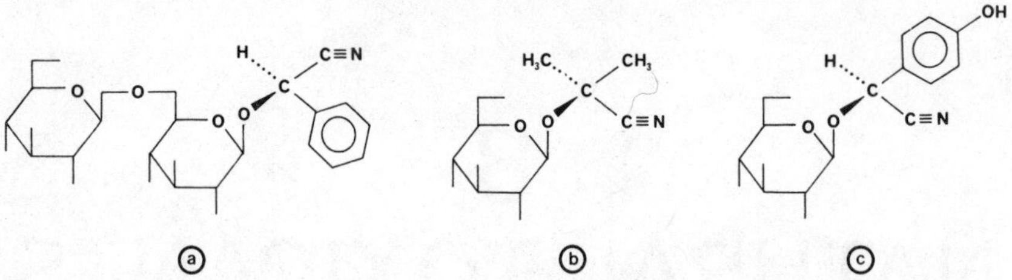

Fig. 8.1. Structures of some common cyanogenic glycosides. (A). Amygdalin, (B). Linamarin, (C). Dhurrin.

ing, sun-drying, soaking in water, or fermentation commonly practiced in the tropical regions, eliminates most of the toxicity.

Chemical Structure

Cyanogenic glycosides consist of a sugar moiety (usually D-glucose or disaccharides such as gentiobiose) linked to a cyanohydrin. Figure 8.1 shows the structure of some of the best known cyanogenic glycosides (1).

Mechanism of Toxicity

The toxicity of these glycosides is due to the release of HCN caused by enzymatic actions. Two enzymes are involved, (1) the β-glucosidase, which hydrolyzes the molecule into the corresponding cyanohydrin and sugar, (2) the hydroxynitrile lyase, which dissociates the cyanohydrin to yield the corresponding aldehyde or ketone and HCN (Eq. 8-1) (2). Both enzymes are found in plants containing the cyanogenic glycosides. Reactions occur when plant tissues are crushed, as in processing or ingestion, to allow the enzyme and substrate to come in contact.

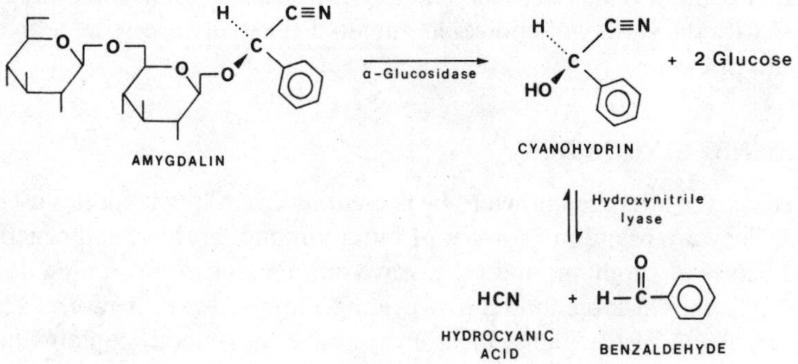

Eq. 8-1

The lethal level of HCN taken orally in a single dose in humans is 0.5–3.5 mg/kg body weight. The primary action of HCN is the inhibition of cytochrome oxidase, resulting in the interruption of cellular respiration. Small amounts of HCN can be detoxified by the enzyme rhodanese in the liver (Eq. 8-2) (2). The reaction requires a supply of thiosulfate and the product is thiocyanate, which is goitrogenic.

$$CN^- + S_2O_3^= \longrightarrow SCN^- + SO_3^= \qquad \text{Eq. 8-2}$$

GLYCOALKALOIDS

Glycoalkaloids are toxic compounds found in the *Solanaceae* plants, notably the cultivated potato (*Solanum tuberosum* L.). There have been recorded cases of poisoning in humans involving the consumption of potato glycoalkaloids.

Chemical Structure

Glycoalkaloids are alkaloid (the aglycone portion) glycosylated with a carbohydrate moiety. The two major glycoalkaloids found in potatoes are the α-solanine ($\sim 40\%$) and α-chaconine ($\sim 60\%$). Both have the same alkaloidal aglycone, solanidine, but contain β-solatriose and β-chacotriose, respectively, as the carbohydrate moiety (Fig. 8.2)

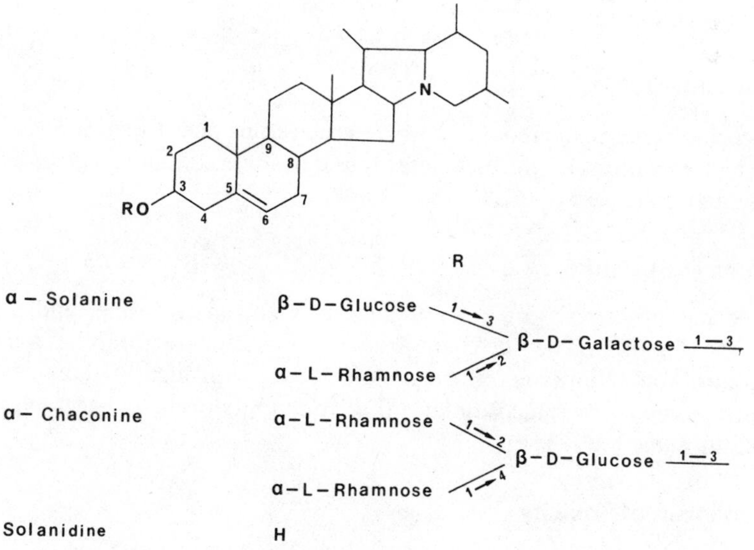

	R
α – Solanine	β–D–Glucose $\xrightarrow{1\rightarrow3}$ β–D–Galactose $\xrightarrow{1\rightarrow3}$; α–L–Rhamnose $\xrightarrow{1\rightarrow2}$
α – Chaconine	α–L–Rhamnose $\xrightarrow{1\rightarrow2}$ β–D–Glucose $\xrightarrow{1\rightarrow3}$; α–L–Rhamnose $\xrightarrow{1\rightarrow4}$
Solanidine	H

Fig. 8.2. Structures of α-solanine and β-chaconine.

Mechanism of Toxicity

Potatoes usually contain glycoalkaloids at the level of approximately 10 mg total/100g fresh weight. The highest concentration of glycoalkaloids are found in the peel and sprouts. Differences also exist among different cultivars with as high as 30 mg/100 g (3). The synthesis of glycoalkaloids can be induced by light or mechanical damage. Injury such as cutting and slicing results in increasing synthesis of glycoalkaloids by the plant. "Greening" occurs when potato is exposed to light. Associated with the increasing accumulation of chlorophyll is an increase in glycoalkaloids. To safeguard against poisoning, potato cultivers with total glycoalkaloid levels over 20 mg/100 g are not commercially acceptable (4). Since potato is a major staple food, greening can cause considerable losses to the potato industry.

Glycoalkaloids are strong inhibitors of cholinesterase and the cause of neurological disorder symptoms. Other toxic action includes disruption of the cell membranes in the gastrointestinal tract. Absorption of glycoalkaloids is usually enhanced by alkaline pH conditions where binding with sterols in cell membranes causes extra disruption (5). Lethal doses for humans range from 3 to 6 mg/kg body weight, although susceptibility varies considerably among individuals. Doses of greater than 2 mg/kg are normally considered toxic. Symptoms of poisoning include vomiting, diarrhea, abdominal pain, apathy, weakness, and unconsciousness. Animals are generally less susceptible to the glycoalkaloid. Birth abnormalities including cranial defects, neural defects, high fetal mortality rate, and resorption of fetus have been implicated in the consumption of high doses of glycoalkaloids in animal studies.

GLUCOSINOLATES

Glucosinolates occur predominantly in cruciferous plants (genus *Brassica*). The concentrations of glucosinolates found in some common vegetables are presented in Table 8.1 (6).

Chemical Structure

Glucosinolates consist of a β-thioglucose, a sulfonate oxime, and a side chain R (Fig. 8.3). The side chain R and the O-sulfonate group has a *trans*-configuration. Numerous glucosinolates have been identified in various plants. Some of the glucosinolates that commonly occur in vegetables are listed in Table 8.2 (7).

Mechanism of Toxicity

Hydrolysis of the glucosinolates is catalyzed by thioglucoside glycohydrolase located separately from the subtrates in intact tissues. Hydrolysis oc-

Table 8.1.   Glucosinolate Content in Edible Parts of Plants

PLANTS	GLUCOSINOLATE (μg/g VEGETABLE)	
	RANGE	MEAN
Cabbage		
Red	410–1090	760
White	260–1060	530
Chinese	170–1360	540
Cauliflower	270–830	480
Turnip	210–600	420
Radish		
Red	90–130	110
White	70–210	140
Mustard		
White	45,100–8 2,300	64,100
Black	32,800–59,800	46,300

Reprinted with permission from Fenwick, G. R., Heaney, R. K., and Mullin, W. J. 1983. Glucosinolates and their breakdown products in food and food plants. *CRC Crit. Rev. Food Sci & Nutr.* **18**, 153, Copyright, CRC PRESS, Inc., Boca Raton, Fl.

curs only when the plants are crushed. The enzyme exhibits optimum activity near neutral pH and is activated markedly by the addition of ascorbic acid.

The unstable aglycone produced from the enzymatic hydrolysis of the glucosinolates forms predominantly the isothiocyanate in a process similar to the Lossen rearrangement (8). Thiocyanates and nitrile may also form to a limited extent (Eq. 8-3). Thiocyanate lacks the pungent flavor of isothiocyanate and has been implicated in the off-flavors in milk. The formation of nitrile is enhanced at acid pH and by metal ions. The reaction forming thiocyanate may involve an enzyme-mediated rearrangement.

Indole glucosinolates (e.g., glucobrassicin) form isothiocyanates that degrade into 3-hydroxymethylindole. The product can then condense or, in the presence of ascorbic acid, form a complex compound, ascorbigen (Eq. 8-4). The aglycone from glucosinolates containing a β-hydroxy (e.g., 2-

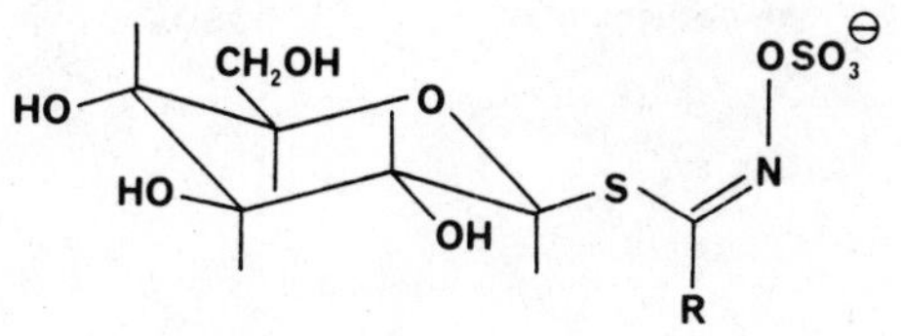

Fig. 8.3.   General structure of glucosinolate.

Table 8.2. Various Glucosinolates in Vegetables

COMMON NAME	R GROUP	OCCURRENCE
Sinigrin	allyl-	cabbage, brussels sprouts, cauliflower, mustard greens
Glucobrassicin	3-indoylmethyl-	cabbage, brussel sprouts, cauliflower, broccoli
Progoitrin	(R)-2-hydroxy-3-butenyl	cabbage, Chinese cabbage, turnips, rutabaga
Gluconapin	3-butenyl-	cabbage, Chinese cabbage, brussels sprouts, cauliflower, mustard spinach
Neoglucobrassicin	N-methoxy-3-inodylmethyl	brussels sprouts, cauliflower, broccoli, rutabagas, radishes
Gluconasturtiin	2-phenylethyl-	cabbage, Chinese cabbage, brussels sprouts, cauliflower, broccoli, mustard greens, mustard spinach, turnips
Glucotropaeolin	benzyl-	mustard greens, cabbage, mustard spinach
Glucobrassicanapin	4-pentenyl-	Chinese cabbage, mustard spinach, mustard greens
Glucoalyssin	4-methylsulfinylbutyl-	cabbage, brussels sprouts, broccoli
Glucoiberin	3-methylsulfinylpropyl-	cabbage, brussels sprouts, cauliflower

Reprinted with permission from Van Etten, C. H., Daxenbichler, M. E., and Wolff, I. A. Natural glucosinates (thioglucosides) in foods and feeds. *J. Agric. Food Chem.* **17**, 484. Copyright 1969 American Chemical Society.

$$R-N=C=S + HSO_4^{\ominus} \qquad\qquad R-S-C\equiv N + HSO_4^{\ominus}$$
Thiocyanate

Lossen Rearrangement   *Enzyme mediated*

$$R-C\overset{S-\beta-D-GLUCOSE}{\underset{NOSO_3^{\ominus}}{\Big\langle}} \xrightarrow[\text{THIOGLUCOSIDASE}]{H_2O} R-C\overset{SH}{\underset{NOSO_3^{\ominus}}{\Big\langle}} + \text{GLUCOSE} \qquad \text{Eq. 8-3}$$

$$\downarrow Fe^{++} \text{ or } H^{+}$$

$$R-C\equiv N + HSO_4^{\ominus} + S$$
Nitrile

hydroxy-3-butenyl glucosinolate) can cyclize to the corresponding oxazolidine-2-thione (Eq. 8-5)(6).

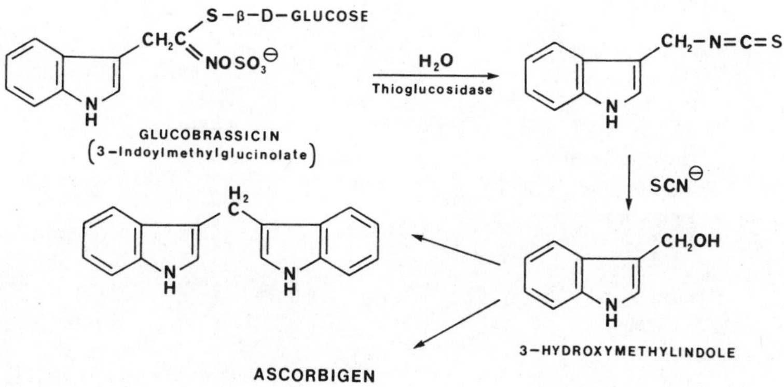

Eq. 8-4

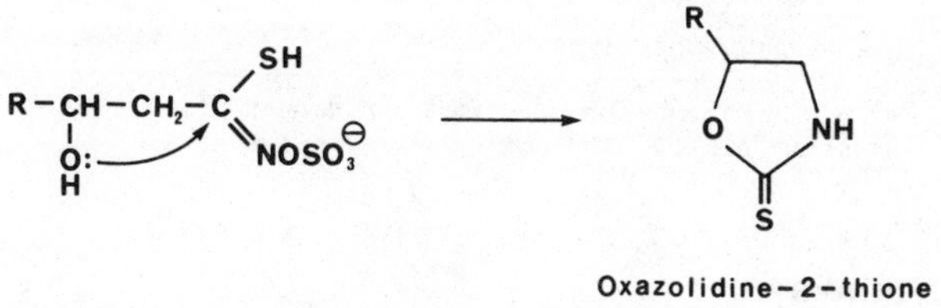

Eq. 8-5

Oxazolidine-2-thione

Isothiocyanate is metabolized in vivo to thiocyanate, which is a goitrogen. The goitrogenic effect of thiocyanate, resulting from its competition with iodine, is shown only under iodine-deficient diets. Oxazolidine-2-thione, however, acts by interfering with the synthesis of thyroxine. The antithyroid effect is not alleviated by supplementing with iodide. The compound 5-vinyloxazolidine-2-thione present in rape, rutabaga, and cabbage seeds is one of the most potent goitrogens reported.

METHYLXANTHINES

The best known methylxanthine is caffeine, with the chemical structure 1,3,7-trimethylxanthine. The most familiar sources of caffeine are coffee, cocoa beans, cola nuts, and tea. The amounts of caffeine present in various beverages, including soft drinks, are shown in Table 8.3. Although coffee contains only caffeine, tea and cocoa contain other methylxanthines besides caffeine, notably theophylline and theobromine (Fig. 8.4).

Table 8.3. Caffeine Contents in Various Drinks

TYPE OF DRINK	CAFFEINE (mg)
Coffee (5-oz cup)	
Drip method	110–150
Precoated	64–124
Instant	40–108
Decaffeinated	2–5
Tea (5-oz cup)	
1-min brew	9–33
3	20–46
5	20–50
Instant	12–28
Iced tea	22–36
Cocoa	
Mix	6
Milk chocolate (1 oz)	6
Baking chocolate (1 oz)	35
Soft drink (12 oz)	
Colas and pepper-type	30–60

Reprinted with permission from Roberts and Barone (9), copyright 1983 by Institute of Food Technologists.

Metabolic Pathway

More than 99% of the caffeine ingested is absorbed rapidly from the gastro-intestinal tract, and its concentration in the blood plasma rises to a peak level within 30 min. Once in the blood stream, caffeine effectively penetrates through all body tissues. Because of the high permeation of caffeine through biological membranes, caffeine must be transformed into metabolites for effective excretion and elimination from the body.

The half-life of caffeine (time required for the body to eliminate one-half of the plasma caffeine) varies from hours to days, depending on age, sex,

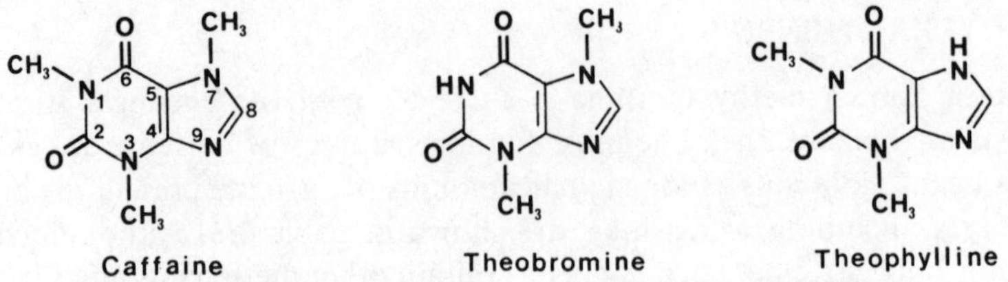

Fig. 8.4. Structures of some common methylxanthines.

medication, and health conditions. Newborns lack the many enzymes required to metabolize caffeine, and the half-life is 3–4 days. Smokers have a shorter half-life (3 hr) than nonsmokers (3–7 hr). Pregnant women require 18 or more hours, and patients with liver problems also have a long half-life (*10*).

Metabolically, caffeine can undergo (1) initial demethylation to the dimethylxanthines—theophylline, theobromine, and paraxanthine (1,7-dimethylxanthine); (2) oxidation at C8 to form 1,3,7-trimethyuric acid; and (C) hydration and ring cleavage at C8 and N9 to form dimethyluracil (Eq. 8-6). The dimethylxanthines are further demethylated and metabolized via reactions similar to (B) and (C) (*11*).

6-Amino-5-[N-methylformylamino]-
1,3-dimethyluracil

Trimethyldihydrouric acid

Trimethyluric acid

Hydration   Oxidation

CAFFEINE   Eq. 8-6

Demethylation

DIMETHYLXANTHINE
(Theophylline, Theobromine, Paxananthine)

DIMETHYLDIHYDROURIC ACID   MONOMETHYLXANTHINE

AMINOMETHYLURACIL   MONOMETHYLURIC ACID

Mechanism of Toxicity

Caffeine is listed in the GRAS list and is widely used in soft drinks and medicine, although it is generally considered to be a stimulant at low doses. One cup of coffee, amounting to the consumption of 1–2 mg/kg body weight, gives a peak plasma level of 50–10 μM. Excessive consumption (> 50 μM) leads to symptoms of "caffeinism"—anxiety, restlessness, sleep latency, diarrhea, muscular tension, heart palpitation. The oral LD_{50} for humans is 150–200 mg/kg (~0.75–1 mM plasma concentration), which amounts to a single consumption of over 75 cups of strong coffee (*12*).

At the molecular level, methylxanthines are known to inhibit the enzyme cyclic adenosine 3′, 5′-monophosphate (cAMP) phosphodiesterase, which catalyzes the hydrolysis of cAMP (Eq. 8-7). Cyclic AMP mediates the action of many hormones, for example, calcitonin, epinephrine, glucagon, norepinephrine, vasopressin, thyroid-stimulating hormone, lipotropin, parathyroid hormone, and corticotropin. However, the concentrations required to inhibit phosphodiesterases are substantially higher than the dose levels for neurophysiological responses.

Eq. 8-7

The effect of methylxanthines may also be attributed to their involvement in blocking certain receptor sites on cell membranes for adenosine (13). Since endogenous adenosine is generally inhibitory on neurophysiologic reactions, it is prostulated that methylxanthins exert a stimulatory action by blocking the receptor sites.

AMINO ACIDS, PEPTIDES, AND PROTEINS

Nonprotein Amino Acids

There are more than 250 nonprotein amino acids found in plants and microorganisms. Many of these nonprotein amino acids are structurally related to the protein amino acids. They are usually homologs, isomers, or products of substitution of protein amino acids. The structural relationships between a few protein and nonprotein amino acids are illustrated in Fig. 8.5.

Mechanism of Toxicity

Not all the nonprotein amino acids are toxic. For example, homoarginine is enzymatically converted to lysine and urea in rat liver. The mechanism of toxicity of nonprotein amino acids (14) is usually due to (1) competition with structurally similar protein amino acids for enzymes, forming inactivated complexes; (2) incorporation into protein, resulting in functionally defective proteins; and (3) interference with protein synthesis in the transfer of protein amino acid to transfer RNA (Eq. 8-8).

$$\text{Amino acid} + \text{tRNA} + \text{ATP} \longrightarrow \text{Amino acid-tRNA} + \text{AMP} + \text{P} \sim \text{P} \quad \text{Eq. 8-8}$$

Seeds of the legume *Lathyrus sativus* are included in diets in certain areas, especially during famine. *Lathyrus* species induces toxic symptoms known as lathyrism, characterized by skeletal defects (osteolathyrism) and damage

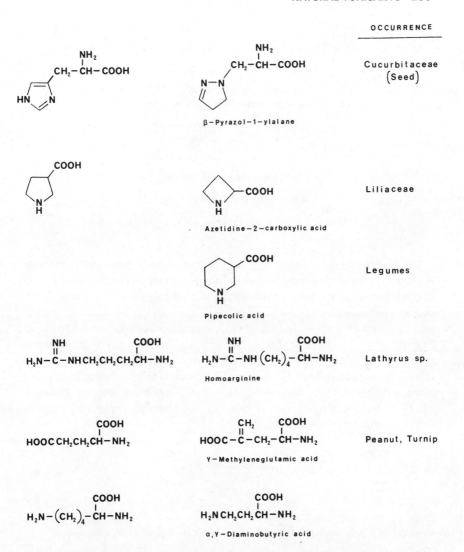

OCCURRENCE

β−Pyrazol−1−ylalane — Cucurbitaceae (Seed)

Azetidine−2−carboxylic acid — Liliaceae

Pipecolic acid — Legumes

Homoarginine — Lathyrus sp.

γ−Methyleneglutamic acid — Peanut, Turnip

α,γ−Diaminobutyric acid

Fig. 8.5. Structural relationships between protein and nonprotein amino acids.

in the nervous system (neurolathyrism). The neurological effect is attributed to the oxalyldiaminopropionic acids (Fig. 8.6) present in the legume. One of the ODAPs, γ-glutamylaminopropionitrile, has been shown to inhibit the enzymes for the cross-linkages in collagen synthesis.

Another compound, 2,4-diaminobutyric acid, has been found to be toxic by inhibiting the enzyme ornithine transcarbamylase. Disruption of the urea cycle induces ammonia toxicity.

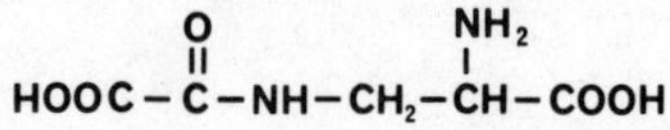

$$\underset{\text{2-Amino-3-oxalylamino propionic acid}}{HOOC - \overset{\overset{\text{O}}{\|}}{C} - NH - CH_2 - \overset{\overset{\text{NH}_2}{|}}{CH} - COOH}$$

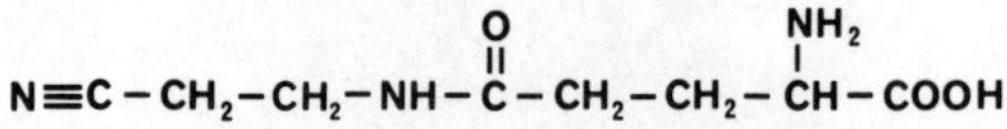

$$\underset{\text{4-Glutamyl-3-aminoproionitrite}}{N\equiv C - CH_2 - CH_2 - NH - \overset{\overset{\text{O}}{\|}}{C} - CH_2 - CH_2 - \overset{\overset{\text{NH}_2}{|}}{CH} - COOH}$$

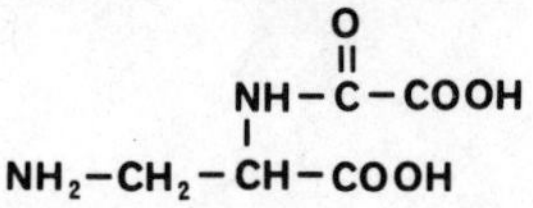

$$\underset{\text{2-Oxalylamino-3-aminopropionic acid}}{NH_2 - CH_2 - \overset{\overset{\overset{\overset{\text{O}}{\|}}{NH - C - COOH}}{|}}{CH} - COOH}$$

Fig. 8.6. Oxalyldiaminopropionic acids.

Unripe fruit of the tropical tree akee (*Blighia sapida*) contains hypoglycin A (β-methylenecyclopropylalanine), which causes severe vomiting and hypoglycemia. In the body, hypoglycine A is metabolized to α-(methylene-cyclopropyl) acetate. The metabolite interferes with the transacylation reaction in which long-chain fatty acid CoA molecules are transferred to carnitine to form acyl carnitine (Eq. 8-9). The reaction is crucial to the effective oxidation of long-chain fatty acids, since only the acyl carnitine can diffuse across the mitochondrial membrane to the matrix where β-oxidation takes place. Hypoglycine A, therefore, interrupts β-oxidation so that glycogen has to be metabolized for energy instead. The resulting depletion of carbohydrate causes hypoglycemia.

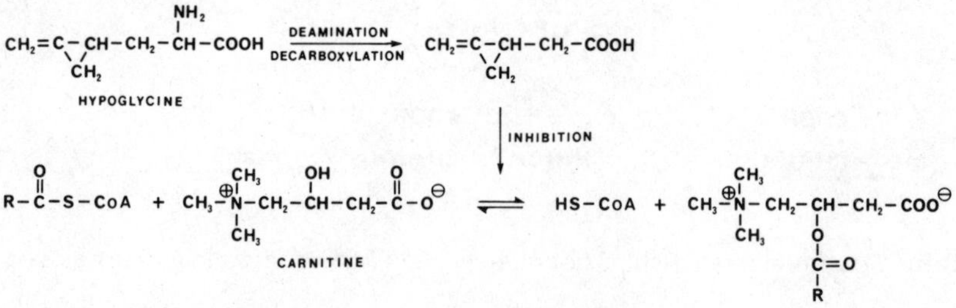

Eq. 8-9

Toxic Peptides

The most notable toxic peptides are the phallotoxins and amatoxins of wild mushrooms (genus *Amanita*). Both phallotoxins and amatoxins are cyclic peptides; the former contain a thioether bridge and the later a sulfoxide group (Fig. 8.7).

Phallotoxins act to inhibit synthesis in liver cells, and amatoxins cause fragmentation of the nucleoli of liver cells. Alpha-amanitin has been shown

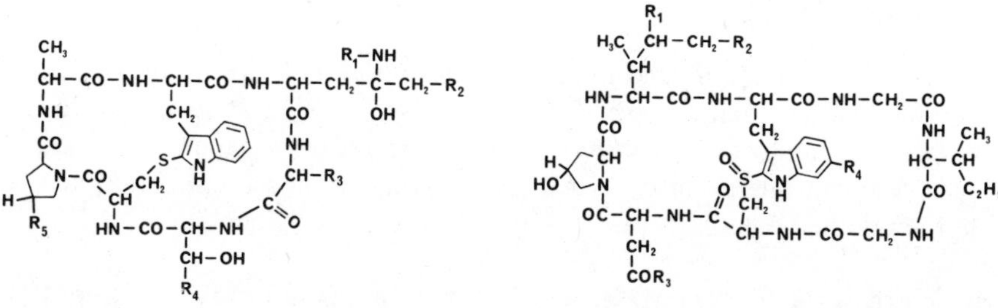

PHALLOTOXIN	R$_1$	R$_2$	R$_3$	R$_4$	R$_5$
PHALLOIDIN	OH	H	CH$_3$	CH$_3$	OH
PHALLOIN	H	H	CH$_3$	CH$_3$	OH
PHALLISIN	OH	OH	CH$_3$	CH$_3$	OH
PHALLICIDIN	OH	H	CH(CH$_3$)$_2$	COOH	OH

AMATOXIN	R$_1$	R$_2$	R$_3$	R$_4$
α–AMANITIN	OH	OH	NH$_2$	OH
β–AMANITIN	OH	OH	OH	OH
γ–AMANITIN	H	OH	NH$_2$	OH
AMANIN	OH	OH	OH	H
AMANULLIN	H	H	NH$_2$	OH

Fig. 8.7. Structures of phallotoxins and amatoxins. (Reprinted with permission from Wieland (15). Copyright 1968 by the AAAS.)

to inhibit DNA-dependent RNA polymerase II. Fragmentation begins 15 hr after administration of α-amanitin to rats, while the cytological effects of phallotoxin occurs in 1–2 hr (15).

The commercial cultivated mushrooms, although they contain no phallotoxins or amatoxins, are found to contain up to 0.04% fresh weight of agaritine (β-N-[γ-L(+)-glutamyl]-4-hydroxymethyl-phenylhydrazine). The enzyme γ-glutamyltransferase, found in mushroom sporophores, hydrolyzes agaritine to L-glutamate and 4-hydroxymethylphenylhydrazine (Eq. 8-10), which is further converted to 4-hydroxymethyl-benzenediazonium ion. Hydrolysis of agaritine with hot, dilute acid also results in the liberation of L-glutamate and the hydrazine (16,17). The diazonium ion is chemically reactive and can couple with aromatic compounds to form azo derivatives.

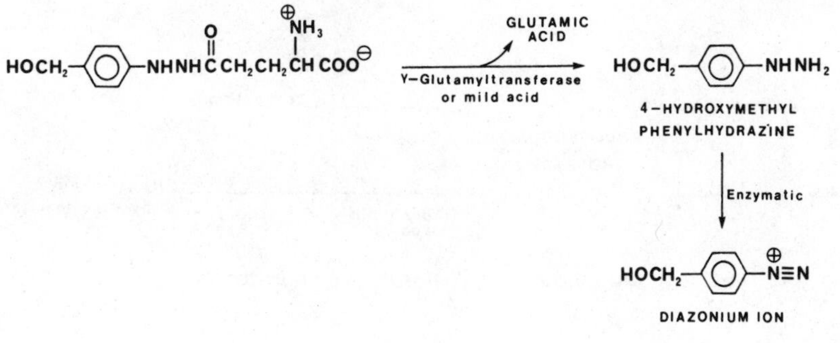

Eq. 8-10

Proteins

The notorious toxins found in *Clostridium botulinum* are proteins. These are nine distinct toxins, types A, B, C, C_1, C_2, D, E, F, and G, produced by various strains of *C. botulinum*. A single bacterium is capable of producing more than one toxin type. All these toxins are neurotoxins that act at cholinergic nerve endings to block the release of acetylcholine.

Structure of the Protein Molecule. All botulinum neurotoxins (except C_2) are synthesized intracellularly as an inactive precursor as a single polypeptide with a molecular weight of ~150,000. The molecule is cleaved by an endogenous protease to yield an active double-chain molecule composed of a heavy chain (MW ~ 100,000) and a light chain (~50,000) linked by an intrachain disulfide bond. This so-called nicking process is necessary for the activation of all the toxins (Fig. 8.8) (*18*).

The C_2 toxin, however, is synthesized as two separate polypeptide chains with MW of 100,000 and 50,000, with no interchain linking. Individual chains are not active but are extremely toxic when combined.

The Mechanism of Action. The botulinum toxin must penetrate into the cholinergic nerve cell before it can exert its blocking effect on the release of acetylcholine. The first step is believed to be the binding of the toxin molecule to specific receptor sites on the plasma membrane of the cholinergic nerve. The binding is directed by the carboxyl terminal segment of the heavy chain.

A membrane translocation step occurs by receptor-mediated endocytosis. The endocytic vesicle (or receptosome) migrates toward the lysosome, with a gradual buildup of pH gradient across the membrane. The fall in pH causes a conformational change in the toxin molecule, allowing the amino terminal segment of the heavy chain to penetrate across the membrane, creating a channel for the light chain to enter into the cytoplasm (Fig. 8.9) (*19*).

Exactly how the light chain interacts to block the release of acetylcholine remains largely unknown. The toxin exerts no direct effects on the enzyme

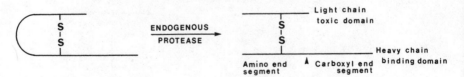

Fig. 8.8. Activation of *C. botulinum* neurotoxin.

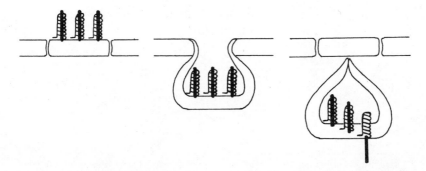

Fig. 8.9. Schematic representation of *C. botulinum* toxin entry into cell by receptor mediated endocytosis. (Reproduced, with permission from the Annual Review of Pharmacology and Toxicology, Volume 26, © 1986 by Annual Reviews Inc.)

choline acetyltransferase that transfers acetate from acetylcoenzyme A with choline to form acetylcholine. It does not affect the ability of loading and storage of acetylcholine in the membrane vesicles, the ability of acetylcholine receptors, or the ability of acetylcholinesterase to degrade acetylcholine.

The binary toxin C_2, however, has been shown to consist of ADP-ribosylating activity. It is speculated that the neurotoxins also possess certain enzymatic activity in the light chain, which reacts with substrates related to the calcium regulation of acetylcholine release.

AMINES

Amine compounds are naturally occurring in food plants at relatively high levels. Foods subjected to bacterial contamination or fermentation processes may also have certain amino acids converted to amines that are vasoactive. Most amines found in animal products are the result of bacterial action, since natural amines are rarely found in significant concentrations in animal tissues.

The commonly known amines found in food plants include tyramine, dopamine, and serotonin (5-hydroxytryptamine) (Table 8.4) (20). All the amines are biosynthesized by the decarboxylation and, in some, hydroxylation of the corresponding amino acid precursors.

The amines tyramine and histamine found in cheese, dairy products, alcoholic beverages, and putrefying meat products (Table 8.5) result from the actions of many bacteria containing amino acid decarboxylases that can convert amino acids to amines. Histamine, formed from bacterial decar-

Table 8.4. Content of Some Common Amines Found in Fruits (μg/g)

FRUIT	SEROTONIN	TYRAMINE	DOPAMINE
Banana	28	7	8
Tomato	12	4	0
Plum (red)	10	6	0
Avocado	10	23	4–5
Pineapple	20	—	—
Orange	0	10	0

Reprinted with permission from Lovenberg, W. Psycho- and vasoactive compounds in food substances. *J. Agric. Food Chem.* **22**, 24. Copyright 1974 American Chemical Society.

boxylation of histidine, has caused food poisoning in fish consumption. Enterobacteria, especially *Proteus* species, are the most important histamine-producing bacteria in fish. *Proteus morganii* and *Enterobacter aerogenes* are among some of the known examples.

Histamine is a vasodilator and causes characteristic skin flushing. Less common symptoms include vomiting, nausea, diarrhea, and hypertension. Fish also contain trimethylamine, formed from the reduction of trimethylamine-*N*-oxide (TMAO) by bacteria. Another amine that gives the characteristic fish smell is dimethylamine, which is the result of enzymatic breakdown of TMAO.

Metabolism

Under normal conditions, amines are rapidly detoxified once absorbed into the body. The mitochondrial amine oxidases catalyze the oxidative deamination of amines to the corresponding aldehyde, which is inactive.

For example, histamine is metabolized by diamine oxidase to imidazole

Table 8.5. Amines from Bacterial Action

PRODUCT	TYRAMINE (μg/g)	HISTIDINE (μg/g)
Cheese (fermented aged)	0–2200	0–2500
Beer and ale	1.8–11.2	—
Wine	0–25	0.3–30
Yeast extract	0–2250	250–2830
Fish	0–470	10–300
Meat extract	95–304	—
Beef liver (stored)	274	65

Reprinted with permission from Smith (*21*), courtesy of Elsevier Applied Science Publishers, Ltd, Essex, England.

acetaldehyde and the acid. Alternatively, the histamine may be N-methylated by the enzyme histamine-N-methyl-transferase before conversion to N-methylimidazole acetic acid by monoamine oxidase (Eq. 8-11) (22). Problems arise only with patients taking drugs (e.g., tranylcypromine) that are inhibitors of the enzyme. In these cases, amines can accumulate in the blood sufficiently to cause severe hypertension.

Eq. 8-11

MYCOTOXINS

Mycotoxins are toxic fungal metabolites that may contaminate peanuts, rice, soybeans, wheat, barley, corn, sorghum, cottonmeal, and dairy products. Many mycotoxins now known are largely attributed to species of *Aspergillus* and *Penicillium*.

Chemical Structure

The most extensively studied aflatoxins are a group of toxic metabolites of *Aspergillus flavus* and *parastiticus*. These are furanocoumarin compounds, consisting of a coumarin nucleus fused to a furan and a lactone. The two major types are aflatoxins B and G (Fig. 8.10).

Mechanism of Toxicity

Aflatoxin contamination is widespread in many foods and has prompted numerous investigations into every aspect of aflatoxin B_1, the most notorious mycotoxin in the group, which is known to cause liver cancer (oral LD_{50} for rat = 5.5–7.5 mg/kg).

Aflatoxin B_1 is biologically metabolized to form various products (23). Microsomal detoxification occurs when aflatoxin B_1 undergoes hydroxyla-

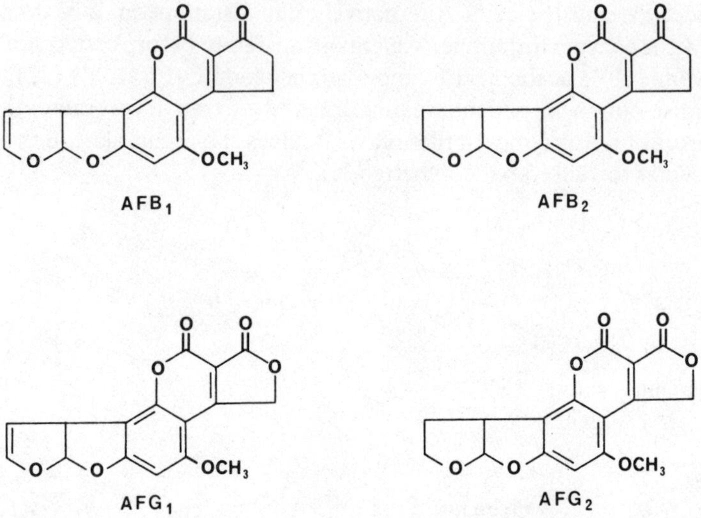

Fig. 8.10. Chemical structures of aflatoxins.

tion or demethylation resulting in aflatoxin M_1, Q_1, B_{2a}, and P_1 (Eq. 8-12). However, aflatoxin B_1 can also undergo epoxidation to the reactive aflatoxin B_1-8,9-oxide, which forms covalently linked conjugates with DNA and proteins (Eq. 8-13). The major aflatoxin-DNA conjugate initially formed has been identified to be 8,9-dihydro-8-(N^7-guanyl)-9-hydroxy-aflatoxin B_1 (AFB$_1$-N^7-GUA). In vivo kinetic studies show that AFB$_1$-N^7-GUA undergoes a rapid decrease in the liver due to the spontaneous hydrolysis of the imidazole ring. One of the identified products, 8,9-dihydro-8-(N^5-formyl-2',5',6'-triamino-4'-oxo-N^5-pyrimidyl)-9-hydroxyl-aflatoxin B_1 (AFB$_1$-FAPY) has been shown to be the persistent DNA conjugate that accumulates in the liver (24).

The aflatoxin B_1-epoxide may also undergo hydrolysis to AFB$_1$-8,9-dihydrodiol. The product, dialdehyde phenolate ion, is stabilized by resonance (Eq. 8-14). The aldehyde groups can then form a Schiff base with the amine groups of proteins. The 8-hydroxy aflatoxins (AFB$_{2a}$, AFG$_{2a}$) also can rearrange to yield dialdehyde phenolate ions (25).

In spite of the many efforts of prevention, aflatoxin contamination remains a problem to be resolved. Cows fed with aflatoxin B_1-contaminated feeds produce milk containing aflatoxin M_1. Although less toxic, aflatoxin M_1 is still considered undesirable. The maximum concentration of M_1 allowed in milk in the United States is 0.5 μg/kg.

Various treatments have been studied to remove aflatoxin. Roasting re-

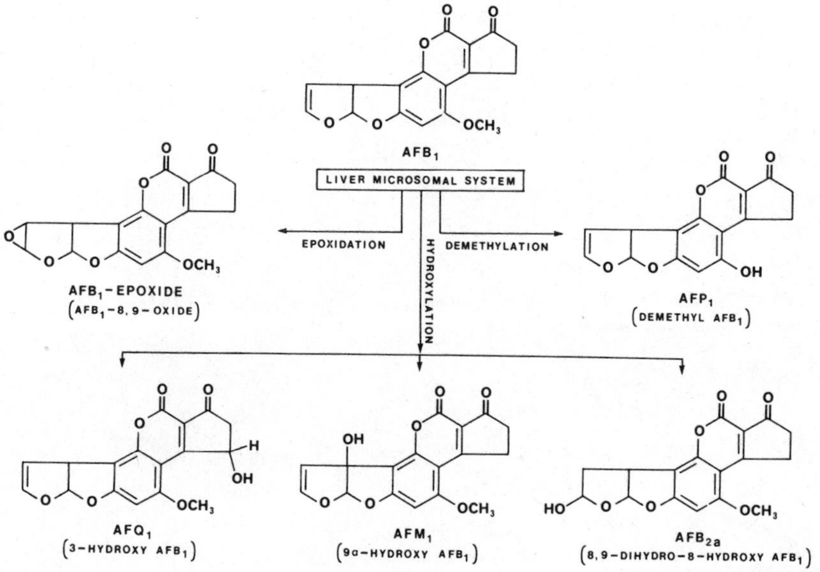

AFB$_1$

LIVER MICROSOMAL SYSTEM

EPOXIDATION   HYDROXYLATION   DEMETHYLATION

AFB$_1$-EPOXIDE
(AFB$_1$-8,9-OXIDE)

OCH$_3$

OH

AFP$_1$
(DEMETHYL AFB$_1$)

Eq. 8-12

AFQ$_1$
(3-HYDROXY AFB$_1$)

OCH$_3$
H
OH

AFM$_1$
(9a-HYDROXY AFB$_1$)

OH
OCH$_3$

AFB$_{2a}$
(8,9-DIHYDRO-8-HYDROXY AFB$_1$)

HO
OCH$_3$

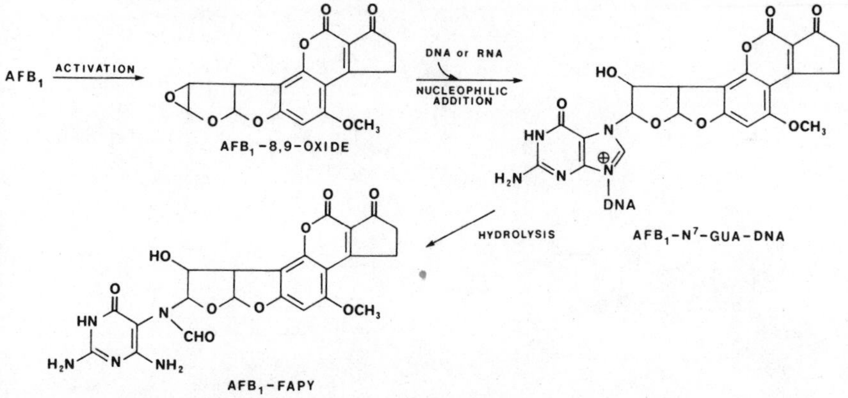

AFB$_1$ →(ACTIVATION)

AFB$_1$-8,9-OXIDE

OCH$_3$

DNA or RNA
NUCLEOPHILIC ADDITION

HO

OCH$_3$

HN
H$_2$N   N
DNA

AFB$_1$-N^7-GUA-DNA

Eq. 8-13

HYDROLYSIS

HO
OCH$_3$

HN
O
N
CHO
H$_2$N   N   NH$_2$

AFB$_1$-FAPY

moves 60–70% of aflatoxin in peanuts, although heat treatment of milk does not cause a decrease in aflatoxin M$_1$. Chemicals such as hydrogen peroxide, sodium hypochlorite, calcium hydroxide, formaldehyde, and lime can destroy aflatoxin. The detoxification process for cattle and chicken feeds in practice is by ammonia treatment under high temperature and pressure (Eq. 8-15). Treatment with ammonia rapidly opens the lactone ring to give the open-ring salt. Loss of ammonia results in the formation of the α-

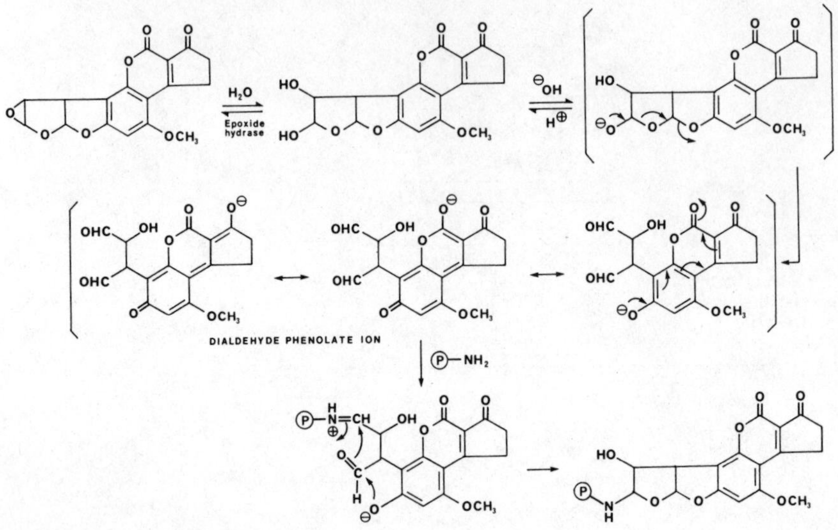

Eq. 8-14

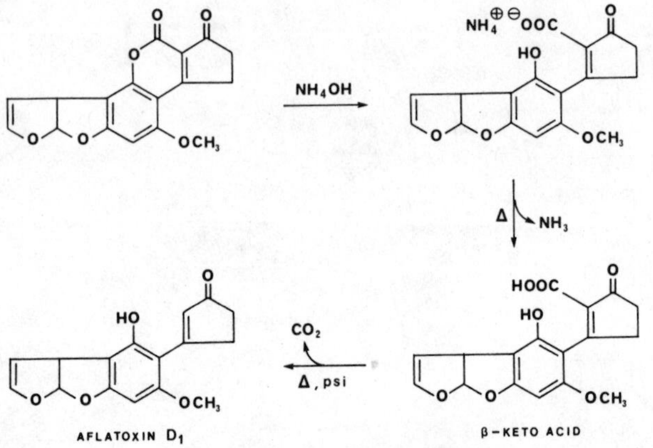

Eq. 8-15

keto acid, which undergoes decarboxylation to aflatoxin D_1 or decomposition to a benzofuran (26).

POLYCYCLIC AROMATIC HYDROCARBONS

Carcinogenic polycyclic aromatic hydrocarbons (PAH) are known to form in grilled meat products. These PAH compounds are formed when fats are pyrolyzed by the high temperature of hot coals. The PAHs in the smoke generated from burned fat are then adsorbed on the meat.

The concentration of PAH depends on the fat content in the meat, the grilling process (contact of melted fat drips with the heat source), and the heat source (amount of smoke formed) and may vary from undetectable amounts to as high as 50 ppm benzo[a]pyrene in some samples. Concentrations of PAH have been reported in sausages, beef, pork, lamb, turkey, chicken, hamburgers, and bacon (27).

Chemical Structure

The four PAHs commonly detected in grilled meat products are benzo[a]pyrene, benzo[a]fluroanthene, benzo[a]anthracene, and chrysene (Fig. 8.11).

Mechanism of Toxicity

Polycyclic aromatic hydrocarbons undergo metabolic activation to reactive intermediates that are responsible for their carcinogenic effects (28). Benzo[a]pyrene is initially converted to the 7,8-epoxide (7,8-dihydro-7,8-epoxybenzo[a]pyrene). The enzyme, aryl hydrocarbon hydroxylase, that catalyzes the conversion is located in the microsomal fraction and requires NADPH. The epoxide is further converted to the dihydrodiol (8-dihydro-7,8-dihydroxybenzo[a]pyrene) by epoxide hydrase. A second epoxidation by the microsomal mono-oxygenase gives the dihydrodiolepoxide (7,8-dihydroxy-9,10-epoxy-7,8,9,10-tetrahydrobenzo[a]pyrene) (Eq. 8-16).

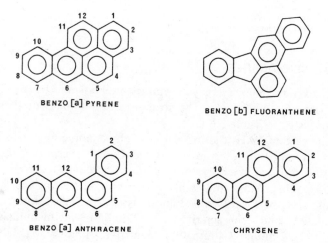

Fig. 8.11. Structures of polycyclic aromatic hydrocarbons.

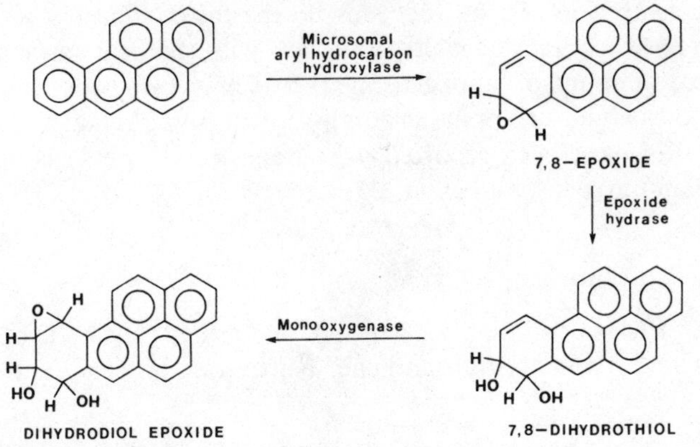

Eq. 8-16

The diol-epoxide of PAH can exist as the syn- or anti-isomers (i.e., the 7-OH is either *cis* or *trans* to the epoxy group), and it has been shown that it is the anti-isomer that is exclusively involved in the binding of DNA (Eq. 8-17) (*29*).

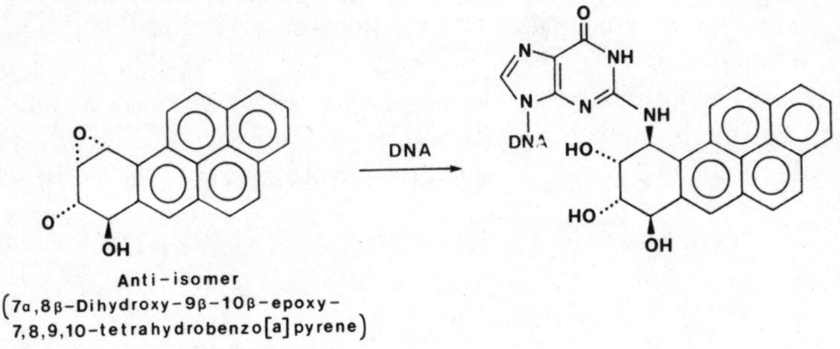

Eq. 8-17

Anti–isomer
(7α,8β–Dihydroxy–9β–10β–epoxy–
7,8,9,10–tetrahydrobenzo[a]pyrene)

HETEROCYCLIC AMINES

A number of mutagenic heterocyclic amines are known to be formed from the pyrolysis of amino acids. The structures and sources of these compounds are listed in Table 8.6 (*30*).

As in the case of PAH, the formation of these amine mutagens in food requires high temperatures in excess of 300°C. However, other mutagenic amines are also found in moderate-temperature cooking (<200°C), such as frying and boiling. Most studies have been on commercial beef extracts, broiled and fried beef, and broiled and roasted fish. These mutagens have been isolated and characterized as 2-amino-3-methylimidazo[4,5-*f*]quinoline (IQ), 2-amino-3,4-dimethylimidazo[4,5-*f*]quinoline (MeIQ), and

Table 8.6. Heterocyclic Amines and Sources

SOURCE	MUTAGEN
Tryptophan	I. Trp-P-1 (3-amino-1,4-dimethyl-5H-pyrido[4,3-b]indole) II. Trp-P-2 (3-amino-1-methyl-5H-pyrido[4,3-b]indole)
Glutamic acid	III. Glu-P-1 (2-amino-6-methyldipyrido[1,2-α:3′, 2′-d]imidazole) IV. Glu-P-1 (2-aminodipyrido[1,2-α:3′,2′-d]imidazole)
Phenylalanine	V. Phe-P-1 (2-amino-5-phenylpridine)

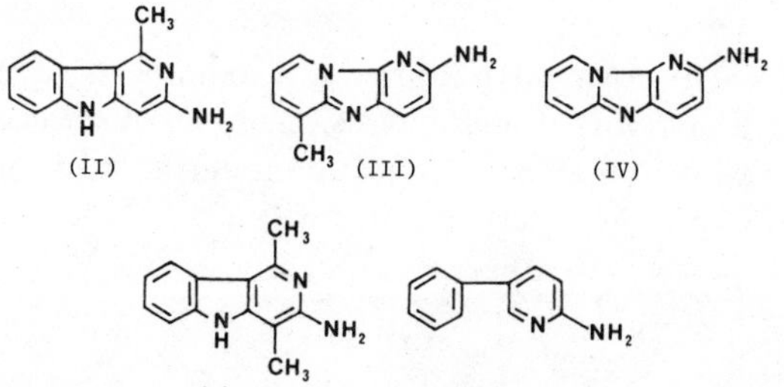

(II)   (III)   (IV)

(I)   (V)

Reprinted with permission from Hashimoto, Y. Shudo, K., and Okamoto, T. Mutagenic chemistry of heteroaromatic amines and mitomycin C. *Acc. Chem. Res.* **17**, 404. Copyright 1984 American Chemical Society.

2-amino-3,8-dimethylimidazo[4,5-f]quinoxaline (MeIQ$_x$). Their chemical structures are presented in Fig. 8.12.

Mutagenicity

The mutagenicity of all the above mentioned heterocyclic amines (to *Salmonella typhimurium* TA98) are compared with other mutagens such as aflatoxin B$_1$ and benzo[a]pyrene in Table 8.7 (*30*).

Mechanism of Toxicity

Metabolically, the heterocyclic amines are enzymatically modified by liver microsomal cytochrome P-450 to the active mutagens, N-hydroxyamines, which are further converted to the corresponding O-acyl derivative. Possible acylations catalyzed by enzymes in cytosol include acetylation, aminoacylation, sulfonation, or phosphorylation. The O-acylated derivative can

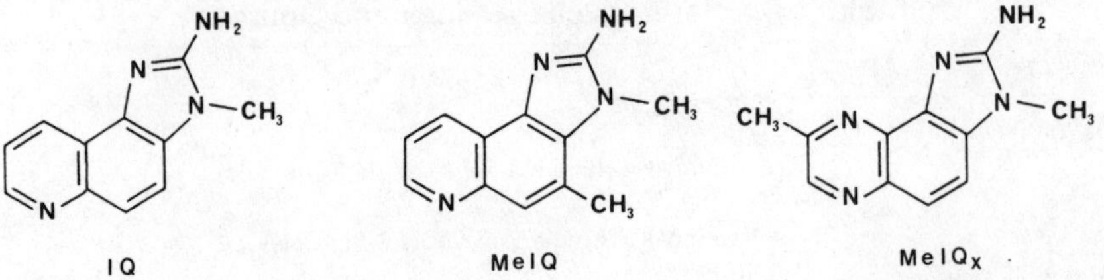

Fig. 8.12. Structures of IQ (2-amino-3-methylimidazo[4,5-f]quinoline), MeIQ (2-amino-3,4-dimethylimidazo[4,5-f]quinoline), and MeIQ$_x$ (2-amino-3,8-dimethylimidazo[4,5-f]quinoxaline).

then undergo efficient covalent binding with guanine base (31). A large substituent at the 8-position of guanine residue in DNA induces mispairing or structural alteration and affects replication and transcription. The metabolic pathway of Trp-P-2 is presented in Eq. 8.18.

NITROSAMINES

Many N-nitrosamines are known carcinogens to animals. The chemistry of the reaction between amines and nitrous acid depends very much on the degree of substitution in the amine. Only secondary and tertiary amines will react with nitrous acid to give nitrosamine products.

Table 8.7. Mutagenic Activity in Various Heterocyclic Amines

COMPOUNDS	REVERTANTS OF TA98/μg
MeIQ	661,000
IQ	433,000
MeIQ$_x$	145,000
Trp-P-2	104,000
Glu-P-1	49,000
Trp-P-1	39,000
Aflatoxin B$_1$	6,000
Glu-P-2	1,900
Benzo[a]pyrene	320
Phe-P-1	41

Reprinted with permission from *Food Technology* **39**(2), 77, copyright 1988, by Institute of Food Technologists.

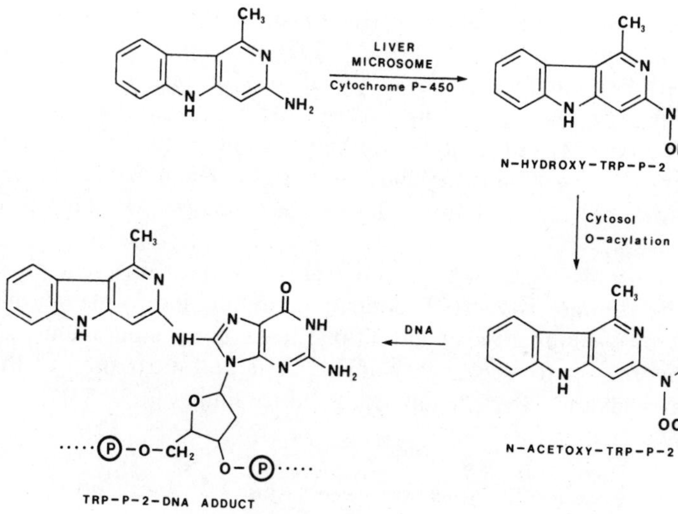

N-HYDROXY-TRP-P-2

Eq. 8-18

N-ACETOXY-TRP-P-2

TRP-P-2-DNA ADDUCT

The General Chemistry of N-nitrosation

Nitrous acid, generated by the action of acid on nitrite salt, is very unstable and decomposes to nitrous anhydride and nitrosonium ion (Eq. 8-19). The initial N-nitrosation reaction is shown by Eq. 8-20. In the case of secondary amines, the N-nitrosonium intermediate loses a proton to form N-nitrosamine (32).

$$NO_2^{\ominus} \underset{pK=\,3.36}{\overset{H}{\rightleftharpoons}} HONO \quad \text{Nitrous acid}$$

$$2\,HONO \rightleftharpoons H_2O + N_2O_3 \,\big(O{=}N{-}O{-}N{=}O\big) \rightleftharpoons NO_2^{\ominus} + {}^{\oplus}N{=}O$$

Nitrous anhydride   Nitrosonium ion

Eq. 8-19

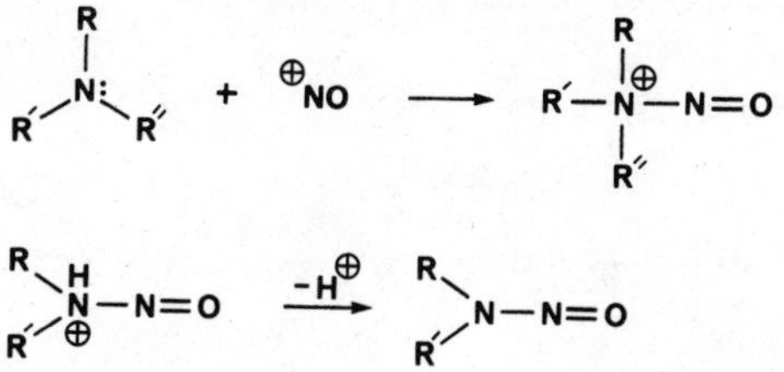

Eq. 8-20

The rate of N-nitrosation [rate $= k(HNO_2)^2(R_2NH)$] depends on pH of the reaction medium and basicity of the amine. The higher the pK_a (i.e., stronger base) of the amine, the lower the nitrosation rate will be, since protonated amine can not react. The effect of pH is twofold. Low pH favors the formation of the unreactive protonated amine but shifts the equilibrium of (nitrite ion $\rightleftharpoons$ nitrous acid) to the right. The nitrosation reaction is, therefore, best carried out with highly basic amines in weakly acidic media (pH ~ 3–3.4).

Tertiary amines also react with nitrous acid, but via a different route forming N-nitrosamine products. The nitrosammonium ion formed in this case undergoes cis-elimination of the nitoxyl to form an immonium ion, which then hydrolyzes to a secondary amine. A second nitrosation of the product then yields the corresponding stable nitrosamine (Eq. 8-21).

Eq. 8-21

Inhibition of Nitrosation

Nitrosation is known to be inhibited by ascorbate. Most curing systems contain 550 ppm sodium ascorbate. Phenolics, thiols and other aromatic compounds can compete with amines for nitrite, forming C- and S-nitroso compounds, respectively (Eq. 8-22) (33). Paradoxically, both p-nitrophenols and S-nitrosocysteine have been shown to catalyze N-nitrosamine formation via transnitrosation. A nitrosating intermediate is proposed for nitrosophenol reacting with nitroso compounds (34). Nucleophilic attack on the intermediate by an amine then produces the corresponding nitrosamine and regenerates the nitrosophenol (Eq. 8-23).

Eq. 8-22

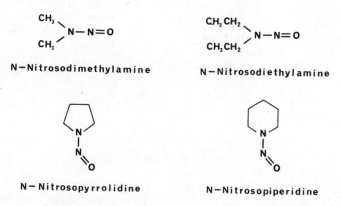

Eq. 8-23

Nitrosamines in Cured meat

A number of nitrosamines have been identified in cured meat and fish products. These include N-nitrosodimethylamine (NDMA), N-nitrosodiethylamine, N-nitrosopyrrolidine (NPYR), and N-nitrosopiperidine (Fig. 8.13). Of these, NDMA and NPYR are found consistently in cooked bacon, and their concentrations are two times higher in the cook-out fat than the rasher (Table 8.8) (35). The nitrosamine NPYR is formed during frying of bacon, largely in the fat phase. During frying, up to 70% of the total NDMA and 50% of the total NPYR appear in the vapor. Since pyrrolidine is not found in raw bacon, free proline is the most likely precursor of NPYR in bacon (36). At high temperatures (>100%C) as in frying, nitrous anhydride (N_2O_3) dissociates to form the $NO_2\cdot$ and $NO\cdot$ radicals. The nitrous oxide radical abstracts the amino proton from proline to give a radical that combines with $NO\cdot$ to yield the nitrosoproline. Further decarboxylation converts the nitrosoproline to NPYR (Eq. 8-24).

N—Nitrosodimethylamine

N—Nitrosodiethylamine

N—Nitrosopyrrolidine

N—Nitrosopiperidine

Fig. 8.13. Nitrosamines found in cured meat.

Table 8.8. Distribution of Nitrosamine in Cooked Bacon

BACON	NDMA (PPB)	NPYR (PPB)
Rasher	3.8	10.4
Cook-out fat	9.9	21.6

Reprinted with permission from Coleman (*35*), courtesy of International Journal of Food Science & Technology, London.

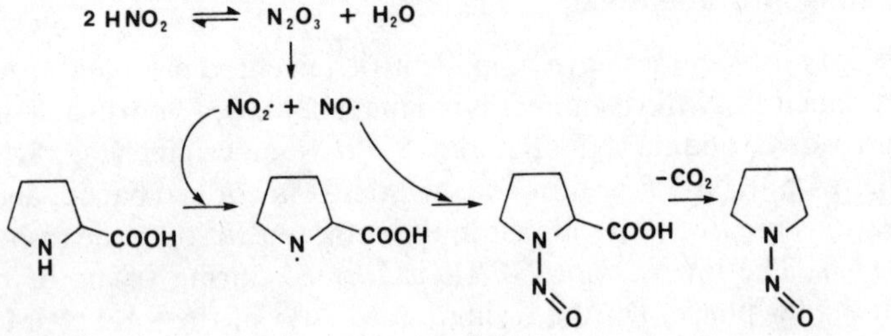

$$2 \, HNO_2 \rightleftharpoons N_2O_3 + H_2O$$

Eq. 8-24

N−NITROSOPYRROLIDINE

In Vivo Nitrosation

It has been shown that nitrate and nitrite are formed endogenously, and the conversion of nitrate to nitrite occurs in the oral cavity and intestine via microbial nitrification. Therefore, *N*-nitroso compounds are likely to form when suitable precursor compounds are present to react with nitrite.

The precursor for nitrosation found in certain brands of soy sauce and fish sauce has been identified to be 1-methyl-1,2,3,4-tetrahydro-β-carboline-3-carboxylic acid (Eq. 8-25) (*38*). The nitroso compound, after reacting with nitrite, is a mutagen (17.4 revertants of TA100 without S9 per μg).

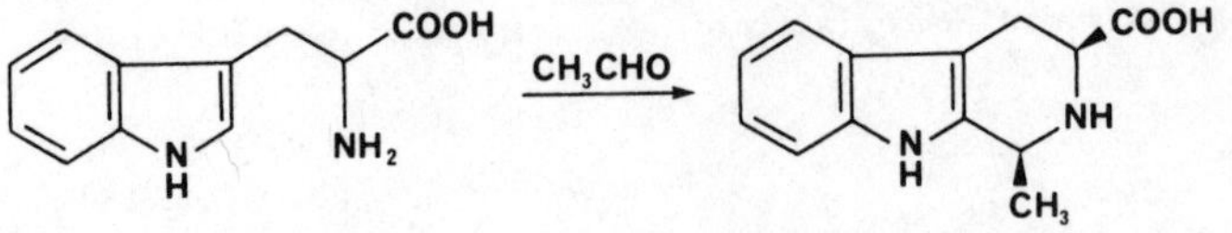

Eq. 8-25

1−Methyl−1,2,3,4−tetrahydro−β−
carboline−3−carboxylic acid

The same compound, along with tetrahydro-β-carboline-3-carboxylic acid, is found in alcoholic beverages. The concentrations range between ~300–400 and 1000–10000 ppm in beer and wine, respectively (*39*). Other tetrahydro-β-carbolines have also been found in alcoholic beverages and other foods, including 6-hydroxy-1-methyl-1,2,3,4-tetrahydro-β-carboline (in beer, certain fruits such as banana and plum, and cheese), 6-hydroxy-1,2,3,4-tetrahydro-β-carboline, and 1-methyl-1,2,3,4-tetrahydro-β-carboline (in alcoholic beverages prepared by fermentation processes). However, these compounds are present at much lower concentrations.

Tetra-β-carbolines are formed by the reaction between tryptamines and carbonyl compounds, and by the subsequent intramolecular cyclization of the Schiff base (Eq. 8-25). These compounds exhibit physical effects including inhibition of monoamine oxidase and of serotonin uptake. Intravenous administration increases voluntary intake of ethanol in rats.

Fava beans, treated with nitrite under gastric conditions, form a mutagenic α-hydroxy-N-nitroso compound (4-chloro-6-methoxyindole-2-hydroxy-1-nitrosoindolin-3-one oxime) that is ~2000 times more potent than the mutagen in soy sauce. The precursor in this case is 4-chloro-6-methoxyindole (Eq. 8-26) (*40*).

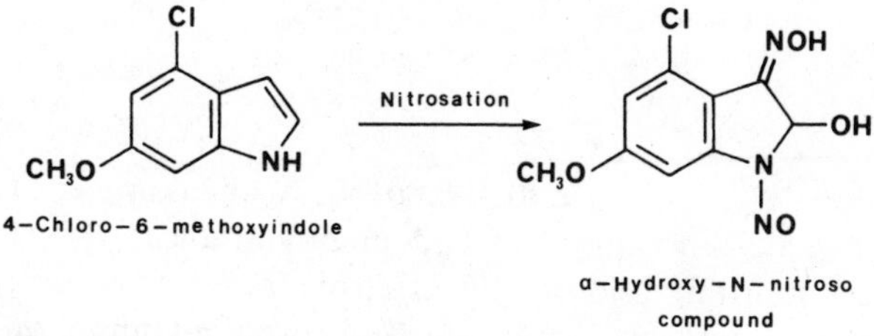

4-Chloro-6-methoxyindole → α-Hydroxy-N-nitroso compound

Eq. 8-26

Chemical Reactions of Nitrosamine

The chemistry of N-nitrosamine is based on the polar resonance structure due to the partial N−N double bond character and the negative charge on the oxygen atom (Fig. 8.14) (*32,37*).

Denitrosation. Nitrosamines undergo denitrosation on heating with mineral acids. While initial protonation occurs at the oxygen atom, it is the N-protonated form that undergoes hydrolysis to the amine and nitrosonium ion (Eq. 8-27).

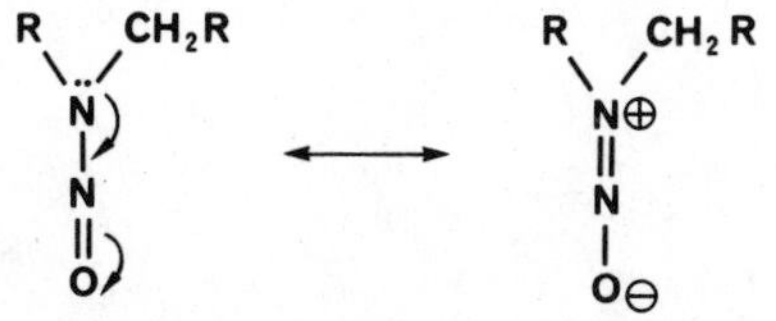

Fig. 8.14. Resonance structure of nitrosamine.

$$\underset{R'}{\overset{R}{>}}N-N=O \; \underset{-H^+}{\overset{+H^+}{\rightleftharpoons}} \; \underset{R'}{\overset{R}{>}}N-N=\overset{\oplus}{O}H \; \rightleftharpoons \; \underset{R'}{\overset{R}{>}}\overset{H}{\underset{\oplus}{N}}-N=O \; \overset{H_2O}{\longrightarrow} \; \underset{R'}{\overset{R}{>}}NH \; + \; \overset{\oplus}{N}O \quad \text{Eq. 8-27}$$

Oxidation and Reduction.   Nitrosamines can be oxidized to the corresponding *N*-nitramines. Oxidizing agents include hydrogen peroxide/nitric acid, and nitric acid/ammonium persulfate (Eq. 8-28). The nitroso group can undergo reduction to *N,N*-disubstituted hydrazines or secondary amines. The most common reductants are zinc dust in acetic acid and lithium aluminum hydride (Eq. 8-29).

$$R_2NNO \; \xrightarrow{\;[O]\;} \; R_2NNO_2 \qquad\qquad \text{Eq. 8-28}$$

$$R_2NNO \; \xrightarrow[CH_3COOH]{\;Zn\;} \; R_2NNH_2 \; + \; R_2'NH \qquad\qquad \text{Eq. 8-29}$$

Photoreaction.   In dilute acid (0.1–$0.001N$), *N*-nitrosamines undergo photoreaction, rapidly generating NO· and aminium radical ($R_2NH·^+$). Excitation causes proton migration to the electrons rich amine nitrogen, in the nitrosamine-acid complex, forming an *N*-nitrosoammonium ion, which decomposes to aminium and nitrous oxide radicals (Eq. 8-30). In aqueous solution, the radical disproportionates to form HNO and the immonium ion. Nucleophilic attack of HNO on the immonium ion yields the *C*-nitroso product, which tautomerizes to the stable oxime (Eq. 8-31). In the absence of an α-hydrogen, tautomerization cannot occur, and reverse elimination of the *C*-nitrosamine will regenerate the more stable immonium ion.

$$\underset{R'CH_2}{\overset{R}{>}}N-N=O \; \xrightarrow{\;dil.\;H^+\;} \; \underset{R'CH_2}{\overset{R}{>}}\overset{\delta^+}{N}\!=\!\!=\!\!N\overset{\delta^-}{=\!\!=}O\cdots HCl \; \xrightarrow{\;\underset{\nu}{h\nu}\;} \; \left\{ \underset{R'CH_2}{\overset{R}{>}}\underset{\oplus}{\overset{H}{N}}-NO \right\}$$

N-Nitrosoammonium ion

Eq. 8-30

$$\downarrow$$

$$\underset{R'CH_2}{\overset{R}{>}}\overset{+·}{N}H \; + \; ·NO$$

$$\begin{array}{c} R \\ > NH \\ R'CH_2 \end{array} \overset{+\cdot}{} + \cdot NO \longrightarrow \begin{array}{c} R \\ > NH \\ R'CH \end{array} \overset{\oplus}{} + HNO \rightleftharpoons \begin{array}{c} R \\ > NH_2 \\ R'CH \\ | \\ N \\ \| \\ O \end{array} \overset{\oplus}{} \longrightarrow \begin{array}{c} R \\ > NH_2 \\ R'C \\ \| \\ N \\ | \\ OH \end{array} \overset{\oplus}{}$$

AMINIUM RADICAL   IMMONIUM ION   C-NITROSAMINE   OXIME   Eq. 8-31

Metabolic Mechanism

Metabolically, nitrosamines undergo enzymatic α-hydroxylation to the α-hydroxynitrosamine. Cleavage of the $C-N$ bond yields the aldehyde and the alkyldiazohydroxide. The latter then breaks down to the diazoalkane or the cationic species (Eq. 8-32). It is believed that the electrophilic intermediates, alkyldiazonium ion and alkyldiazohydroxide, react with DNA to form covalent conjugates (41).

$$\begin{array}{c} R' \\ > N-N=O \\ RCH_2 \end{array} \xrightarrow[\text{Enzymatic}]{\text{Hydroxylation}} \begin{array}{c} R' \\ > N-N=O \\ R-CH \\ | \\ OH \end{array} \xrightarrow{R'CHO} RCH_2N=N-OH \xrightarrow{} RCH=N_2$$

RCH=N$_2$ DIAZOALKANE

α-HYDROXYNITROSAMINE   ALKYLDIAZOHYDROXIDE   Eq. 8-32

$$RCH_2^{\oplus} \longleftarrow RCH_2N_2^{\oplus}$$

CARBONIUM ION   ALKYLDIAZONIUM ION

Numerous nitroso compounds have been tested on animals and are known to be carcinogenic. Nitrosamines are widely distributed in body tissues following oral administration, often producing tumors at various sites, including liver, nasal cavities, kidney, and stomach. However, all these animal tests have been conducted using dosage levels in the range of mg/kg, well above the ppb range found in any food products.

9
ADDITIVES

The use of food additives is fundamentally necessary for an industry that has to meet the demand of a continuous, adequate supply of a substantial variety of food items to a large population of consumers. Chemicals, natural or synthetic, are deliberately added to various foods for the purpose of changing the chemical and physical properties of a food system. A number of additives are discussed in various chapters. These include antioxidants, sequestrants, emulsifiers, colors, flavoring ingredients, preservatives, textural modifiers, and enzymes. In this chapter, only those additives that are of importance, but have not been covered elsewhere, are presented in detail. The chemistry of their functional properties, their fate in the food system, and their metabolic pathway after consumption will help our understanding of why additives are used, how to assure their safe use, and how to establish the conditions for effective utilization.

PHOSPHATES

There are more than 20 commonly available phosphates with widespread applications in most processed food products. Much of the confusion comes from naming the specific compounds within these groups, especially when trade names are commonly used in the food-processing industry.

Chemical Structure

Food phosphates can be conveniently classified into (1) orthophosphates and (2) condensed phosphates. The latter is composed of polyphosphates and metaphosphates. Their general structures are shown in Fig. 9.1. Cyclic phosphates of three and larger rings are shown to exist. Long-chain poly-

314

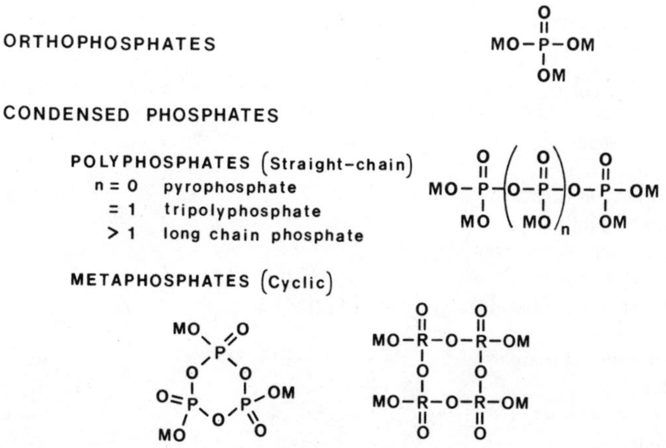

ORTHOPHOSPHATES

$$MO-\overset{\overset{\text{O}}{\|}}{\underset{\underset{\text{OM}}{|}}{P}}-OM$$

CONDENSED PHOSPHATES

POLYPHOSPHATES (Straight-chain)
  n = 0   pyrophosphate
    = 1   tripolyphosphate
    > 1   long chain phosphate

$$MO-\overset{\overset{\text{O}}{\|}}{\underset{\underset{\text{MO}}{|}}{P}}-\left(O-\overset{\overset{\text{O}}{\|}}{\underset{\underset{\text{MO}}{|}}{P}}\right)_n O-\overset{\overset{\text{O}}{\|}}{\underset{\underset{\text{OM}}{|}}{P}}-OM$$

METAPHOSPHATES (Cyclic)

Fig. 9.1. Structures of phosphates.

phosphates commonly exist in glasslike crystals. Table 9.1 tabulates all the common food phosphates with chemical formulas under each of the three groups (*1, 2*).

Sequestering

Phosphates are added to food for technological purposes. One of the most important functions of phosphates is their effectiveness in sequestering metals. A sequestrant is an anion that forms a soluble complex with metal ions in the presence of other complexing or precipitating anions. Based on the amount of the free metal ion, in equilibrium with solutions of sequestrants and of precipitating anions, the complexing ability of the sequestrants for the metal ion can be compared (*3*). Hence, from Fig. 9.2, the long-chain phosphates are stronger anions for iron than the orthophosphates, and sequestration decreases with increasing pH.

Water Holding Capacity

Polyphosphates are used in processed meat, poultry, and seafood largely for controlling fluid loss to give a more tender and juicy product.

The most commonly used polyphosphates in meat, poultry and seafood products include sodium tripolyphosphate, often blended with sodium hexametaphosphate for better effect. The phosphates added are usually hydrolyzed to pyrophosphate which is believed to be the active agent. In meat processing, a combination of 2% salt and 0.3% phosphate is commonly

Table 9.1. Commonly used food phosphates

CHEMICAL NAMES AND SYNONYMS	FORMULA	pH (1% SOLUTION)	NEUTRALIZING VALUE
The Orthophosphates			
1. Phosphoric acid[a] Orthophosphoric acid Monophosphoric acid	H_3PO_4	1.6	—
2. Monosodium dihydrogen monophosphate	NaH_2PO_4 (anhydrous)	4.6	70
Monosodium Phosphate[a]	$NaH_2PO_4 \cdot H_2O$ (monohydrate)		
Monosodium monophosphate	$NaH_2PO_4 \cdot 2H_2O$ (dihydrate)		
Sodium dihydrogen phosphate			
Sodium phosphate, monobasic			
Sodium biphosphate			
Primary sodium phosphate			
3. Disodium monohydrogen monophosphate	Na_2HPO_4 (anhydrous)	9.0	—
Disodium phosphate[a]	$Na_2HPO_4 \cdot 2H_2O$ (dihydrate)		
Disodium monohydrogen phosphate	$Na_2HPO_4 \cdot 7H_2O$ (heptahydrate)		
Sodium phosphate, dibasic	$Na_2HPO_4 \cdot 12H_2O$ (dodecahydrate)		
Neutral sodium phosphate			
Disodium monophosphate			
Disodium phosphate			
Secondary sodium phosphate			
4. Monocalcium phosphate[a]	$Ca(H_2PO_4)_2$ (anhydrous)	4.6	8
Calcium phosphate, monobasic	$Ca(H_2PO_4)_2 \cdot H_2O$ (monohydrate)		
Calcium acid phosphate			
Calcium biphosphate			
Primary calcium phosphate			
5. Dicalcium phosphate[a]	$CaHPO_4 \cdot H_2O$ (dihydrate)	7.5	33
Calcium phosphate, dibasic			
Calcium phosphate, secondary			
6. Tricalcium phosphate[a]	$Ca_{10}(OH)_2(PO_4)_6$	7.3	—
Calcium phosphate, tribasic			
Tertiary Calcium phosphate			
7. Monoammonium phosphate[a]	$NH_4H_2PO_4$	4.6	62

Table 9.1   (Continued)

CHEMICAL NAMES AND SYNONYMS	FORMULA	pH (1% SOLUTION)	NEUTRALIZING VALUE
Ammonium biphosphate	ˮ		
Ammonium phosphate, monobasic			
8. Diammonium phosphate[a]	$(NH_4)_2HPO_4$	8.0	—
Ammonium phosphate, dibasic			
9. Monopotassium phosphate[a]	KH_2PO_4	4.6	—
Acid potassium phosphate			
Potassium phosphate, monobasic			
10. Tripotassium phosphate[a]	K_3PO_4	11.5	—
Basic potassium phosphate			
Potassium phosphate, tribasic			
11. Sodium aluminum phos-	$Na_3Al_2H_{15}(PO_4)_8$	3.4	100
phate	$NaH_{14}Al_3(PO_4)_8 \cdot 4H_2O$		
The Polyphosphates			
1. Sodium pyrophosphate	$Na_2H_2P_2O_7$	4.2–4.8	74
Sodium acid pyrophosphate[a]			
Disodium dihydrogen diphos-phate			
Dibasic sodium pyrophos-phate			
Disodium pyrophosphate			
2. Sodium tripolyphosphate[a]	$Na_5P_3O_{10}$	9.8	—
Pentasodium triphosphate			
Sodium triphosphate			
3. Sodium hexametaphosphate[a]	mixture of various chain lengths	6.9	
Sodium phosphate glass			
Graham's salt	$Na_{n+2}P_nO_{3n+1}$		
Sodium polyphosphate			
The Metaphosphates			
1. Sodium trimetaphosphate	$Na_3P_3O_9$		

[a]Common names for commercial products in the United States.

Reprinted with permission from *Food Phosphates* (2), courtesy of Monsanto Nutrition Chemicals Division, St. Louis, Mo.

used to increase water uptake. Approximately 0.8–1.0 M (4.6–5.8%) sodium chloride is required for maximum swelling, but the addition of phosphate reduces the concentration of chloride needed.

The effect of polyphosphate on meat hydration is due to three factors: (1) increase in pH, (2) increase in ionic strength, and (3) sequestration of

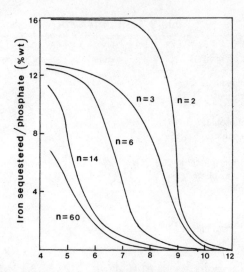

Fig. 9.2. Sequestration of iron (III). (Reprinted with permission from Irani and Morgenthaler (3), courtesy of American Oil Chemists' Society.)

metal ions. The greater the increase of pH from the isoelectric point of meat (pH ~ 5.4), the stronger is the electrostatic repulsion due to increasing negative net charge of the protein. Screening of the positive charges by phosphate anions causes further weakening of electrostatic attractions. The increase in the water-holding capacity is also related to the sequestering property of the phosphate anions. Polyphosphates complex with protein-bound muscle calcium and magnesium, freeing the protein molecules from cross-linking by these cations. The polypeptide chain, once separated, can then give way to electrostatic repulsion. More water molecules can then be immobilized in the loosened protein network. This effect is greatly enhanced in the basic pH range (> pI) or by the addition of salt, which causes increasing electrostatic repulsion, and further loosening of the protein molecule (Fig. 9.3) (4)

It has recently been observed by phase-contrast microscopy that the increase in water-holding capacity is caused by the expansion of the volume of the myofibrils. Based on the structural organization of the myofibril, the swelling must be constrained at the Z-, M-lines, as well as the cross-bridges that tie together the thick and thin filaments. Both chloride and pyrophosphate have been shown to cause (1) the depolymerization of the myosin filament, and (2) dissociation of the actomyosin complex (5). The resulting disruption of the structural constraints allows the interfilament volume to

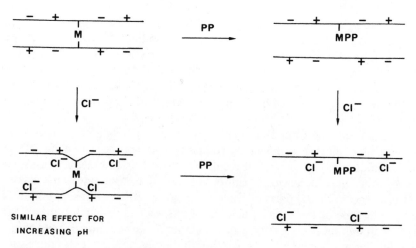

Fig. 9.3. Effect of phosphate ions on meat hydration. From Hamm (4).

expand. Similar dissociation have also been suggested for the cytoskeletal proteins, titin and nebulin.

Stabilizing Emulsion

Phosphates are also employed in the manufacture of processed cheese. They help stabilize the emulsion of butterfat in the protein-water matrix. It is suggested that phosphates, by sequestering the calcium ions from the *para-κ*-casein, expose the solubilizing groups in the milk protein. The resulting cheese has a smooth texture and melts down with no fat separation. The common phosphates used for this purpose are the sodium salts of the ortho-phosphates. Polyphosphates such as sodium acid pyroposphate and sodium hexametaphosphates are also used (2). Alkaline orthophosphates tend to give the cheese product a soft texture with lower melting temperature. Acidic phosphates and polyphosphates increase the melting temperature and hardness of the product. Up to 3% by weight of the finished product of these emulsifying salts can be used.

Leavening

The single most unique application of phosphates is their functioning as leavening acids that react with sodium bicarbonate to release carbon dioxide. The most commonly used phosphates are anhydrous monocalcium phosphate, monocalcium phosphate monohydrate, dicalcium phosphate dihydrate, sodium aluminum phosphate, and sodium acid pyrophosphate. The amount and the types of leavening acids used for a particular product depend largely upon the available acidity and the rate of reaction of the leavener. The term "neutralizing value" (NV) (Table 9.1) and "dough reaction rate" (DRR) are used by the industry to define the properties of leavening acids (6).

The neutralizing value is defined as the parts by weight of sodium bicarbonate that will be neutralized by 100 parts of leavening acid. Thus, NV = (wt. sodium bicarbonate/wt. leavening acid) × 100. In practical terms, the NV corresponds to the weight amount of sodium bicarbonate that will completely react with 100 parts of leavening acid to release all the carbon dioxide, with little of the soda or acid phosphate salt remaining after baking.

The dough reaction rate measures the rate of carbon dioxide release during mixing and in the bench stage (holding period) in a dough under standardized conditions. The DDR allows comparing the reactivity of differing leavening acids in a chemical leavening system.

Monocalcium phosphate monohydrate, one of the first acidic phosphates used, reacts rapidly with soda, with much carbon dioxide released during the mixing stage of dough preparation. Further, it disproportionates in water to form the dicalcium phosphate and phosphoric acid (Eq. 9-1).

$$Ca\left(H_2PO_4\right)_2 \cdot H_2O + H_2O \rightleftharpoons CaHPO_4 \cdot 2H_2O + H_3PO_4 \qquad \text{Eq. 9-1}$$

Dicalcium phosphate dihydrate is slow acting, and only reacts with soda above 140°F, at which temperature the dicalcium phosphate starts to disproportionate to $Ca(H_2PO_4)_2$ and $Ca_5(PO_4)_3OH$.

Anhydrous monocalcium phosphate is hydroscopic and is usually used in a form stabilized by a coating of mixed potassium, aluminum, calcium, and magnesium metaphosphate to slow down dissociation and reaction with soda.

Sodium aluminum phosphate exhibits slow reaction during holding, and the leavening action starts largely when the product is heated.

Sodium acid pyrophosphate comes in several grades, each with a different rate of reaction, ranging from DRR of 20 to 45 (2). Most household baking powders are "double acting"—containing monocalcium phosphate monohydrate, to provide rapid reaction during mixing, and sodium aluminum sulfate, which gives little reaction until the dough or batter is heated in the oven. Cake mixes, frozen doughs, and batter require slow-acting leavening acid.

Buffering

Orthophosphates such as phosphoric acid, monosodium phosphate, disodium phosphate, and sodium acid pyrophosphate are commonly used as buffering and acidifying agents. Long-chain polyphosphates generally have poor buffering capacity.

CITRIC ACID

Citric acid is a multifunctional additive suitable for a wide range of applications. Citric acid is a tribasic acid with four ionizable groups (7), with pK_1 = 3.13, pK_2 = 4.76, and pK_3 = 6.40 for the three carboxyl groups and a pK_4 of 11 or greater for the hydroxyl group. The carboxyl group α to the hydroxyl group is ionized first, followed by the two terminal carboxyl groups, then finally the hydroxyl group.

Acidification

Citric acid is commonly used for acidifying low-acid foods to a final equilibrium pH of the food of 4.6 or below in the canning process. Foods such as bean, carrot, cucumber, cabbage, cauliflower, spinach, pepper, corn, pea, asparagus, tropical fruit, and fish have a finished equilibrium pH higher than 4.6 (Fig. 9.4) and a water activity (a_w) greater than 0.85 (8). If the pH is lowered, the processing time and temperature in canning these foods can be reduced, since a pH of 4.6 or below inhibits the sporation of the spoilage organisms of greatest concern, such as *Clostrium botulinum*.

The fermentation of grapes that are low in total acidity normally results in poor wine quality. Citric acid is often used to make up for the acidity deficiency before fermentation.

Buffering

Citric acid, used in combination with its salts, provides a good buffer that serves to stabilize the pH during various stages of food processing, as well as in the formulation of the finished product.

Flavor Enhancing

Due to its high solubility ($\sim$ 160 g/100 g H_2O at 256°C) and good flavor-blending characteristic, citric acid has long been used by the beverage indus-

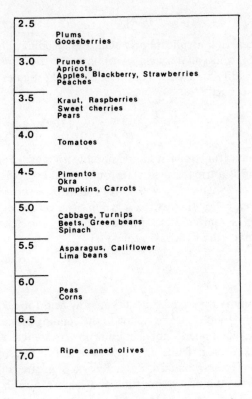

2.5	
	Plums
	Gooseberries
3.0	Prunes
	Apricots
	Apples, Blackberry, Strawberries
	Peaches
3.5	Kraut, Raspberries
	Sweet cherries
	Pears
4.0	
	Tomatoes
4.5	Pimentos
	Okra
	Pumpkins, Carrots
5.0	
	Cabbage, Turnips
	Beets, Green beans
	Spinach
5.5	Asparagus, Califlower
	Lima beans
6.0	
	Peas
	Corns
6.5	
7.0	Ripe canned olives

Fig. 9.4. pH values of various canned foods. (Reprinted with permission from Schmidt (8), courtesy of Miles Laboratories, Inc., Elkhart, In.)

try to impart a pleasant sour taste of fruit flavor, and also to enhance the many natural and artificial flavorings used in various beverages (9). Most instant teas also utilize citric acid to enhance the lemon flavor. Table 9.2 presents the levels of citric used in certain common beverages.

Sequestering

Another important function of citric acid is to sequester metal ions that can accelerate oxidation (rancidity in fats and oil), browning (color deterioration in beverages), and complex formation (turbidity in wine and ice tea). Citric acid has seven possible ligating groups (seven oxygen atoms) which can coordinate with metals (7). The citric ion forms a tridentate chelate with ferrous ion in which one terminal carboxyl, the central carboxyl, and

Table 9.2.   Citric Acid Usage in Beverages

BEVERAGE	% CITRIC ACID (w/v)
Carbonated beverage	
Orange	0.133
Diet orange	0.173
Lemon lime	0.144
Creme soda	0.046
Ginger ale	0.099
Tonic	0.363
Still beverage	
Citrus drink	0.14
Cherry	0.14
Peach	0.14
Orange	0.19

Reprinted with permission from Irwin (*9*), courtesy of Miles Laboratories, Inc. Elkhart, In.

the hydroxyl group are coordinated to a single ferrous ion (Fig. 9.5). The remaining terminal carboxyl group is coordinated to two other ferrous ions. The metal ions are each, in turn, coordinated with other citrate ions, thus forming a chain structure. Adjacent chains of ferrous citrate are linked by extensive hydrogen bonding through water molecules and hexaquoiron counter ions ($Fe(II)(H_2O)_6$) (Fig. 9.6). Magnesium and manganese form similar structures in solution (*10*).

Citric acid is used in fats and oils to increase the effectiveness of the antioxidants, and in wines for protection against haze formation in the finishing process. Trace metals present in fruits and vegetables often cause discoloration in processing. Some known examples are surface darkening

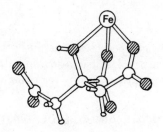

Fig. 9.5.   Molecular structure of citric acid–iron complex. (Reprinted with permission from Clusker, J. P. Citrate conformation and chelation: Enzymatic Implications. *Acc. Chem. Res.* **13**, 347. Copyright 1980 American Chemical Society.)

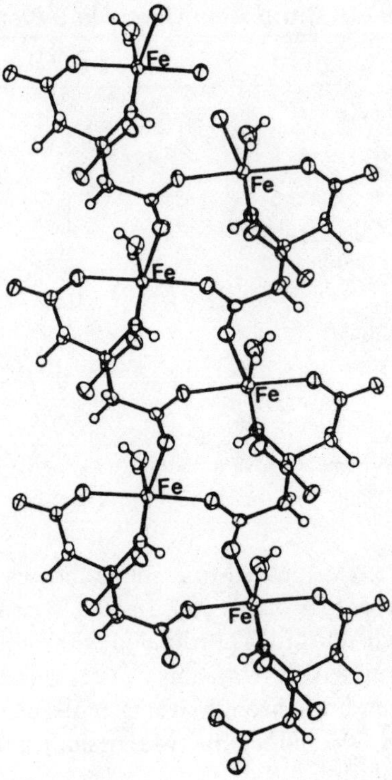

Fig. 9.6. Infinite chain of $Fe(II)C_6H_5O_7(H_2O)$ along a 2_1 axis of the unit cell. (Reprinted with permission from Strouse, J., Layten, S. W., and Strouse, C. E. Structural studies of transition metal complexes of triionized and tetraionized citrate. Models for the coordination of the citrate ion to transition metal ions in solution and at the active site of aconitase. *J. Am. Chem. Soc.* **99**, 563. Copyright 1977 American Chemical Society.)

of cauliflower, kidney beans, potatoes, and mushrooms, and the formation of pink color in canned pears. Citric acid can help control discoloration during fruit and vegetable processing by removing the trace metals.

ANTIMICROBIAL SHORT-CHAIN ACID DERIVATIVES

Included in the category of antimicrobial short-chain acid derivatives are the sodium benzoate, alkyl esters of *p*-hydroxybenzoate (parabens), sorbate and its salts (commonly sodium or potassium), propionate, and sulfite. Sulfite, due to its unique chemistry and functions, is to be discussed in a separate section.

Mechanism of Inhibition

The antimicrobial effect of these short-chain acids comes from the undissociated form (*11*), the concentration of which is determined by the pK_a of the acid and the pH of the medium (Table. 9.3). Ionized species fail to penetrate the cell membrane to any great extent. Maximum usage levels are in the range of 0.1%.

Inhibition of microbial growth is slowly reversible and concentration dependent. The amount needed for inhibition decreases with increasing chain length of the acid. It is generally believed that inhibition of microbial growth by these acid preservatives involves disruption of the cellular membrane transport system, although the exact mechanism is not clear.

It has been shown in *Bacillus subtilis* and *Escherichia coli* that cellular uptake of amino acids, organic acids, and phosphates and other compounds is inhibited with added preservatives (*12*). Similar inhibition of transport activity is also observed in the fungus *Penicillum chrysogenum* (*13*). It has been suggested that the undissociated acids penetrate through the lipophilic cell membrane and then become ionized after reaching the interior cell compartments of low proton concentration. The intracellular concentration of the organic acids upsets the proton or charge gradient involved in energizing membrane transport.

Intracellular accumulation of high concentrations of preservatives has also been shown to occur in yeast. But one particular species, *Saccharomyces bailli,* grows in the presence of 600 mg/L benzoic acid or sorbic acid and at a pH less than the pK_a of the acids. The yeast is able to maintain a relatively low intracellular concentration of preservatives by adapting an inducible energy system of secreting the preservatives from the cell (*14*).

Table 9.3. The pK^a and Undissociated Forms of Selected Acids

NAME	pK^a	UNDISSOCIATED ACID FORM
Benzoate	4.2	⬡–COOH
Paraben	8.47	HO–⬡–COOR
Propionate	4.87	$CH_3CH_2CH_2COOH$
Sorbate	4.8	$CH_3CH=CH\ CH=CHCOOH$

SULFITE

Sulfur (IV) oxospecies are widely used food preservatives which include sulfur dioxide (SO_2) and salts of sulfite (SO_3^{2-}) and bisulfite (HSO_3^-). Gaseous sulfur dioxide generated from burning sulfur solution is commonly applied to preserving dried fruits and vegetables, while solutions of sodium or potassium sulfite are commonly used for liquid foods.

Regardless of the form applied, the equilibrium between the various sulfur oxospecies always exist. However, in the food industry, sulfur dioxide, sulfite, and bisulfite are used indiscriminately. For convenience, "sulfite" will be used as a general term for any or all of these species.

Chemical Equilibrium of the Oxospecies

The oxospecies exist in an aqueous solution of SO_2 described by the following equilibria (Eq. 9-2) (15,16). Dissolved sulfur dioxide exists as SO_2, denoted as $SO_2 \cdot H_2O$, and not the commonly expected sulfurous acid, H_2SO_3.

$$pK_a$$
$$\text{25C, low ionic}$$

$$SO_2\,(gas) + H_2O \rightleftharpoons SO_2 \cdot H_2O\,(solution)$$

$$SO_2 \cdot H_2O \rightleftharpoons H^+ + HSO_3^- \qquad 1.81$$

$$HSO_3^- \rightleftharpoons H^+ + SO_3^= \qquad 7.18$$

Eq. 9-2

In the normal pH range of food, the predominant species will be HSO_3^- as indicated in Eq. 9-2 (16). Increasing ionic strength reduces the pK_a for Eq. 9-2.3, and SO_3^{2-} is therefore expected to be present in greater proportion.

Also, as the total concentration is increased, the bisulfite ions dimerze to form disulfite (metabisulfite, pyrosulfite), as in Eq. 9-3. Dimerization is unimportant at concentrations lower than $0.01M$.

$$2\,HSO_3^- \rightleftharpoons S_2O_5^= + H_2O$$

Eq. 9-3

Inhibition of Nonenzymatic Browning

Nonenzymatic browning consists of the following reactions: (1) Maillard reaction, (2) ascorbic acid browning, and (3) caramelization, as discussed in Chapter 4.

One major action of sulfite on the browning reaction is to interrupt the steps leading to the formation of colored products, by forming sulfonates with carbonyl intermediates such as 3-deoxyosulose and 3,4-dideoxyosolos-

3-ene (*17,18*). In the case of 3-deoxyosulose, the hydroxyl groups β to the enediol grouping are very labile. Replacement of the 4-hydroxy group by sulfur oxoanions yields the sulfonate product, 3-deoxy-4-sulfo-osulose, with inversion at C-4 (Eq. 9-4). However, 3-deoxyosulose can undergo dehydration to 3,4-dideoxyosulo-3-ene, which is an α, β-unsaturated carbonyl. Nucleophilic 1,4-addition of sulfur oxoanions to the double bond also leads to the formation of the same product (Eq. 9-5).

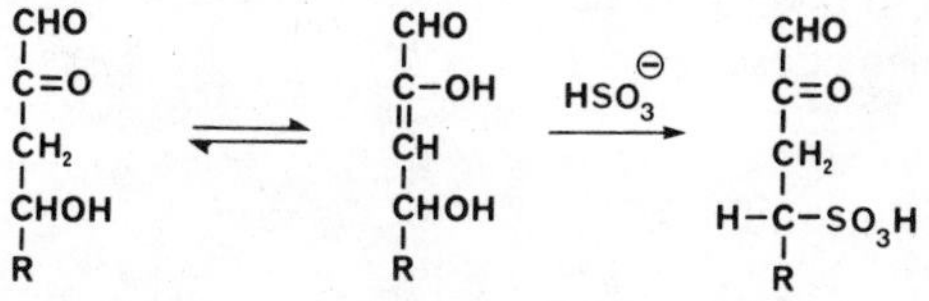

Eq. 9-4

3-Deoxyosulose

Eq. 9-5

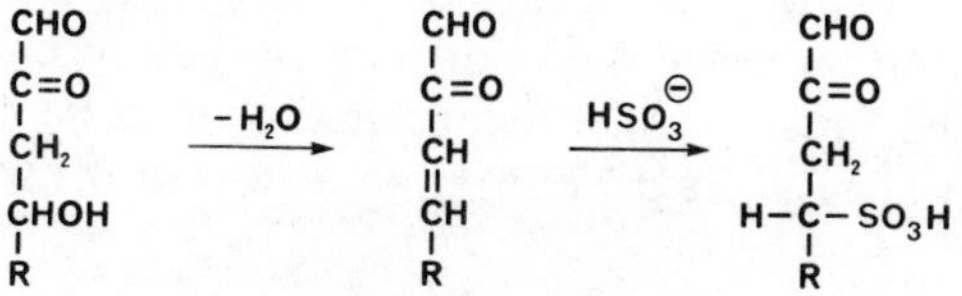

3,4-Dideoxyosulos-3-ene

The mechanism involving the addition of sulfur oxoanions to the α, β-unsaturated intermediate constitutes the major route of inhibition of non-enzymatic browning (Chapter 3). The irreversible formation of the sulfonate product effectively removes the reaction intermediates from the later stages of browning reactions.

Inhibition of Enzymatic Browning

Inhibition of enzymatic browning by sulfur oxoanions is mainly accomplished by removing the *o*-quinones produced from *o*-diphenols by the action of polyphenol oxidase from further oxidation and polymerization. Nucleophilic addition of sulfite to the *o*-quinone leads to the formation of sulfonate (Eq. 9-6). There is also a direct effect of sulfite on the enzyme, but the exact mechanism is not known.

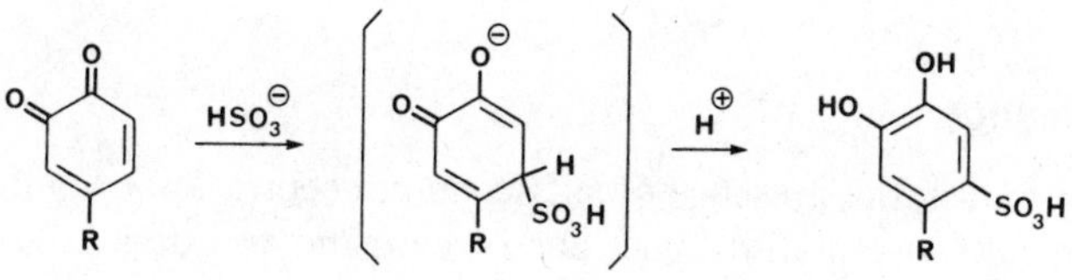

Eq. 9-6

Antimicrobial Action

The antimicrobial activity of sulfite depends on the penetration of molecular SO_2 across the cell membrane. It is suggested that sulfur dioxide activates an ATPase system located in the cell membrane. Sulfite concentration as low as $1mM$, at pH 5.0 or below, has been shown to cause rapid depletion of ATP prior to cellular death in yeast (19).

The SO_2, once inside the cell, may also react with cell components. Sulfite has been shown to react with cofactors and coenzymes, amino acids, pyrimidines, and nucleosides.

Reaction with Pyrimidines

The pyrimidine uracil reacts with $NaHSO_3$ at pH 6–7 to yield an additive compound, 5,6-dihydrouracil-6-sulfonate (Eq. 9-7). The sulfonate is stable at neutral and acidic pH, but hydrolyzed by alkali (pH > 9) to uracil (20). The rate of addition for uracil is proportional to the concentration of sulfite ion and unionized uracil. Uracil has been shown to be converted to the addition product at bisulfite concentration as low as $0.1M$ (21).

Eq. 9-7

5,6-Dihydrouracil-6-sulfonate

Under similar conditions, cytosine forms an addition product with bisulfite, dihydrocytosine-6-sulfonate which undergoes deamination to yield the 5,6-dihydrouracil-6-sulfonate (Eq. 9-8). The optimum rate of deamination occurs at acidic pH < 5 and in a high sodium bisulfite concentration ($>0.5M$). At physiological pH, the rate of deamination declines to only about 1% of that observed at the optimum pH. Furthermore, at low sulfite concentrations compatible with in vivo conditions, the rate can be expected to be extremely low, estimated to be $4.0 \times 10^{-4}/\text{sec}$ for a sulfite concentration of $10^{-3}M$ (22).

As anticipated from the chemistry of deamination reaction, the conversion of cytosine to uracil can cause C-G to T-A site mutation.

Transamination

At neutral pH, in the presence of sufficient concentration of amines, bisulfite catalyzes transmination reactions of cytosine and derivatives with the formation of N-substituted cytosines (Eq. 9-9) (23). Apparently, the amines

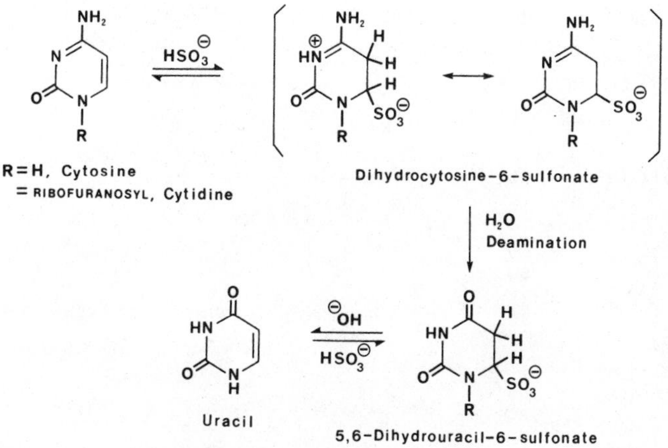

R = H, Cytosine
= RIBOFURANOSYL, Cytidine

Dihydrocytosine-6-sulfonate

H_2O
Deamination

Eq. 9-8

Uracil

$^{\ominus}OH$
$HSO_3^{\ominus}$

5,6-Dihydrouracil-6-sulfonate

$HSO_3^{\ominus}$

$R''R'NH$
Transamination
neutral pH

pH < 5
H_2O
Deamination

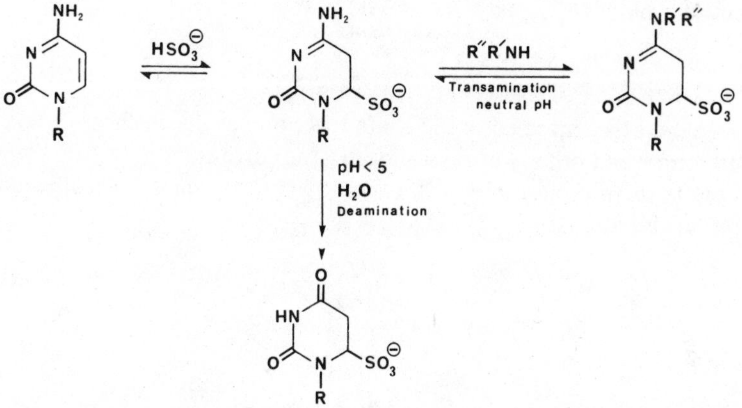

Eq. 9-9

compete successfully with water as nucleophiles for formation of the intermediate dihydrocytosine-6-sulfonate at neutral pH, where the deamination reaction is minimal.

Lysine and polylysine have been shown to cross-link cytosine and polycytidylic acid. Cross-linkage of proteins and nucleic acids by bisulfite was demonstrated in RNA bacteriophage MS2. Therefore, the transamination reaction that occurs optimally at neutral pH raises the possibility of adverse effects at the cellular level.

Free-Radical Reactions

Sulfite can be oxidized to sulfate via a free-radical chain reaction initiated by metal ions, or by photochemical, electrolytic, or enzymatic processes (Eq. 9-10). The sulfite radicals formed undergo a range of propagation reactions. The SO_3^- and SO_5^- radicals act as chain carriers in the oxidation of sulfite in systems of low sulfite concentration, and in which Eqs. 9-10.5 and 9-10.6 are negligible. In strong alkaline solution, SO_5^- and OH^- are the main chain carriers (24).

INITIATION $\quad SO_3^= \longrightarrow SO_3^- + e_{aq}^- \quad (+O_2 \rightarrow O_2^-) \qquad 9-10.1$

PROPAGATION $\quad SO_3^- + O_2 \longrightarrow SO_5^- \qquad\qquad\qquad 9-10.2$

$\qquad\qquad\qquad SO_5^- + SO_3^= \longrightarrow SO_4^- + SO_4^= \qquad 9-10.3$

$\qquad\qquad\qquad SO_4^- + SO_3^= \longrightarrow SO_4^= + SO_3^- \qquad 9-10.4$

At strong alkali pH, $\qquad\qquad\qquad\qquad\qquad\qquad\qquad$ Eq. 9-10

$\qquad\qquad\qquad SO_4^- + OH^- \longrightarrow SO_4^= + OH\cdot \qquad 9-10.5$

$\qquad\qquad\qquad OH\cdot + SO_3^= \longrightarrow OH^- + SO_3^- \qquad 9-10.6$

TERMINATION $\quad \left.\begin{array}{l} SO_5^- + SO_5^- \\[1ex] SO_4^- + SO_4^- \end{array}\right\} $ Non-radical products

The free radicals generated in the chain reaction have been shown to oxidize methionine and other sulfide analogs to the corresponding sulfoxide at neutral pH in the presence of Mn^{2+}, O_2, and SO_3^{2-} during the aerobic oxidation of sulfite ion (Eq. 9-11) (25).

$$R-S-R' + HO\cdot \longrightarrow R-\overset{\oplus}{S}-R' + OH^\ominus$$

$$\downarrow HO\cdot \qquad\qquad\qquad\qquad \text{Eq. 9-11}$$

$$R-\overset{\oplus}{\underset{OH}{S}}-R' \longrightarrow R-\overset{O}{\underset{\parallel}{S}}-R' + H^\oplus$$

In the presence of Mn^{2+}, O_2, and glycine, sulfite catalyzes the destruction of β-carotene at the pH of food systems (Eq. 9-12) (26). Experiments indicate the reaction is rapid and more than 90% of the carotene loss occurs in one minute (Table 9.4). Destruction can be effectively inhibited by free-radical scavengers, such as α-tocopherol and BHT.

$$Mn^{++} + O_2 \longrightarrow Mn^{+++} + O_2^-$$

$$Mn^{+++} + SO_3^= \longrightarrow Mn^{++} + SO_3^-$$

$$\beta\text{-Carotene} + O_2^- + H^+ \longrightarrow \beta\text{-Carotene}\cdot + H_2O_2 \qquad \text{Eq. 9-12}$$

$$\beta\text{-Carotene} + SO_3^- \longrightarrow \beta\text{-Carotene}\cdot + H^+ + SO_3^=$$

$$\beta\text{-Carotene}\cdot + O_2 \longrightarrow \beta\text{-Carotene}-OO\cdot$$

$$\beta\text{-Carotene}-OO\cdot \longrightarrow \text{Oxidation products}$$

Eq. 9-12
(cont.)

Sulfite has also been shown to induce lipid oxidation in emulsified linoleic acid (27). The formation of hydroperoxide correlates with the loss of sulfite (Eq. 9-13).

$$SO_3^- + LH \longrightarrow L\cdot + H^+ + SO_3^=$$

$$L\cdot + O_2 \longrightarrow LOO\cdot$$

$$LOO\cdot + LH \longrightarrow LOOH + L\cdot$$

$$LOO\cdot + H^+ + SO_3^= \longrightarrow LOOH + SO_3^-$$

Eq. 9-13

The sulfite free radicals also can initiate reactions in nucleic acids. Sulfite at concentration of 10mM catalyzes the transformation of 4-thiouridine, a nucleoside in bacterial transfer RNA, to uridine-4-sulfonate in the presence of oxygen, at neutral pH (Eq. 9-14) (28). The thiosulfate formed in the initial reaction is unstable toward light or acid and decomposes into uridine-4-sulfonate. A high concentration of sulfite inhibits the overall reaction due to the competition between the substrate (4-thiouridine) and sodium bisulfite for the sulfite radical.

The 4-sulfonate group of the product is susceptible to attack by various nucleophiles. Reactions with ammonia or alkyl amines results in the substi-

Table 9.4. Destruction of β-Carotene by Sulfite at pH 5.7 and 9.2

	RELATIVE β-CAROTENE CONCENTRATION[a]	
TIME	pH 5.7	pH 9.2
0	1.00	1.00
10	0.76	0.15
20	0.51	0.10
30	0.26	0.07
40	0.12	0.05
50	0.09	0.05

[a]Initial concentration: β-carotene = 80μM, sulfite = 100μM, MnCl₂ = 52μM, glycine = 21mM, in 0.025 Mbuffer.

Reprinted with permission from Wedzicha and Lamikanra (26), courtesy of Elsevier Applied Science Publishers, Ltd., Essex, England.

Eq. 9-14

4−THIOURIDINE
R = β−D−ribofuranosyl

URIDINE−4−
THIOSULFATE

URIDINE−4−
SULFONATE

tution of amino and alkyl amino groups, forming cytidine and its derivatives. Sulfite at 10mM concentration also reacts with DNA, leading to the cleavage of the nucleotides. The reaction is oxygen dependent and occurs rapidly at pH 7, in the presence of Mn^{2+} ions.

Mutagenicity

The mutagenicity of sulfite is anticipated from the chemistry of all the possible reactions with the many biological molecules. Although most studies done on viruses, bacteria, and yeast employ 1M or higher concentrations of bisulfite, in a few cases, relatively low concentrations of bisulfite (0.01M) at neutral pH have been shown to promote mutation in *Micrococcus aureus* and *Saccharomyces cerevisiae.* In mammalian cell cultures, low concentrations of sulfite (less than 0.01M) inhibits DNA, RNA, and protein synthesis; prevent mitosis; reduce cell growth; and cause chromosomal abnormalities (*29*).

In spite of sulfite-induced chromsomal abberations in in vitro experiments, no adverse effects have been observed in chronic sulfite feeding studies, unless sulfite is administered at significantly higher doses. The effect of high doses can be attributed to the indirect toxicity due to the destruction of thiamine and other vitamins in the feeding diet (*30*).

Studies on the effect of sulfite on three generations of rats for a two-year period suggests a noneffect level equivalent to an intake dose of 72 mg SO_2 equivalent/kg/day (*31*). Applying a hundred-fold safety factor, the estimated maximum acceptable daily intake becomes 0.7 mg/kg, which amounts to a safety level of 42 mg per average adult (60 kg body weight). The estimated total sulfite as SO_2 in some sulfited foods is listed in Table 9.5 (*32*).

Table 9.5. Estimated Total SO_2 as Consumed for Some Sulfited Foods (ppm)

FOOD[a]	ppm AS SO_2 EQUIVALENT[b]
≥ 100 ppm	
Dried fruit (excluding dark raisins & prunes)	1200
Lemon juice (nonfrozen)	800
Salad bar lettuce	400–950
Lime juice (nonsugar)	160
Wine	150
Molasses	125
Sauerkraut juice	100
50–90.9 ppm	
Dried potatoes	35–90
Grape juice (white, white sparkling, pink sparkling, red sparkling)	85
Wine vinegar	75
Gravies, sauces	75
Fruit topping	60
Maraschino cherries	50
< 50 ppm	
Pectin	≤ 10–50
Shrimp (fresh)	≤ 10–40
Corn syrup	30
Sauerkraut	30
Pickled pepper	30
Pickles/relishes	30
Cornstarch	30
Fresh mushrooms	13

[a]The use of sulfite was banned from raw fruits and vegetables on July 9, 1986. Labels are required in packaged foods with detectable amounts of sulfite (10 ppm of SO_2), effective on Jan 9, 1987.
[b]SO_2 equivalent measured by Monier-Williams Assay.

Reprinted with permission from Taylor and Bush (32), copyright 1986 by Institute of Food Technologists.

Metabolic Pathway

The principal defense against toxicity of sulfite in mammalian systems is its reduction to sulfate by sulfite oxidase (Eq. 9-15), a hemoprotein containing molybdenum located in the intermembrane spaces of mitochondria. Electrons in the oxidation can be transferred to the oxidative phosphorylation system.

The natural function of the enzyme is to participate in the metabolism of sulfur-containing amino acids. Endogenous sulfite produced in the metabo-

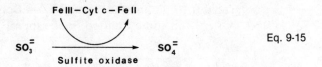

Eq. 9-15

lism of sulfur-containing compounds is estimated to be considerably higher than the estimated daily intake of exogenous sulfite from the diet (average 6 mg SO_2 equivalent/capita). Animals deficient in sulfite oxidase are found to be more susceptible to a lower dose of sulfite (*33*). Certain asthmatic patients, especially those dependent on steroid drugs, are more sensitive to sulfite.

10
VITAMINS

The degradation of vitamins during food processing has always been of great concern and the subject of particular interest to food scientists. The chemical changes that occur to a vitamin in a complex food system are difficult to study. The assessment of losses in vitamins in a food system is further hampered by the lack of systematic studies and by the variables in treatment conditions. Most of our present knowledge in this respect results from extrapolating experiments done in model systems. While the chemical degradation of vitamins can be used as an index for the possible nutritional loss of the particular food, the fate of the degradative products and their possible reactions remain largely unknown.

Another area of interest that deserves greater attention concerns the biochemical mechanisms of the vitamins—especially how the biochemical reaction mechanisms can be understood in terms of known mechanisms of organic reactions. This understanding is particularly important in view of the general misconception regarding vitamins as "natural" miracle treats. An appreciation of the chemical nature of vitamins can help not only to remove some of the myths, but to understand their functional and physiological properties on the basis of good sense and science.

Vitamins are commonly classified into two groups, the fat-soluble vitamins A, D, E and K, and the water-soluble vitamin C and vitamins of the B complex.

VITAMIN A

Vitamin A (retinol) is a diterpene alcohol consisting of a trimethylcyclohexenyl ring with a side chain of isoprene units.

The main source of vitamin A is β-carotene (Fig. 10.1), which is one

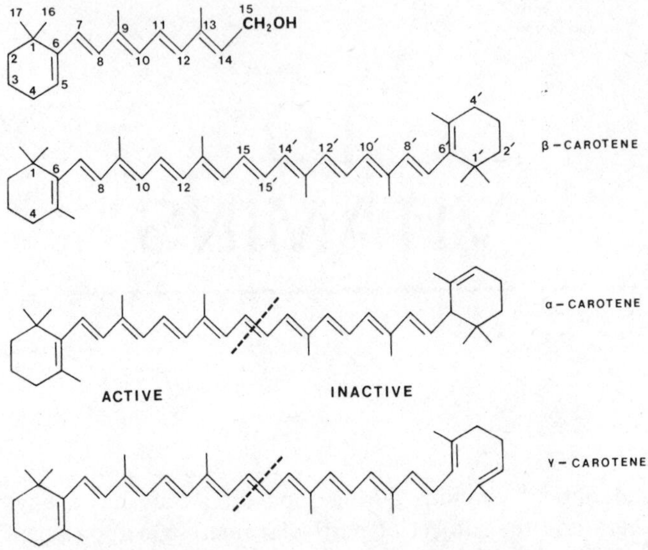

Fig. 10.1. Activity of various carotenes.

of the vitamin A precursors found in carotenoid. In general, a carotenoid provitamin A compound requires an unsubstituted β-carotene to be 100% vitamin A activity, α- and γ-carotene (Fig. 10.1) possess approximately 50% vitamin A activity.

Biological Function

Biosynthesis of vitamin A from precursor compounds occurs in the intestinal muscosa. The molecule is cleaved either by (1) symmetric or asymmetric fission or (2) terminal cleavage to yield vitamin A aldehyde (retinal), which is then reduced to the alcohol (retionl) (Eq. 10-1). The enzyme 15, 15'-dioxygenase has been isolated and demonstrated to cleave precursor compounds in the presence of oxygen to yield retinal.

Vitamin A plays a role in development and growth, regulation of stability of biological membranes, maintenance of mucus-secreting cells of epithelia, biosynthesis of glycoprotein, and prevention of keratinization. The mechanism of many of these physiological functions remains unclear, except its specific role in vision. The retina consists of photoreceptor cells, known as rods and cones. In the rod cells are photosensitive pigments called rhodopsin. Rhodopsin is opsin (a protein) with 11-*cis*-retinal attached to the ε-amino group of a lysine residue via a Schiff-base linkage (*1*). One result of

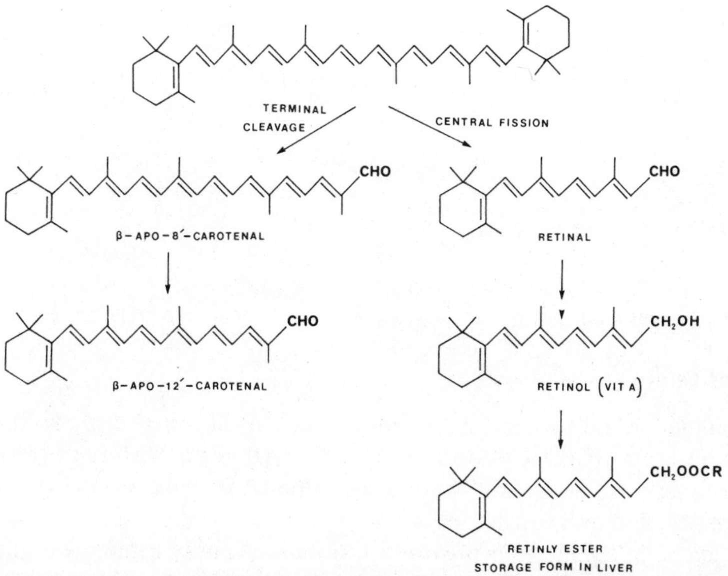

Eq. 10-1

the Schiff-base formation is a marked shift of λ_{max} to a longer wavelength (red). Absorption of light by rhodopsin causes isomerization of 11-*cis*-retinal to the all-*trans* configuration. This photochemical event is followed by a series of dark reactions that leads to hydrolysis. The all-*trans* retinal released is reconstituted to the *cis*-configuration catalyzed by the enzyme retinal isomerase (Eq. 10-2).

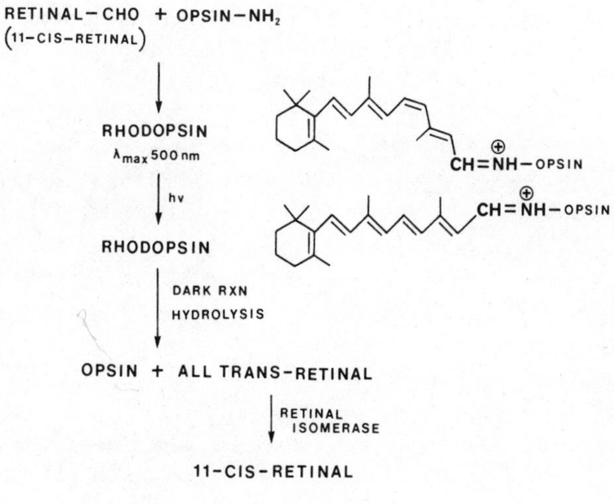

Eq. 10-2

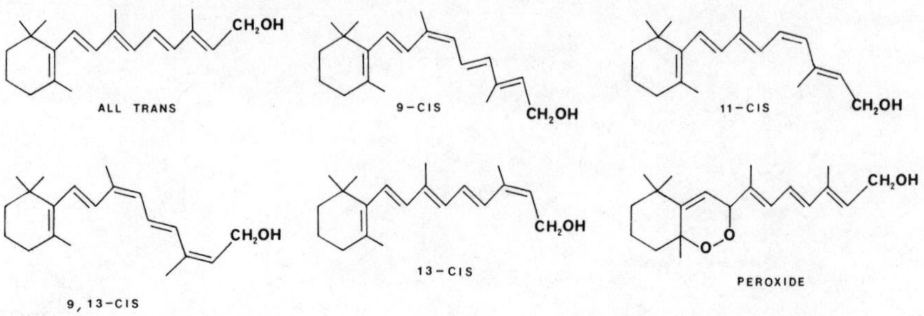

Fig. 10.2. Isomeric mixture of vitamin A in the presence of dilute acid.

Action of Dilute Acid

The carotenes found in natural sources are in the all-*trans* form. In the presence of dilute HCl, the all-*trans* isomerizes, yielding a mixture of *trans* and *cis* isomers. Isomerization can also be initiated by light or heat (refer to the next section and to Chapter 4).

In an isomeric equilibrium mixture of vitamin A, only a few sterically favored isomers exist (Fig. 10.2), although vitamin A with 5 carbon-carbon double bonds can theoretically exist in 32 stereoisomeric forms. Among the isomers found to exist include 9-*cis*, 13-*cis*, and 9,13-*cis*.

Photochemical Reactions

Photodimerization. The formation of dimers is expected from a Diels-Alder reaction between the vitamin A molecules. The unsaturated side chains of vitamin A are *trans* to each other and diequatorial. The two CHOX groups attached to C14 and C14′ are *cis* to each other (Eq. 10-3) (*2*).

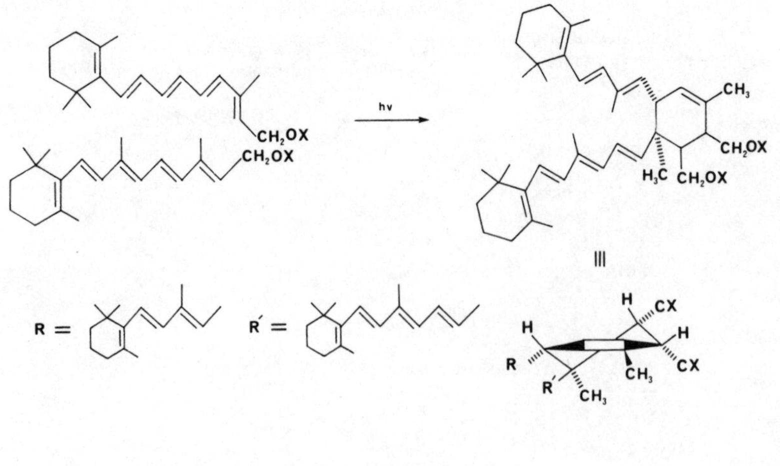

Eq. 10-3

Photoisomerization.   Irradiation of all-*trans* vitamin A in hexane yields a mixture of all-*trans* (50%), 13-*cis* (45%), 9-*cis* (8%), and peroxide (5%). Depending on the solvent used, 11-*cis* is also found.

Photosensitized Oxidation.   When vitamin A is irradiated with ultraviolet light in the presence of a sensitizer, peroxide is formed, analogous to the adducts formed with dienophiles in the Diels-Alder reaction (Eq. 10-4).

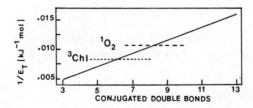

EXOPEROXIDE

Eq. 10-4

Quenching. β-Carotene is very efficient in physical quenching of $^1O_2$, with a quenching rate constant (K_Q) of 1.3×10^{10} $M^{-1}s^{-1}$ (in benzene), close to that expected for diffusion-controlled reactants and 10^4 times the rate (K_R) of irreversible reaction with $^1O_2$ (Eq. 10-5).

$$^1\text{Carotene} + {}^1O_2 \xrightarrow{K_Q} {}^3\text{Carotene} + {}^3O_2$$

$$\downarrow K_R \qquad\qquad\qquad \downarrow$$

$$\text{Products} \qquad\qquad {}^1\text{Carotene}$$

Eq. 10-5

The energy difference between $^1O_2$ and $^3O_2$ is estimated to be 94 KJ mol^{-1}. A linear plot of the inverse triplet energy ($1/E_T$) against the number of conjugated double bonds in the carotenoid indicates that only carotenoids with more than nine conjugated double bonds can quench the energy of $^1O_2$ (Fig. 10.3) (*3*.).

Fig. 10.3. The relationship between the reciprocal of carotenoid triplet energy levels and the number of conjugated double bonds in the pigment. (Reprinted with permission from Krinsky, N.I. Carotenoid protection against oxidation. Pure & Appl. Chem. *51*, 655. Courtesy of International Union of Pure and Applied Chemistry, Oxford, England.)

Fragmentation.   Singlet oxygen can react with carotene and other carotenoids via (1) the ene-reaction, (2) the dioxetane mechanism, and (3) the peroxy epoxide mechanism (Eq. 10-6). Various aliphatic and monocyclic isoprenoids are formed by fragmentation of the oxidative products. Beta-13-, β-14-, and β-15-apo-carotenone have been identified in simulated commercial processing systems.

Eq. 10-6

VITAMIN B$_1$

Thiamin is present in small quantities in almost all foods; whole cereal grains and organ meat such as liver, heart, and kidney are good sources.

Biochemical Mechanism

The biochemical mechanism of thiamin has been thoroughly studied. Thiamin (as pyrophosphate) participates as a coenzyme in the reaction involving (1) α-keto acid decarboxylases, α-keto acid oxidases, (3) transketolases, and (4) phosphoketolase. The conversion of pyruvate to acetyl CoA is used to illustrate the chemistry behind the action of thiamin. Here, thiamin is the cofactor of the pyruvate dehydrogenase complex, which is comprised of

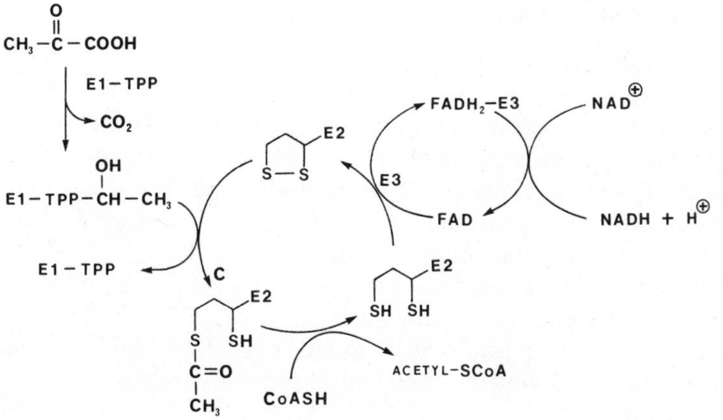

Fig. 10.4. Scheme of reactions catalyzed by the pyruvate dehydrogenase complex.

pyruvate decarboxylase (E1), dihydrolipoyl transacetylase (E2), and dihydrolipoyl dehydrogenase (E3) (Fig. 10.4).

The basic chemistry of the reaction is a benzoin-type condensation. The thiamin molecule consists of a pyrimidine and a thiazolium ring: the C4′ amino group acts as a weak base to pull off the proton at C2 in the thiazolium, resulting in the formation of a "ylid." The process is likely enzyme oriented. The "ylid" carbanion condenses with a ketal donor, such as pyruvate in this case. The thiazolium ion acts as an "electron sink" to accept electrons from the $C-C$ bond cleaved in decarboxylation. The enolic intermediate acts as a nucleophile and acylates lipoic acid (linked to the ϵ-amino of lysine of enzyme lipoyl transacetylase) before being transferred to CoA (Eq. 10-7).

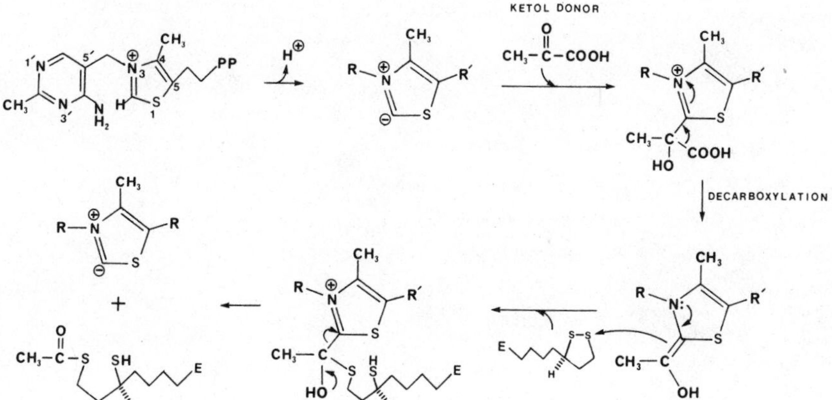

Eq. 10-7

Action of Alkali and Acid

In alkaline solution, the hydroxide ion is added to the C2 of the thiazolium ring, forming an unstable pseudobase. Disruption of the $-C-S-$ bond yields a colorless thiol form. In solution of high pH (>11), the amino group of the pyrimidine reacts intramolecularly with the C2 in the thiazolium ring to form a tricyclic intermediate, dihydrothiachrome. The ionization of the dihydrothiachrome with opening of the thiazolium ring yields a yellow thiol form, which, in time, converts to the colorless thiol form (Eq. 10-8). Degradation of the thiol form yields hydrogen sulfide and a variety of breakdown products (4). Heating of thiamin solution at pH 6.0 or below results in cleavage at the methylene bridge to give 2-methyl-4-amino-5-hydroxy-methyl pyridine and 4-methyl-5-(β-hydroxyethyl)thiazole.

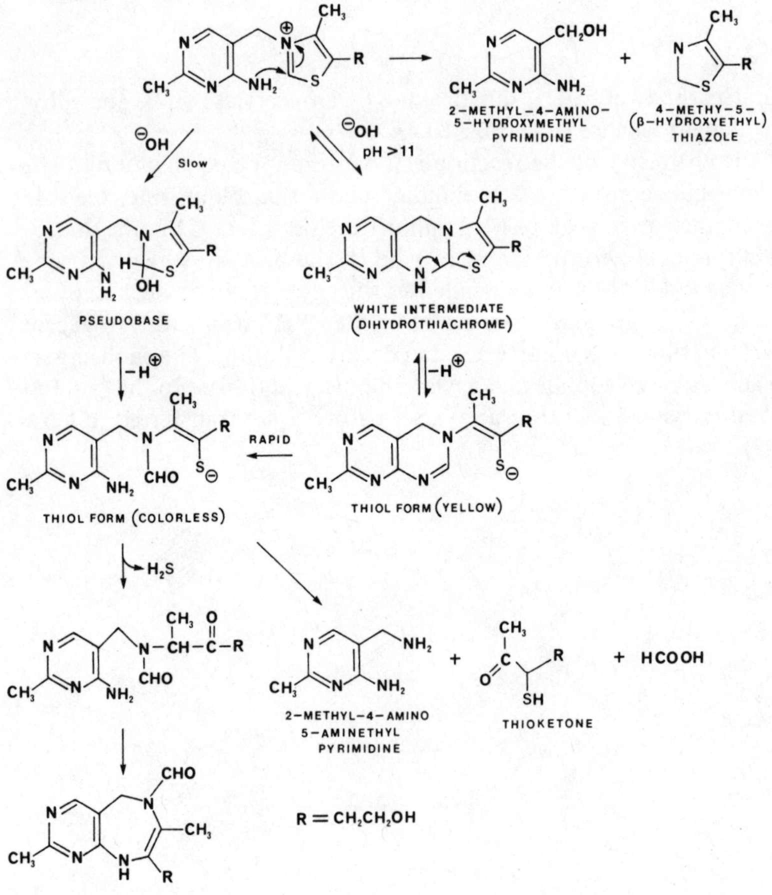

Eq. 10-8

OXIDATION AND REDUCTION

Both the yellow thiol form and the intermediate precursor can be oxidized to thiochrome with an intense blue fluorescense. The colorless thiol can be oxidized, or it reacts in sulfhydryl-disulfide exchange to form dimers (Eq. 10-9). Oxidizing agents such as H_2O_2 and I_2 accelerate the reactions in neutral or alkaline solutions. The thiochrome is not found if the thiamin stays in alkaline solution long enough for the yellow thiol form to disappear before oxidizing agent is introduced.

Eq. 10-9

REACTION WITH BISULFITE

Thiamin is unstable and undergoes cleavage in bisulfite solution. The first step is a bisulfite addition to the 6-position in the pyrimidine ring, followed by a nucleophilic substitution at the methylene carbon by a second bisulfite ion via a S_N2 mechanism (5). It is also suggested that the addition of the first bisulfite ion is followed by the elimination of the leaving group (thiazole) to give a stabilized carbocation intermediate. Addition of a second bisulfite ion with the loss of the first sulfite ion yields the final sulfonate product (Eq. 10-10) (6).

Eq. 10-10

PHOTOLYSIS

Light irradiation of thiamin causes cleavage at the $C-N$ and $C-S$ bonds, forming 3-mercapto-5-hydroxy-2-pentanone, which is the key intermediate in the formation of various degradation products. Cleavage at the α- and β-position of the intermediate yields hydrogen sulfide, formaldehyde, and acetaldehyde. Elimination of sulfur from the intermediate followed by formylation results in the formation of 3-formyl-5-hydroxy-2-pentanone, which cyclizes readily to an equilibrium mixture of 2-methyl-3-formyl-4,5-dihydrofuran and 3-acetyl-4,5-dihydrofuran.

The intermediate compound may cyclize to form 2-hydroxy-2-methyl-3-mercapto-furan. Dehydration and oxidation of the latter gives a disulfide, *bis*(2-methyl-4,5-dehydrofuran-3-yl)disulfide. Or it reacts with formaldehyde or acetaldehyde to form the characteristic thiamin odor, 1-methylbicyclo[3.3.0]-4-thia-2,8-dioxaoctane, or 1,3-dimethylbicyclo[3.3.0]-4-thia-2,8-dioxaoctane, respectively. The thresholds of these two thiamin odor compounds are $1/10^7$ parts water and $1/10^9$ parts of water, respectively (7) (Eq. 10-11).

VITAMIN B$_2$

Riboflavin is 7,8-dimethyl-10-(1′-D-ribityl)isoalloxazine. The biochemically functional coenzyme is flavin adenine dinucleotide (FAD) (Fig. 10.5).

Flavoenzymes catalyze a diverse range of reactions; some typical examples are given in Fig. 10.6. The reaction substrates include amines, amino acids, alcohols, hydroxy acids, dithiols, aldehydes, ketones, and acids.

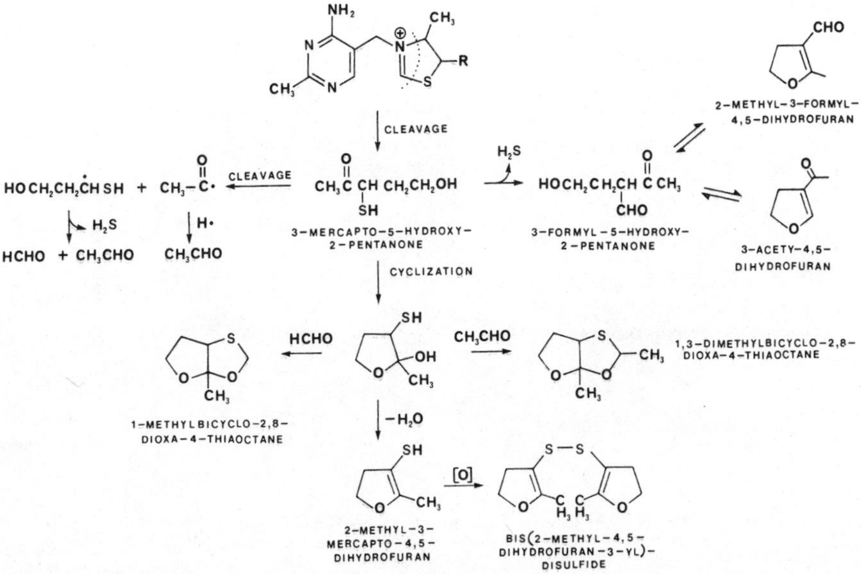

Eq. 10-11

Biochemical Mechanism

Flavin has a unique property of undergoing both $2e^-$ and $1e^-$ reactions. The intermediate semiquinones are stable radicals, with the unpaired electrons delocalized through the conjugated isoalloxazine ring (Eq. 10-12).

Most flavins bind strongly but noncovalently with the enzyme. The equilibrium constant K is on the order of $10^{-8}M$, indicating a strong binding

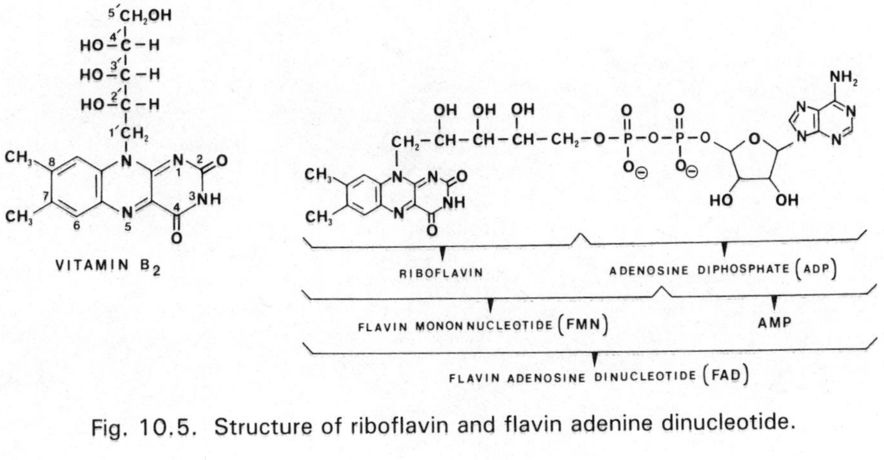

Fig. 10.5. Structure of riboflavin and flavin adenine dinucleotide.

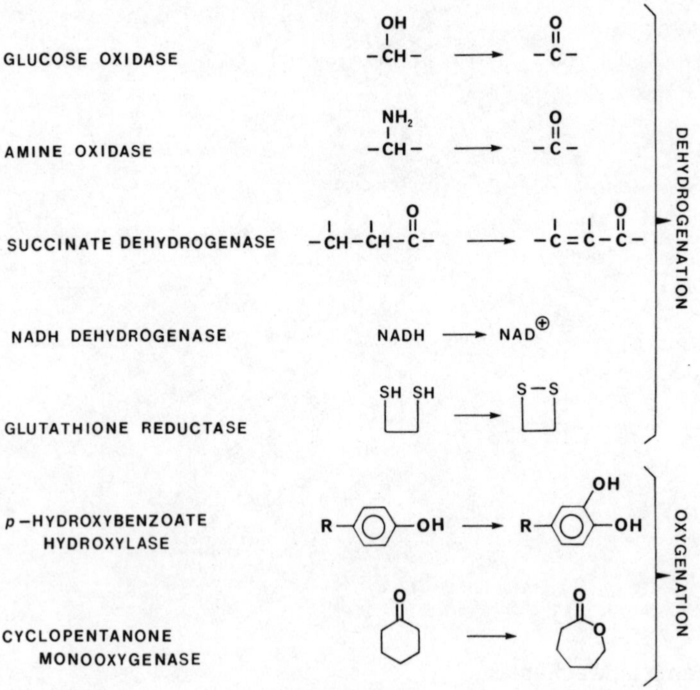

Fig. 10.6. Reactions catalyzed by flavoenzymes.

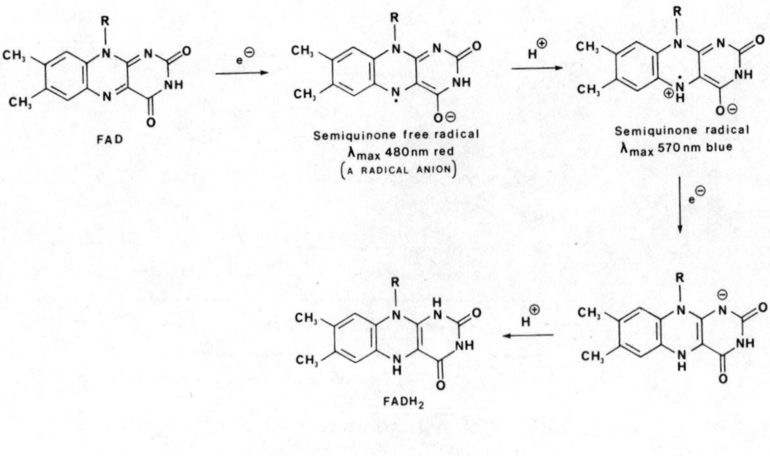

Eq. 10-12

(Eq. 10-13). The flavin coenzyme must complete the redox cycle (two half-reactions) before the enzyme can carry out the next cycle of catalysis (Eq. 10-14.1, 10-14.2). The electron acceptor Y in the oxidation half-reaction is commonly molecular oxygen, which undergoes a $2e^-$ reduction to H_2O_2 or a $4e^-$ reduction to H_2O by successive $1e^-$ reductions (Eq. 10-15).

$$\text{PROTEIN} - \text{flavin} \underset{}{\overset{K}{\rightleftharpoons}} \text{PROTEIN} + \text{Flavin} \qquad \text{Eq. 10-13}$$

REDUCTION

$$\text{ENZYME} - \text{FAD} + \text{XH}_2 \rightleftharpoons \text{ENZYME} - \text{FADH}_2 + \text{X} \qquad 10-14.1$$

Eq. 10-14

OXIDATION

$$\text{ENZYME} - \text{FADH}_2 + \text{Y} \rightleftharpoons \text{ENZYME} - \text{FAD} + \text{YH}_2 \qquad 10-14.2$$

$$\tfrac{1}{2}O_2 \xrightarrow{e^{\ominus}} \tfrac{1}{2}O_2^{\bar{}} \xrightarrow[H^{\oplus}]{e^{\ominus}} \tfrac{1}{2}H_2O_2 \xrightarrow{e^{\ominus}} \cdot OH \xrightarrow[H^{\oplus}]{e^{\ominus}} H_2O \qquad \text{Eq. 10-15}$$

SUPEROXIDE   HYDROXY RADICAL

E^o -330 mV $+94$ mV $+136$ mV $+233$ mV

Monoamine oxidase is the flavoenzyme that is involved in the inactivation of catecholamine neurotransmitters by converting the amines to aldehydes. The overall reaction is represented in Eq. 10-16.

Eq. 10-16

The reoxidation of flavin is mediated by a radical mechanism with the oxygen molecule to produce the superoxide anion ($O_2^-\cdot$) and the semiquinone radical (flavin-H·) (Eq. 10-14). The superoxide anion and flavin-H· recombine to form flavin hydroperoxide, which decomposes to the products H_2O_2 and oxidized flavin (Eq. 10-17).

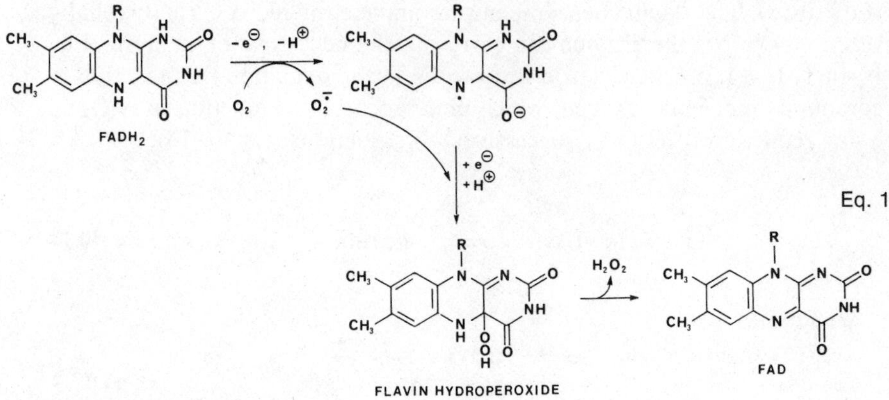

FADH$_2$

FLAVIN HYDROPEROXIDE

FAD

Eq. 10-17

In the reoxidation of enzyme-FADH$_2$, oxygen undergoes a $2e^-$ reduction to H$_2$O$_2$. Most flavoenzymes involved in the $4e^-$ reduction of oxygen to H$_2$O are of bacterial origin and therefore will not be discussed.

Alkaline Degradation

In alkaline solution, riboflavin is hydrolyzed to urea and oxocarbonic acid (Eq. 10-18). The hydrolysis occurs even at room temperature.

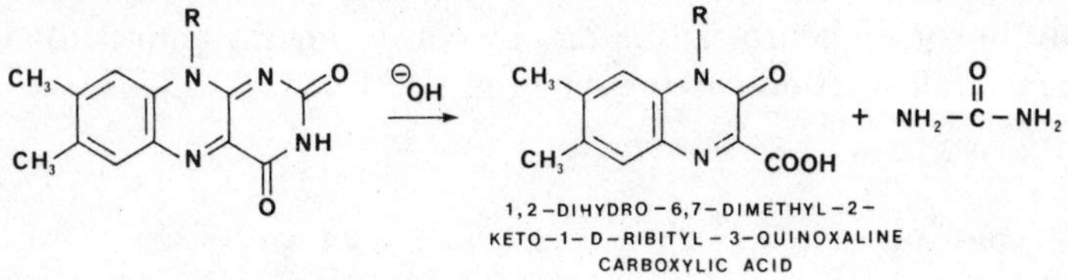

1,2−DIHYDRO−6,7−DIMETHYL−2−
KETO−1−D−RIBITYL−3−QUINOXALINE
CARBOXYLIC ACID

Eq. 10-18

Reaction with Sulfite

Flavoenzymes (particularly the oxidases) and, to a much lesser extent, the flavins, react with sulfite at the N-5 position to form an addition complex (8). Acid hydrolysis under anaerobic conditions yields the reduced flavin and sulfite (Eq. 10-19).

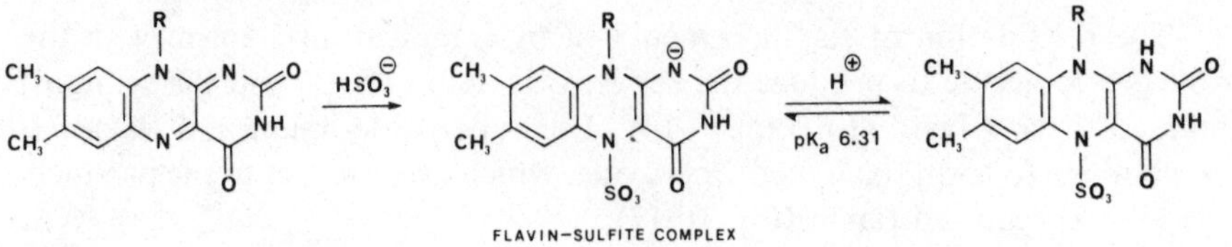

FLAVIN−SULFITE COMPLEX

pK$_a$ 6.31

Eq. 10-19

Photochemical Reaction

The photochemistry of flavin can be divided into four categories: photoreduction, photoalkylation, photoaddition, and photosensitized reactions (9).

Photoreduction. Under anaerobic conditions, flavin undergoes a singlet to triplet excitation, following by hydrogen abstraction from the ribityl C-2′. The diradical intermediate undergoes disproportionation to form deuteroflavin. Alternately, the diradical formed by the hydrogen abstraction of the triplet riboflavin undergoes a proton exchange and forms a N1-N10 bridged leucodeuteroflavin. Subsequent oxidation yields the deuteroflavin (Eq. 10-20). (Lumichrome is an alloxazine, and riboflavin and deuteroflavin are isoalloxazines, as indicated in Fig. 10.7).

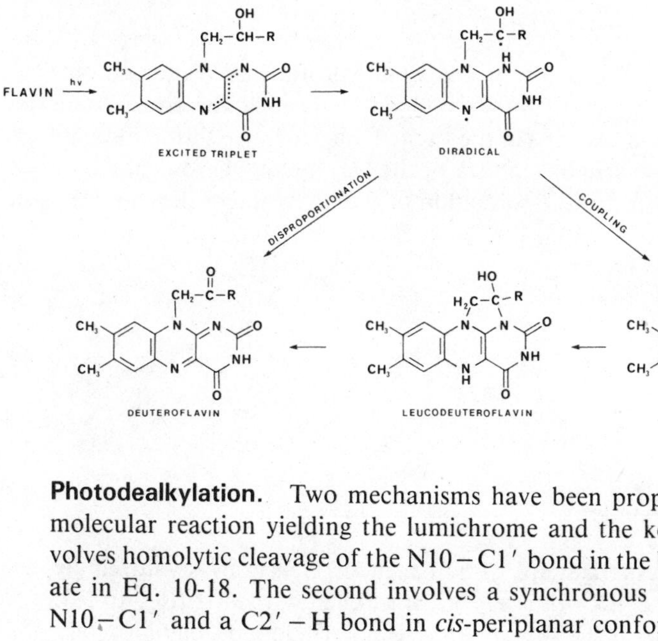

Eq. 10-20

Photodealkylation. Two mechanisms have been proposed for the intramolecular reaction yielding the lumichrome and the ketone. The first involves homolytic cleavage of the N10 − C1′ bond in the biradical intermediate in Eq. 10-18. The second involves a synchronous fragmentation of a N10 − C1′ and a C2′ − H bond in *cis*-periplanar conformation with direct

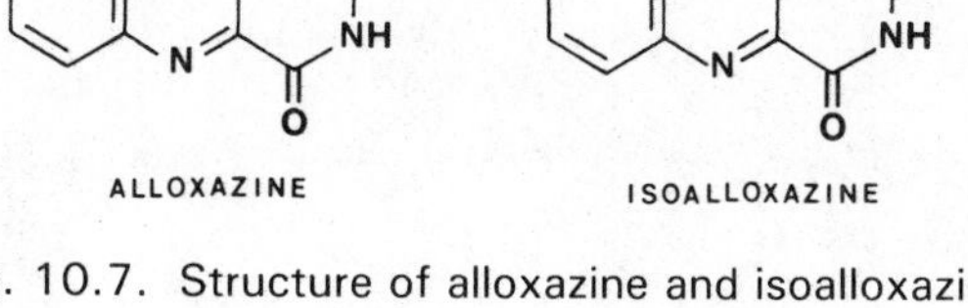

Fig. 10.7. Structure of alloxazine and isoalloxazine.

proton transfer. The second mechanism proceeds without a biradical inter-mediate (Eq. 10-21).

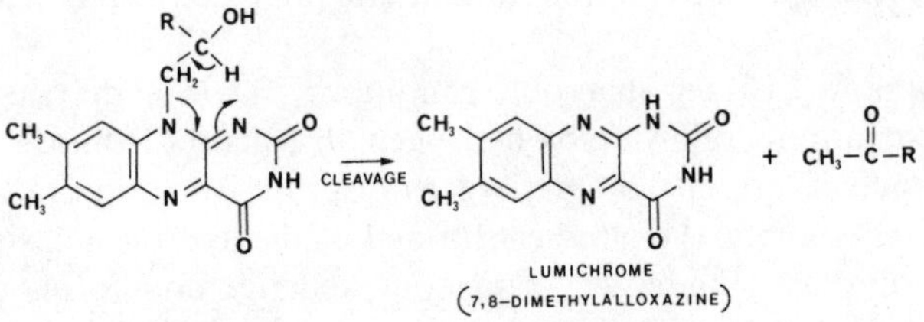

LUMICHROME
(7,8-DIMETHYLALLOXAZINE)

Eq. 10-21

Photoaddition. Photoaddition follows the general equation:

$$Fl_{ox} + ROH \xrightarrow{h\nu} RO-Fl_{red}H$$

The addition product is hydroxy or alkoxy dihydroflavins. While photore-duction and dealkylation occur predominantly through the triplet state $^3Fl^*_{ox}$, photoaddition proceeds via the excited singlet $^1Fl^*_{ox}$. Intramolecular addition involves a nucleophilic attack of the C2' hydroxyl group at C9. A 9 α,5-proton shift leads to the formation of 9-alkoxyl-flavin (Eq. 10-22).

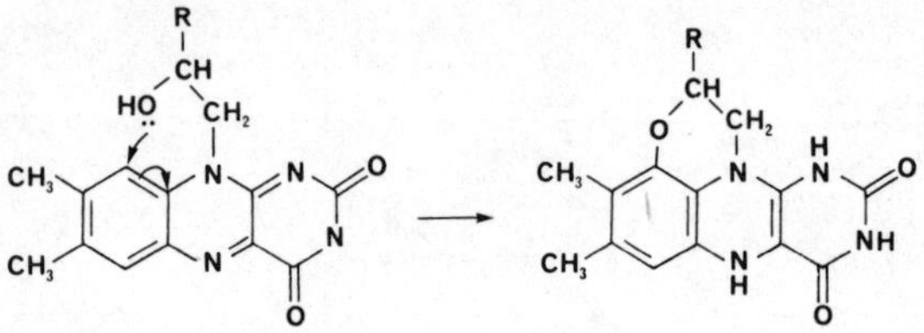

Eq. 10-22

Flavin-sensitized Reactions

Photosensitized Isomerization. Photochemical reactions sensitized by flavins generally proceed with a triplet-triplet energy transfer (Eq. 10-23). Flavins can photosensitize the *cis-trans* isomerization of olefinic com-pounds. Retinol, cinnamic acid, and stilbene 4-carboxylic acid are specific examples.

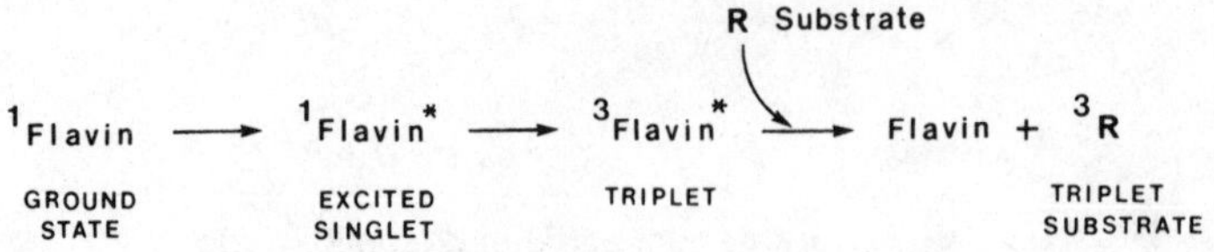

R Substrate

$$^1Flavin \longrightarrow {}^1Flavin^* \longrightarrow {}^3Flavin^* \longrightarrow Flavin + {}^3R$$

Eq. 10-23

GROUND STATE   EXCITED SINGLET   TRIPLET   TRIPLET SUBSTRATE

Photosensitized Oxidation. Flavin-sensitized oxidation usually yields complex products. In the case of aliphatic amino compounds, the products are carbon dioxide and an aldehyde (Eq. 10-24).

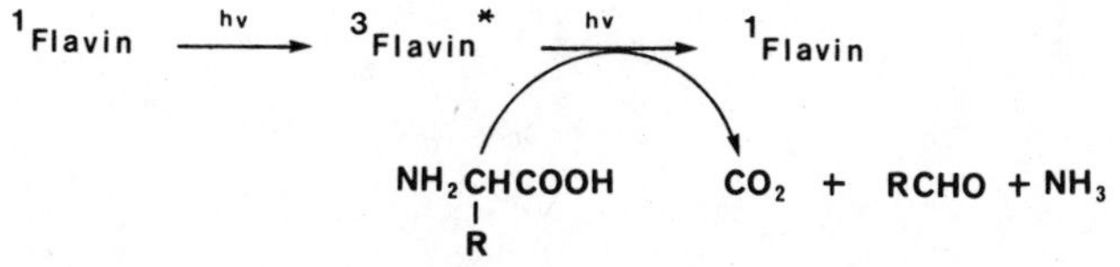

$$^1\text{Flavin} \xrightarrow{h\nu} \,^3\text{Flavin}^* \xrightarrow{h\nu} \,^1\text{Flavin}$$

$$\begin{array}{ccc} & \text{NH}_2\text{CHCOOH} & \text{CO}_2 + \text{RCHO} + \text{NH}_3 \\ & | & \\ & R & \end{array}$$

Eq. 10-24

VITAMIN B₆

Vitamin B_6 consists of multiple forms. Pyridoxol (pyridoxine), which is 4,5-di-(hydroxymethyl)-3-hydroxy-2-methylpyridine, was first isolated, followed by the discoveries of pyridoxal and pyridoxamine (Fig. 10.8).

In the liver, pyridoxol is converted to the biochemically active cofactor pyridoxal phosphate (Eq. 10-25).

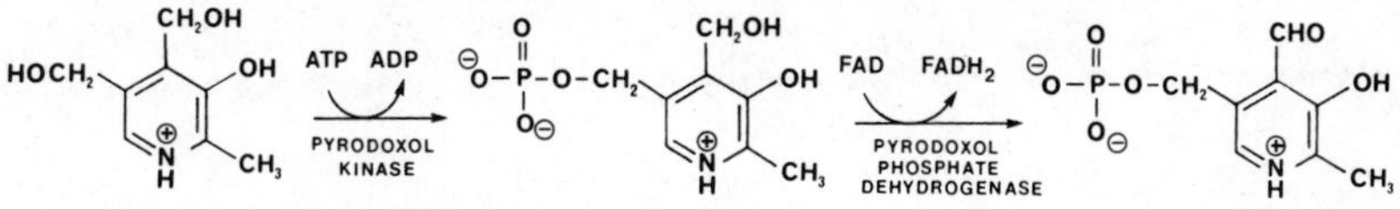

Eq. 10-25

Biochemical Mechanism

Pyridoxal phosphate is the coenzyme that catalyzes enzymatic reactions such as (1) recemization, (2) transmination, (3) elimination, (4) decarboxylation, and (5) reverse condensation (10).

All the reactions require Schiff-base formation between the aldehyde of the pyridoxal and the amino group of the substrate. The initial step, dissociation of the α-H, yields the carbanion intermediate. Reprotonation at the α-carbon results in racemization of the amino acid. An aldimine-ketimine conversion results in transamination. Electron shift toward the β-carbon

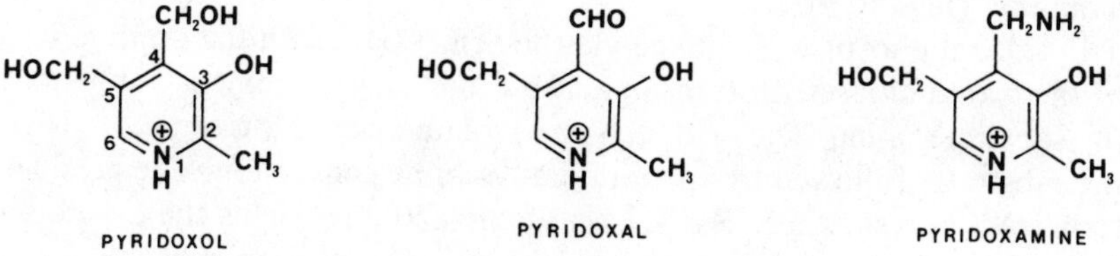

Fig. 10.8. Multiple forms of vitamin B_6.

causes β-elimination of the substituent group of the amino acid. In the case that the carbanion intermediate is formed by the shifting of the bonding between the α- and the carboxyl carbon, decarboxylation occurs (Eq. 10-26).

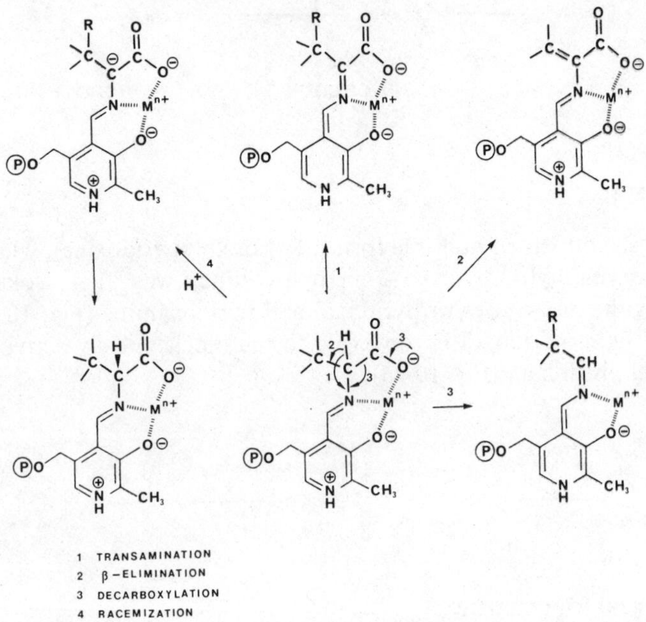

Eq. 10-26

1 TRANSAMINATION
2 β-ELIMINATION
3 DECARBOXYLATION
4 RACEMIZATION

In all cases, hydrolysis as the aldimine yields an amine and pyridoxal, while hydrolysis as the ketimine gives keto acid and pyridoxamine. The metal ions help stabilize the conjugation system between the Schiff base and the pyridine ring, and accelerate the rate of reaction by acting as a general acid catalyst.

To illustrate the biochemical mechanism of pyridoxal phosphate, the reaction of transamination serves as a typical example. In the degradation of amino acid, the α-amino group of many amino acids is transferred to α-ketoglutarate to form glutamate, which is oxidatively deaminated to yield ammonia (Fig. 10.9).

Pyridoxal phosphate is linked via the Schiff base with the ϵ-amino lysine of the transaminases. In transamination, the enzyme-imine is converted to the substrate-imine. The next step involves an abstraction of the α-H from the substrate, followed by an aldimine-ketimine conversion. The product is stabilized by resonance. Hydrolysis of the ketimine yields the α-keto acid and the pyridoxamine phosphate. The pyridoxamine phosphate formed then reacts with α-ketoglutarate (an α-keto acid) to yield glutamate (an

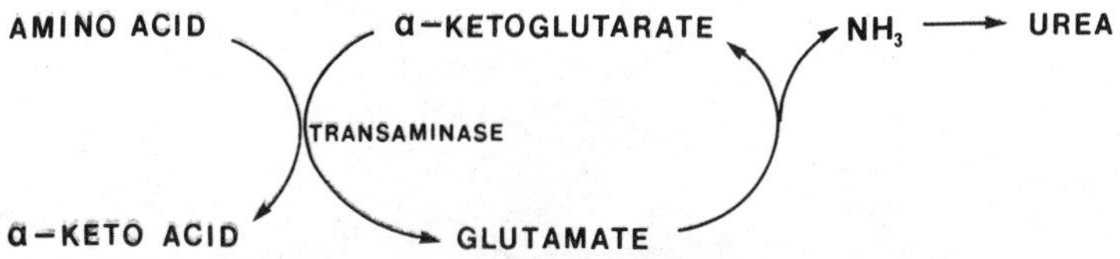

Fig. 10.9. The oxidative deamination cycle.

α-amino acid) in a reverse mechanism to complete the reaction cycle (Eq. 10-27).

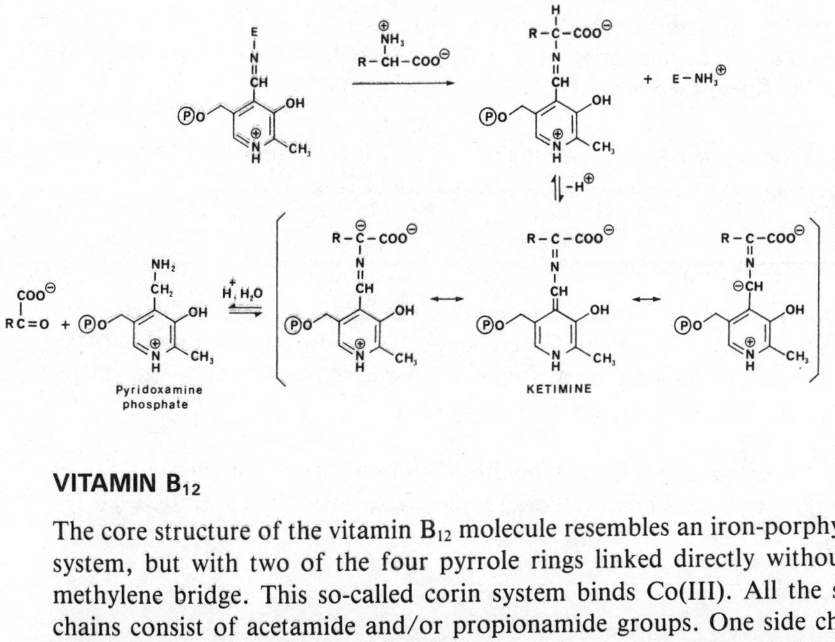

Eq. 10-27

VITAMIN B$_{12}$

The core structure of the vitamin B$_{12}$ molecule resembles an iron-porphyrin system, but with two of the four pyrrole rings linked directly without a methylene bridge. This so-called corin system binds Co(III). All the side chains consist of acetamide and/or propionamide groups. One side chain consists of an amide-phosphate-ribose bonded to a dimethylbenzimidazole group which coordinates with the Co(III). The corrin ring is numbered clockwise, starting from ring A involved in the direct linking of the pyrrole ring (Fig. 10.10A).

The top axial ligand in the vitamin B$_{12}$ isolated initially from natural sources is cyanide ion. However, the cyanide was only picked up during isolation and purification steps in the early investigations. The compound has since then also been known as cyanocobalamin (Fig. 10.10B). Cyanocobalamin is the form used in dietary supplements. Cyanocobalamin is not biochemically active, and the ingested vitamin must be enzymatically modi-

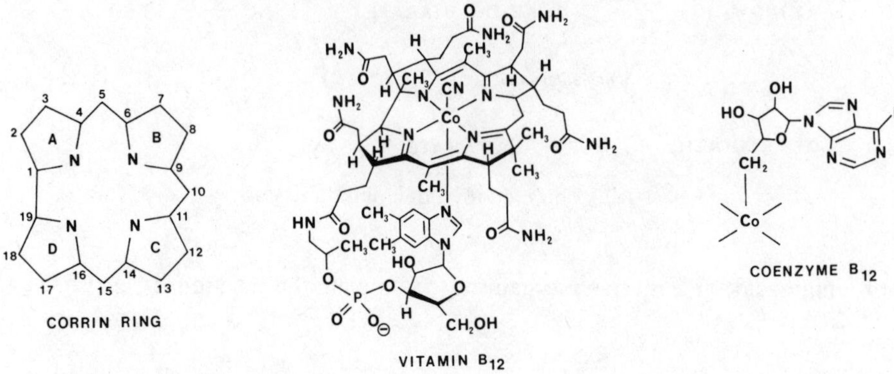

Fig. 10.10. Structure of (A) corrin ring, (B) vitamin B_{12}, and (C) coenzyme B_{12}. (Reprinted with permission from *Acc. Chem. Res.* **9**, 114. Copyright 1976 by American Chemical Society.)

fied to coenzyme B_{12}, by replacing the cyano with a 5′-deoxyadenosyl group (hence known as 5′-deoxyadenosylcobalamin) (Fig. 10.10C).

Biochemical Mechanism

Coenzyme B_{12} is a cofactor for various enzymatic reactions, the reaction mechanism involving the 1,2-interchange of a hydrogen atom and another substituent. In the case of methylmalonyl-coenzyme A mutase rearrangement, the following steps are depicted (*11*):

1. Enzyme-induced homolytic dissociation of the cobalt-carbon bond to generate cobalamin (II) and 5′-deoxyadenosyl radical (Eq. 10-28.1).

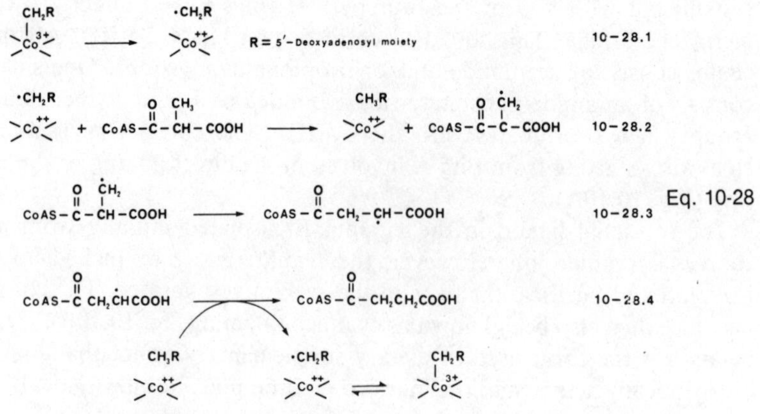

2. Abstraction by the 5′-deoxyadenosyl radical of a hydrogen atom from the substrate to yield a substrate radical and 5′-deoxyadenosine (Eq. 10-28.2).
3. Rearrangement of the substrate radical with 1,2-migration of the thioester group (Eq. 10-28.3).
4. Transfer of a hydrogen atom from 5′-deoxyadenosine to the radical (Eq. 10-28.4).

The homolytic fission of the $Co-C$ bond in Eq. 10-27.2 is generally well established. The normal cobalt-alkyl bond is stable with a bond dissociation energy of 18–25 kcal/mol. The binding of substrate to the enzyme induces a conformational change of the apoprotein that, in turn, distorts the cobalt coordination sphere of the coenzyme and labilizes the $Co-C$ bond. The equilibrium in Eq. 10-28 is displaced by a factor of $>10^6$ in favor of bond breaking.

Many enzyme systems other than the methylmalonyl CoA mutase are known to require coenzyme B_{12}. These systems include α-methyleneglutarate mutase, glutamate mutase, diol dehydrase, glycerol dehydrase, L-β-lysine mutase, D-α-lysine mutase, and ornithine mutase. The rearrangement mechanisms of these enzymatic reactions remain largely unclear.

Photolytic Degradation

Light irradiation causes the homolytic fission of the cobalt-carbon bond. Under anaerobic conditions, the 5′-deoxyadenosyl radical cyclizes rapidly. Under aerobic conditions, the products are the hydroxocobalamin and 5′-aldehyde of adenosine (Eq. 10-29) (*12*).

Eq. 10-29

Oxidation

Oxidation of vitamin B_{12} under mild alkaline conditions yields dehydrovitamin B_{12}. The acetamide side chain at C-7 in the corrin B ring cyclizes to form a γ-lactam (cyclic amide). In the presence of an oxidizing agent (e.g., iodine), a γ-lactone (cyclic ester) is formed. Both are biologically inactive (Fig. 10.11).

BIOTIN

The chemical structure of biotin consists of fused rings of ureido and tetrahydrothiophene, with an aliphatic side chain covalently bound to the ϵ-amino group of a lysine residue of the enzyme (Fig. 10.12). Biotin is a cofactor of a number of carboxylases that catalyze the carboxylation of some important metabolic acids (Eq. 10-30).

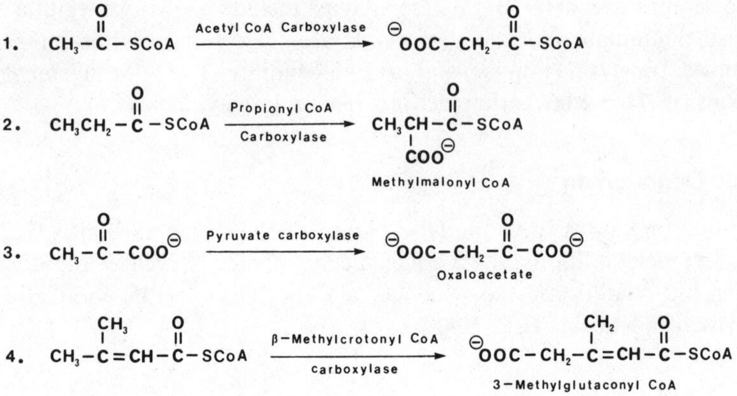

Eq. 10-30

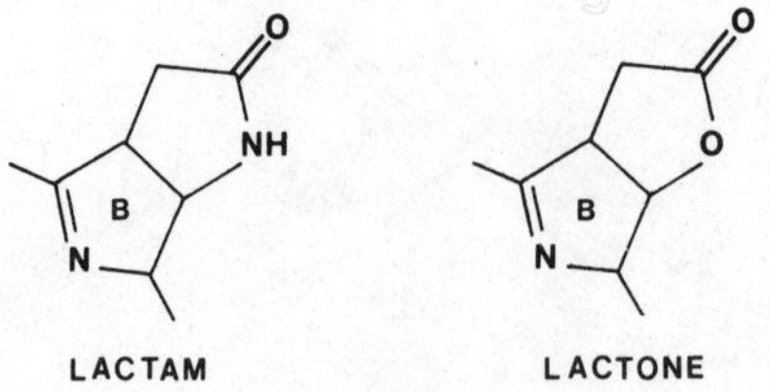

Fig. 10.11 Chemical structures of lactam and lactone.

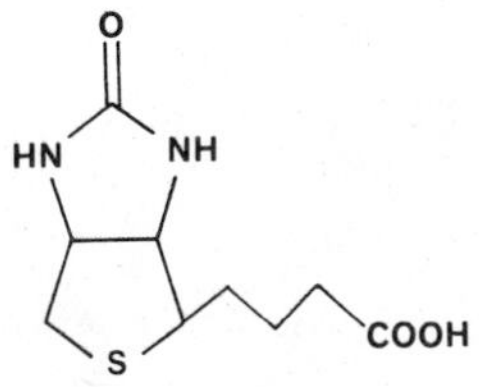

Fig. 10.12. Structure of biotin.

Biochemical Mechanism

The overall reaction can be represented by two half-reactions which occur at different subunits of the enzyme (Eq. 10-31).

CARBOXYLATION

$$\text{E-Biotin} + \text{ATP} + HCO_3^{\ominus} \xrightarrow{Mg^{++} \text{ or } M^{++}} \text{E-Biotin-}CO_2^{\ominus} + \text{ADP} + P_i$$

Eq. 10-31

TRANSCARBOXYLATION

$$\text{E-Biotin-}CO_2^{\ominus} + \text{Acceptor} \longrightarrow \text{E-Biotin} + \text{Accepter-}CO_2^{\ominus}$$

The carboxylation half-reaction involves the formation of *N*-carboxybiotin via *O*-phosphobiotin. The free energy generated by the cleavage of ATP is utilized to form the *O*-phosphobiotin intermediate (Eq. 10-32) (*13*).

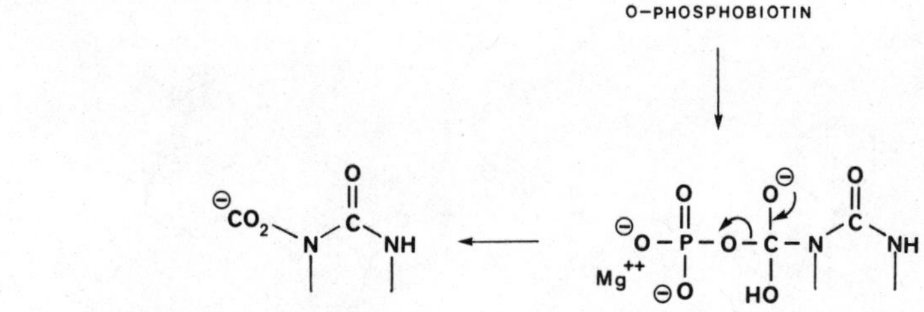

Eq. 10-32

The transcarboxylation reaction proceeds via a stepwise mechanism, in which the proton removal from the acceptor forming the carbanion and the carboxyl addition occurs in separate steps (Eq. 10-33) (*14*).

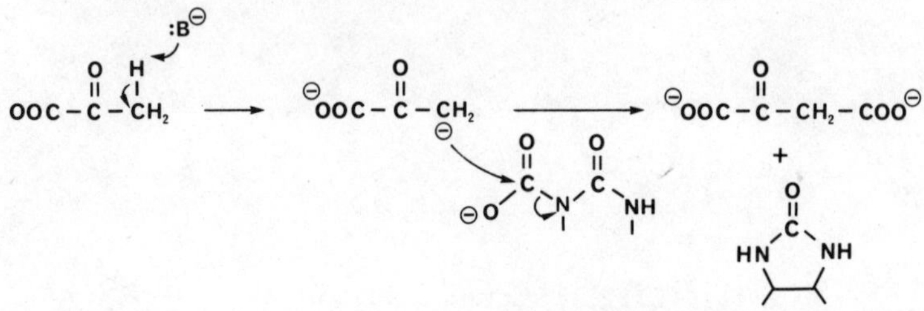

Eq. 10-33

Biotin-deficiency cases are uncommon. Most clinical studies relate to in-born deficiency in biotin-dependent enzymes that are involved in the key steps in tricarboxylic acid cycle and in the metabolism of amino acids. The immediate result is acidosis accompanied with various neurological symptoms such as hypotonia, delayed motor development, muscle atropy, tongue fibrillation, seizure, and mental retardation.

NIACIN

Niacin (nicotinic acid) and nicotinamide are the commercial form of the vitamin. The active form is nicotinamide adenine dinucleotide (NAD^+) (Fig. 10.13), which is the coenzyme involved in dehydrogenase-catalyzed reactions.

Biochemical Mechanism

Dehydrogenase-catalyzed reactions typically involve hydride transfer between the C6 of the coenzyme and the carbonyl group of the substrate (Eq. 10-34). The reaction is stereospecific and only one (R or S) isomer is formed. The enzyme positions the substrate and the coenzyme so that the asymmetric reaction ensures the correct configuration of the product.

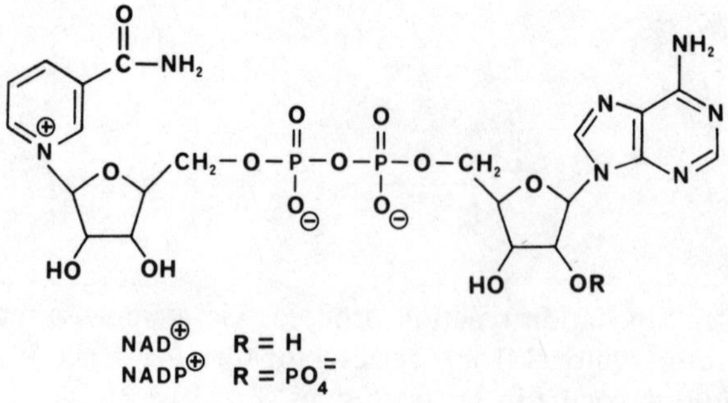

$$NAD^{\oplus} \quad R = H$$
$$NADP^{\oplus} \quad R = PO_4^{=}$$

Fig. 10.13. Structure of nicotinamide adenine dinucleotide.

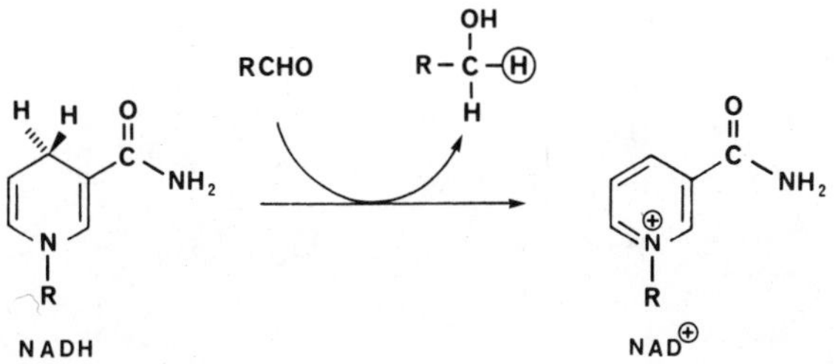

Eq. 10-34

VITAMIN C

Fresh fruits and vegetables have been known for centuries to prevent and cure scurvy. The pure crystalline substance finally isolated is L-ascorbic acid (AH_2). L-Ascorbic acid is a lactone with an enediol group (cyclic ester of a hydroxy carboxylic acid). The oxidized product of ascorbic acid, dehydroascorbic acid (A), is a 2,3-diketal which exists predominantly as a bicyclic hydrated form in solution (Eq. 10-35). Both AH_2 and A possess biological activity. The common substitute for L-ascorbic acid in most food uses is D-isoascorbic acid, which possesses essentially no biological activity.

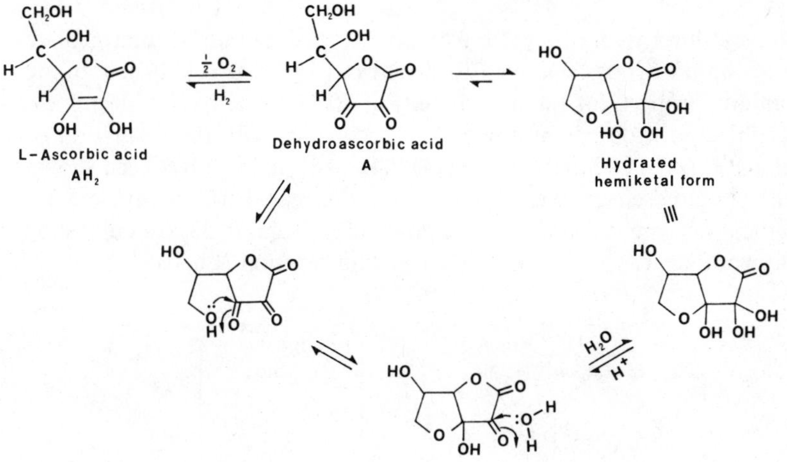

Eq. 10-35

Biochemical Mechanism

Acidity. L-Ascorbic acid is dibasic, with the first $pK_a = 4.0$ and second $pK_a = 11.3$. The acidity of ascorbic acid is mainly due to the monoanion formed from the dissociation of the proton at the 3-hydroxyl. In contrast, loss of proton from the 2-hydroxyl yields a monoanion without resonance

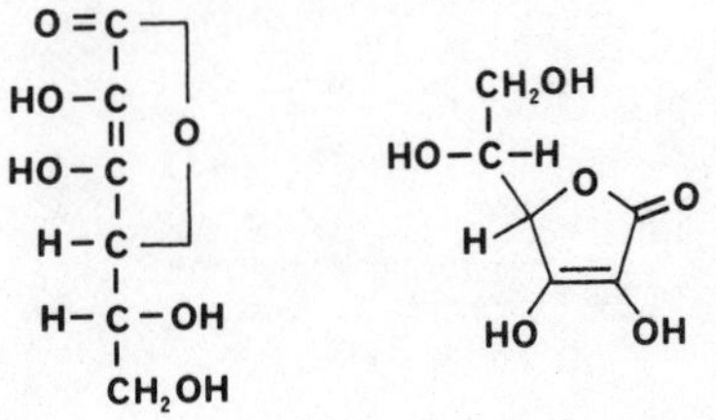

Fig. 10.14. Structure of D-isoascorbic acid.

stabilization (Eq. 10-36). The second dissociation is less favorable, as suggested by the higher pK value. Here, a proton is pulled from the 2-hydroxyl that is hydrogen bonded, and from a negatively charged molecule to form a completely dissociated dianion.

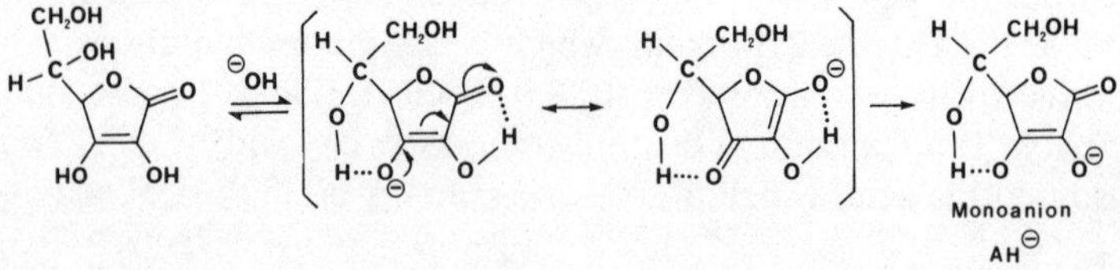

Eq. 10-36

The Radical Intermediate. The formation of a free-radical intermediate in the oxidation of ascorbic acid to dehydroascorbic acid is probably the most unique property of the ascorbate (Eq. 10-37). The oxidation process is reversible and proceeds in a two-step mechanism with the formation of radical anion $(A^-\cdot)$ as intermediate (15). The radical $(A^-\cdot)$ has been shown with the unpaired electron spread over the conjugated tricarbonyl system. The radical anion is very stable and nonreactive, and it decays chiefly by reacting with itself, terminating the free-chain reaction (Eq. 10-37).

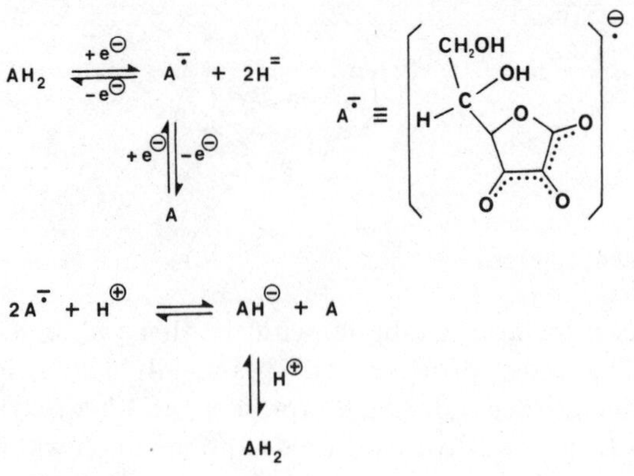

Eq. 10-37

Biological Function.   Vitamin C plays a vital role in (1) collagen synthesis, (2) conversion of 3,4-dihydroxyphenylethylamine to noradrenaline, and (3) protection against free-radical damage.

The enzyme involved in the hydroxylation of proline to hydroxyproline in collagen synthesis, protocollagen proline hydroxylase, requires O_2, Fe^{3+}, α-ketoglutarate, and ascorbate (Eq. 10-38). An abnormal collagen precursor is formed in the absence of ascorbic acid. Ascorbic acid maintains the activity of enzyme hydrolase by the iron atom in the reduced ferrous state. Collagen insufficiently hydroxylated has a lower melting point. This abnormal development of the collagen causes the skin lesions and fragile vessels evident in scurvy.

Eq. 10-38

The enzyme dopamine hydroxylase involved in the conversion of 3,4-dihydroxyphenylethylamine to noradrenaline also requires ascorbic acid and oxygen. (Eq. 10-39.)

Eq. 10-39

Although the chemical mechanisms are not yet clearly elucidated, ascorbic acid is suggested to be involved in the metabolism of iron and calcium, and the hydroxylation reactions in the metabolism of a number of proteins and steroids. All these indicate the importance of the redox balance ($AH_2 \rightleftharpoons A$) of ascorbic acid.

The protective role of ascorbic acid in biological systems via free-radical reaction has been extensively studied. The general reaction in the oxidation of ascorbate by an organic compound is represented by Eq. 10-40. In a biological system, vitamins E and C act synergistically. Vitamin E, being lipophilic, is considered probably to be the primary antioxidant, especially in lipid peroxidation in cell membranes. Vitamin C reacts with a vitamin E radical to regenerate vitamin E, and the resulting ascorbic acid radical can be reduced back to vitamin C by NADH (*16*) (Fig. 10.15).

$$R \cdot + AH^- \longrightarrow RH + A^{\cdot -}$$ Eq. 10-40

Loss of Vitamin C

Vitamin C is one of the least stable vitamins. Retention of vitamin C is most affected by processing, handling, and storage. Enzymes found in fruits and vegetables that oxidize vitamin C are ascorbic acid oxidase, cytochrome oxidase, and peroxidase. However, in food processing, losses of vitamin C due to enzymatic destruction are minimal. Losses are mainly due to nonenzymatic reactions—oxidative and nonoxidative. Nonoxidative reaction is comparatively slow and can be accelerated by lowering pH.

Nonoxidative Degradation.   The nonoxidative degradation of ascorbic acid to 2-furaldehyde and carbon dioxide is acid catalyzed. The initial step involves cleavage of the lactone ring to form 2,3-enediol, which then undergoes dehydration, followed by rearrangement to form a 2,3-diketo acid (Eq. 10-41). Decarboxylation of the diketo acid yields the 3-deoxy-L-pentosulose (corresponding to the 3-deoxyglycosulose in the Maillard reaction) (*17*).

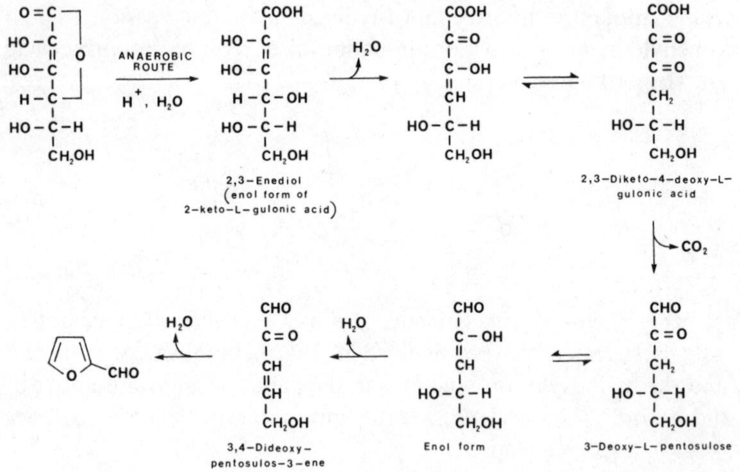

Eq. 10-41

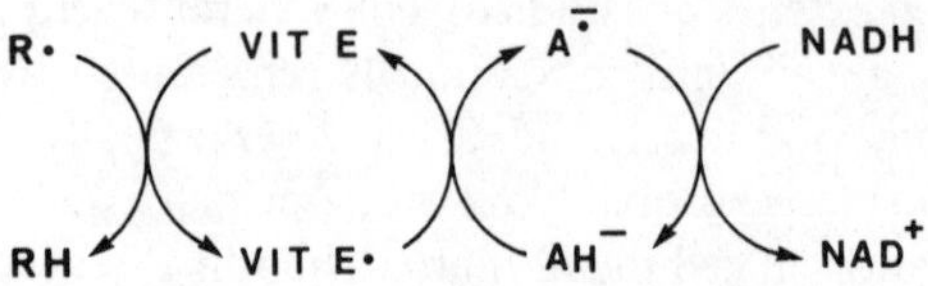

Fig. 10.15.   Interaction between vitamin E radical and ascorbate.

Oxidative Degradation. Oxidative browning in an acid medium proceeds by way of dehydroascorbic acid. Cleavage of the lactone ring yields the 2,3-diketogulonic acid, which decarboxylates to pentosulose (L-xylosone) (Eq. 10-42). 2,3-Enolization and dehydration of the xylosone yields a tricarbonyl compound, 3-keto-4-deoxypentosulose. 3,4-Enolization of 2,3-diketogulonic acid, followed by dehydration and decarboxylation, also gives the same product. It is suggested that the route via 2,3-diketogulonic acid constitutes the major pathway (18).

Eq. 10-42

Strecker Degradation. Dehydroascorbic acid has been shown to be involved in the discoloration of foods, especially oranges, lemon juice, and dehydrated foods, through the browning reaction with amino acids. The initial Strecker degradation of an α-amino acid in the presence of dehydroascorbic acid produces scorbamic acid, which undergoes oxidation to yield dehydro-L-scorbamic acid. Reaction of the two acids yields the condensation product, 2,2'-nitrilodi-2(2')-deoxy-L-ascorbic acid ammonium salt, which is postulated to be the intermediate compound for further polymerization (Eq. 10-43) (19).

Reaction with Metal Ions. Ascorbic acid can be oxidized by metal ions such as Fe^{3+} and Cu^{2+} in a two-sequential one-electron transfer mechanism. The formation of an ascorbate-M^{n+} complex allows a one-electron transfer to yield the ascorbic radical anion-$M^{(n-1)+}$ complex, which dissociates readily. The resulting radical anion then undergoes a second one-electron trans-

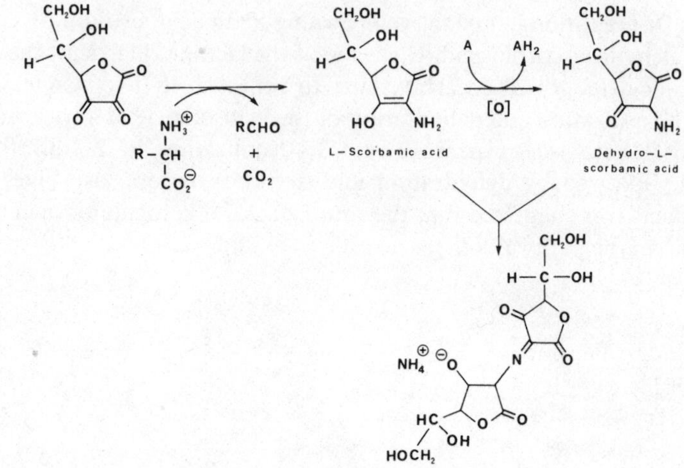

Eq. 10-43

fer by complexing with another metal ion (M^{n+}) to give the final product, dehydroascorbic acid (20) (Eq. 10-44).

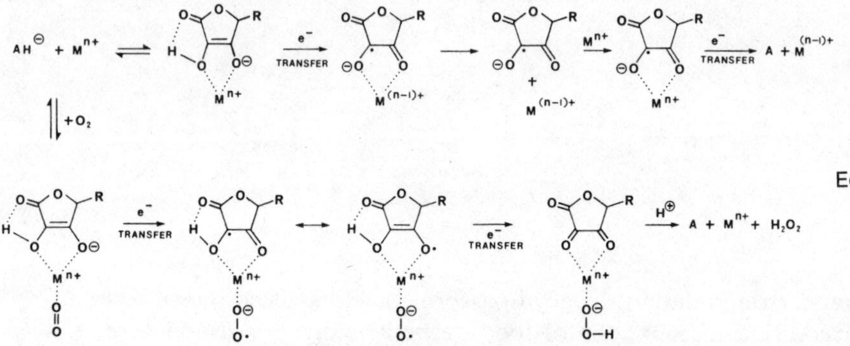

Eq. 10-44

Metal ions can also catalyze the autoxidation of ascorbic acid. In the reaction, molecular oxygen is reduced by a two-electron transfer to peroxide, while the oxidative state of the metal remains unchanged. The first step involves the formation of an ascorbate-metal-dioxygen complex intermediate which allows the transfer of one electron to yield resonance forms in which the M^{n+} is coordinated to the ascorbate radical and superoxide anion. A second electron transfer yields a complex of dehydroascorbic acid-M^{n+}-hydroperoxide that rapidly dissociates to dehydroascorbic acid and hydrogen peroxide. It is also possible that a one-step two-electron transfer may occur to give the same products.

Uses of Vitamin C

1. Prevention of browning in fruit and vegetables. In the presence of ascorbic acid, the o-quinone-type compounds are reduced back to the o-phenolic forms. At the exhaustion of the ascorbic acid in the system, the o-quinone compounds accumulate and polymerize to browning products (refer to Chapter 5, polyphenol oxidase).

2. Inhibition of oxidation in beer, wine, vegetable oil, milk and dairy products.

3. Stabilization of meat color; color fixing in curing of meat. Ascorbic acid acts as a reductant, similar to sulfhydryl compounds and NADH-flavin found in meat, to generate nitrous oxide and nitrosylmyoglobin (refer to Chapter 4, Fig. 4.40).

4. Improvement of dough. During dough mixing, the flour protein gluten undergoes sulfhydryl-disulfide exchange with low-molecular sulfhydryl peptides. The exchange is suggested as one reason for the decrease in dough stability. Dehydroascorbic acid acts as an improver by competing with the gluten in the oxidation of the thiol (Fig. 10.16) (21).

VITAMIN D

The addition of vitamin D to milk in the United States and the consequent eradication of rickets have been attributed to the antirachitic property of the vitamin.

The multiple chemical structures of the D vitamins need clarification. Practically, D vitamins of natural sources are the vitamin D_2 (ergocalciferol) from plants, and vitamin D_3 (cholecalciferol) present in animal tissues. Other vitamin D's with antirachitic activity are also known.

The Chemical Structures

Vitamin D_2 is formed by irradiation of the provitamin D_2 (ergosterol) (Eq. 10-45.1) and similarily, D_3 from provitamin D_3 (7-dehydrocholesterol) (Eq.

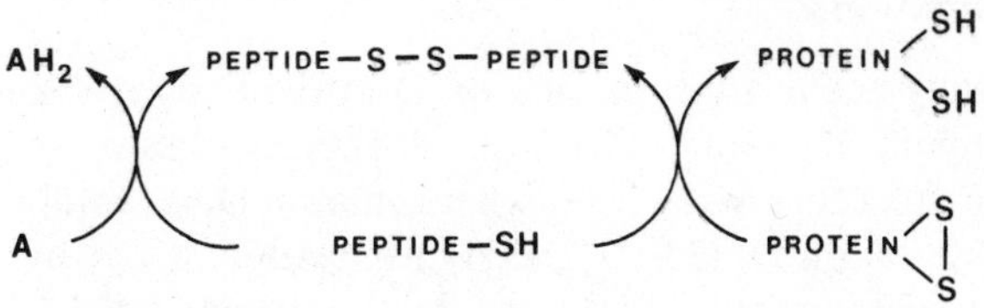

Fig. 10.16. Competition for thiol oxidation between vitamin C and flour protein.

10-45.2). The only difference between the vitamin D_2 and D_3 is exclusively in the side chain attached at C17. Vitamin D_2 has an unsaturated side chain (double bond) between C22 and C23, with one extra methyl group at C24.

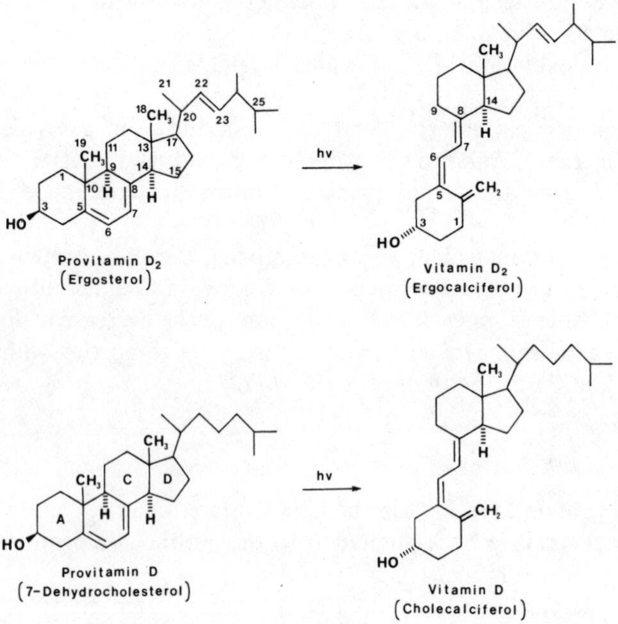

Provitamin D_2
(Ergosterol)

Vitamin D_2
(Ergocalciferol)

10-45.1

Provitamin D
(7-Dehydrocholesterol)

Vitamin D
(Cholecalciferol)

10-45.2

Eq. 10-45

The photochemical reaction involves electrocyclic ring opening of the cyclohexadiene (B ring) of provitamin D. The product (previtamin D) undergoes [1,3] sigmatropic reaction to form vitamin D (Eq. 10-46).

The same reaction is used to convert plant ergosterol into ergocalciferol, the vitamin D_2 that is added to milk: (ergosterol $\xrightarrow{h\nu}$ pre-ergocalciferol $\longrightarrow$ ergocalciferol.

Biological Functions

The biologically active form of vitamin D in the body is the steroid 1,25-dihydroxyvitamin D_3 (1,25 $(OH)_2D_3$). The biosynthesis of 1,25$(OH)_2D_3$ from vitamin D_3 occurs in the liver, where vitamin D_3 is rapidly metabolized to 25-hydroxyvitamin D_3 (25-OHD). Hydroxylation at C-1 of 25-OHD occurs only in the kidney and is catalyzed by the enzyme renal 1-hydroxylase. The synthesis of 1,25$(OH)_2D_3$ is controlled by the calcium demand of the system and the parathyroid hormone PTH (Fig. 10.17) (22).

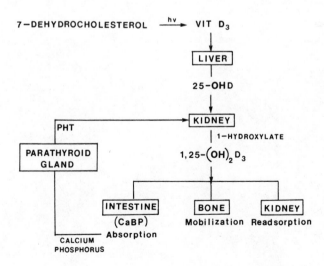

Eq. 10-46

The physiological functions of vitamin D are (1) absorption of calcium and phosphorus in the intestine, possibly involving vitamin D-dependent calcium-binding protein (CaBP), (2) mobilization of calcium and phosphorous from bone, and (3) renal readsorption of calcium and phosphorus.

Vitamin D differs from other common steroids in that the B ring is open, resulting in the conformational flexibility of the A ring. To illustrate this, the structure of $1,25(OH)_2D_3$ is presented in the chain form (Eq. 10-47) (23).

Fig. 10.17. Metabolic pathway of vitamin D_3.

Notice the interconversion of the equatorial and axial positions of the two hydroxyl groups in the A ring. The two chair conformers are in equilibrium, and the interconversion is many thousands of times per second. It is postulated that the 25-OH group functions to steer the molecule of $1,25(OH)_2D_3$ in a position for the receptor protein in the target cell to interact with the A-ring. The resulting $1,25(OH)_2D_3$-receptor complex "freezes" the A-ring into a fixed position.

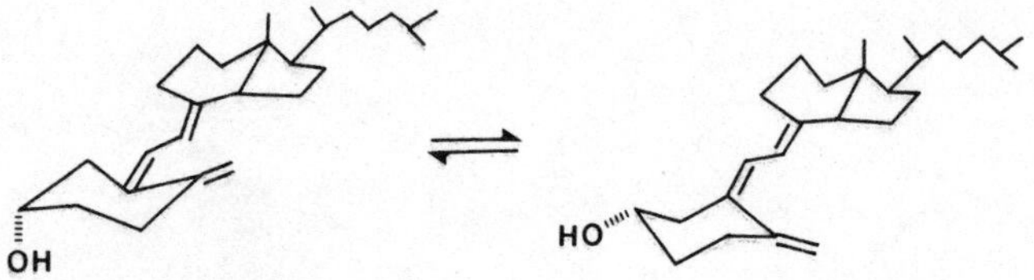

Eq. 10-47

Side Reactions of Irradiation

Two products from side reactions of irradiation of ergosterol are known: tachysterol and lumisterol. Both reactions proceed from a reversible photochemical side reaction of pre-ergocaliferol. The most effective wavelength for vitamin D production is 280 nm. Irradiation at higher wavelengths favors the formation of lumisterol and at lower wavelengths increases the yield of tachysterol. Lumisterol is also formed irreversibly from tachysterol (Eq. 10-48) (24).

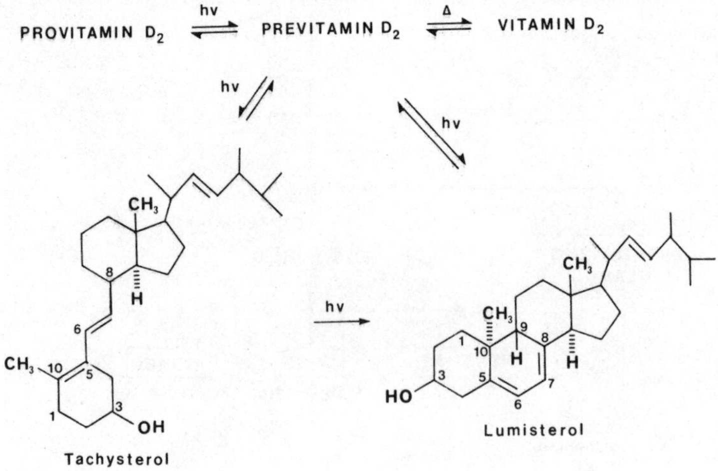

Eq. 10-48

Tachysterol

Lumisterol

Overexposure to irradiation produces suprasterol (which is biologically inactive) from ring closure of the conjugated triene system (Eq. 10-49). A number of other compounds are also formed by ring closure between other carbons, followed by various fragmentation or dimerization.

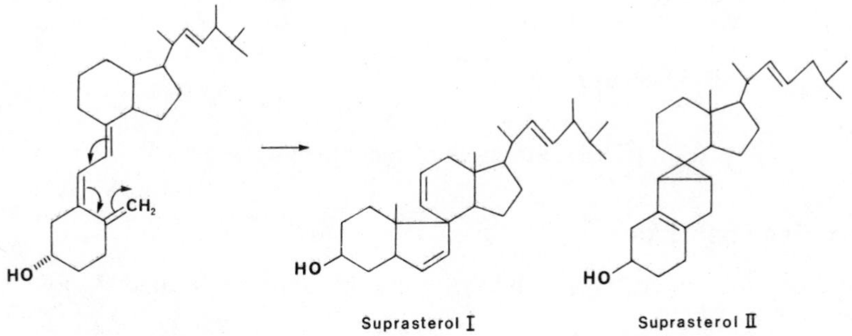

Eq. 10-49

Suprasterol I   Suprasterol II

Thermal Reaction

When vitamin D_2 is heated at 180–190°C, pyro-ergocalciferols are formed. (Fig. 10.18). Both of these, like lumisterol, are isomers (*cis* 9,10) of ergocalciferol and are biologically inactive.

VITAMIN E

Vitamin E consists of a chroman ring (chroman-6-ol) with a saturated isoprenoid C_{16} side chain. Depending on the position of substitution of methyl groups in the chroman ring, the compounds are referred to as α, β, γ, δ (Fig. 10.19).

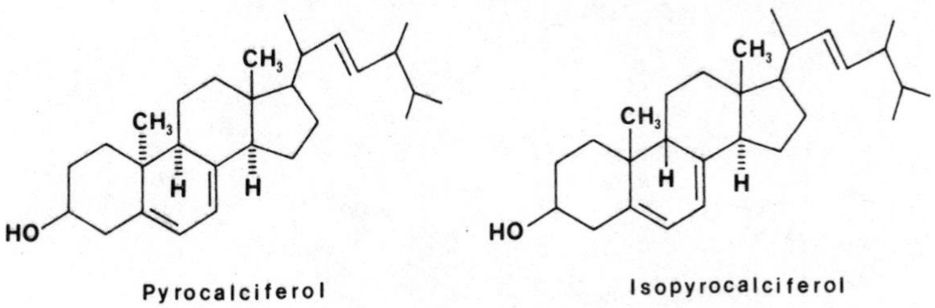

Pyrocalciferol   Isopyrocalciferol

Fig. 10.18. Structure of pyro-ergocalciferols.

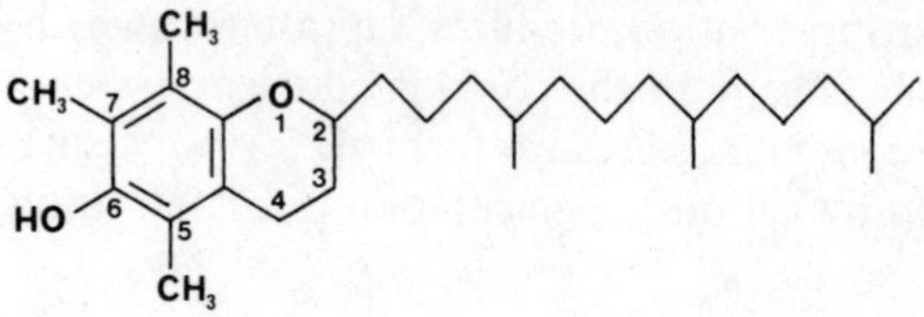

Substitution	Tocopherol
5,7,8 – Trimethyl–	α–
5,8 – Dimethyl–	β–
7,8 – Dimethyl–	γ–
8 – Methyl–	δ–

Fig. 10.19. Structure of tocopherols.

Thermal Decomposition

Pyrolysis of α-tocopherol yields the hydroquinone and unsaturated C_{19} hydrocarbon (Eq. 10-50).

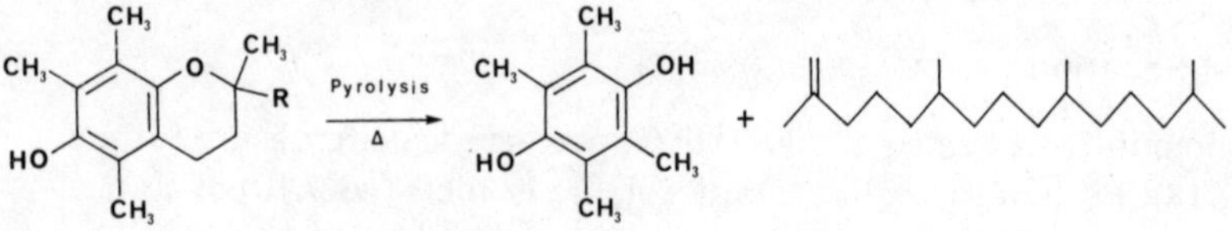

Eq. 10-50

$$R = CH_2-(CH_2-CH_2-CH-CH_2)_3 H$$

Chemical Oxidation

In many chemical oxidative degradation studies of α-tocopherol, the main product formed is α-tocopherol quinone (Eq. 10-51).

Tocopherol quinone

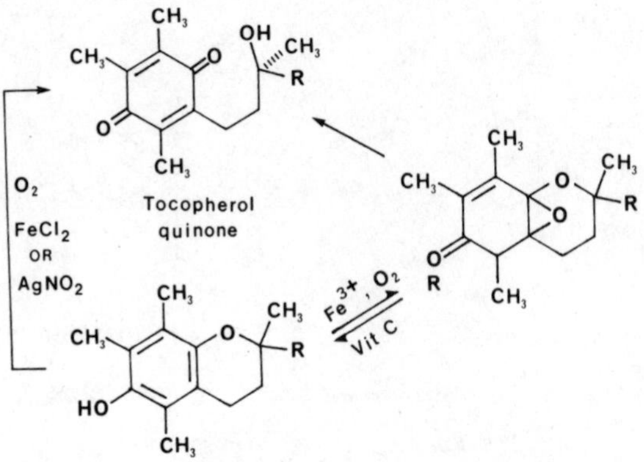

Eq. 10-51

Free-Radical Reactions

α-Tocopherol undergoes free-radical oxidation to form α-tocopherol quinone via an intermediate, α-tocopherol quinone methine (25). The first two steps involve abstractions of two hydrogen atoms, forming an unstable quinone methine, which reacts with a proton to produce an oxonium ion (Eq. 10-52). The oxonium ion then rearranges to a carbonium ion, which then adds a water molecule to form the quinone. Alternatively, it has also been suggested that the tocopherol radical reacts with the second peroxy radical to form the 4-(alkylperoxy)cyclohexadienone, which upon acid hydrolysis yields the tocopherol quinone. In both cases, the tocopherol molecule consumes two peroxy radicals to form a nonradical product. Schematically, the reaction sequence can be represented by Eq. 10-53.

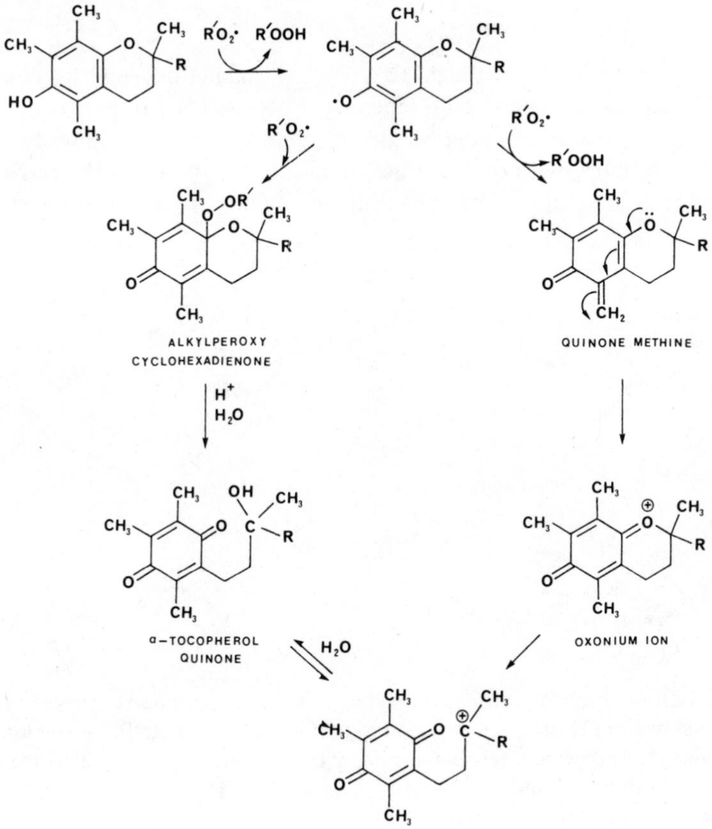

ALKYLPEROXY CYCLOHEXADIENONE

QUINONE METHINE

Eq. 10-52

α-TOCOPHEROL QUINONE

OXONIUM ION

$$R\cdot \ + \ O_2 \ \longrightarrow \ ROO\cdot$$

$$ROO\cdot \ + \ E\text{--}OH \ \longrightarrow \ ROOH \ + \ E\text{--}O\cdot \qquad\qquad \text{Eq. 10-53}$$

$$E\text{--}O\cdot \ + \ ROO\cdot \ \longrightarrow \ \text{Non--radical products}$$

The high efficiency of tocopherols as a chain-breaking antioxidant is largely attributed to the 6-hydroxychroman system. The alkyl substitutions at both ortho and meta positions help to reduce the energy of activation for the transition state, and hence increase the formation of phenoxyl radical. The para-alkoxy group is held in an orientation that the unpaired electron of the phenoxyl radical can delocalize with the oxygen p orbital (Fig. 10.20) (26). In contrast, 2,3,5,6-tetramethyl-4-methoxyphenol, although very similar in structure, shows no enhancement in the formation of phenoxyl radical since the methoxy is out of the plane of the aromatic ring.

Quenching of Singlet Oxygen

Tocopherols are also highly efficient in quenching singlet oxygen by a combination of chemical and physical quenching processes (Eq. 10-54). Among the two processes of scavenging singlet oxygen, the physical quenching process predominates, with the $^1O_2$ quenching rate 15–100 times the rate of the irreversible reaction. The physical quenching rate constant for

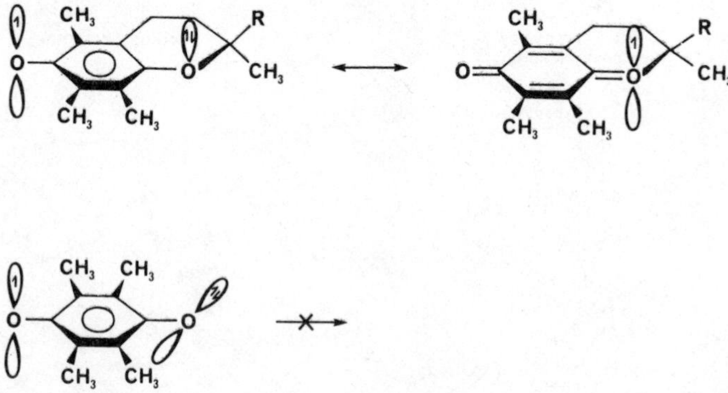

Fig. 10.20. Orientation of the para-alkoxy group in α-tocopherol. (Reprinted with permission from Burton, G. W., and Ingold, K. U. Autoxidation of biological molecules. 1. The antioxidant activity of vitamin E and related chain-breaking phenolic antioxidants in vitro. *J. Am. Chem. Soc.* **103**, 6476. Copyright 1981 American chemical society.)

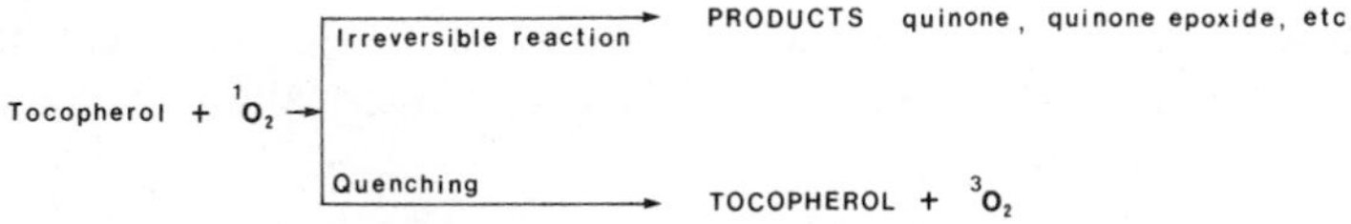

$$\text{Tocopherol} + {}^1O_2 \rightarrow \begin{cases} \text{Irreversible reaction} \longrightarrow \text{PRODUCTS} \quad \text{quinone, quinone epoxide, etc} \\ \\ \text{Quenching} \longrightarrow \text{TOCOPHEROL} + {}^3O_2 \end{cases} \qquad \text{Eq. 10-54}$$

α-tocopherol is determined to be $2.5 \times 10^8\ M^{-1}s^{-1}$ (in benzene). Alpha-tocopherol is one of the most reactive naturally occurring singlet oxygen acceptors and is the most reactive among the four tocopherols. The singlet oxygen reactivity of α, β, γ, and δ correlates with the vitamin E activity.

$$\begin{array}{cccc} \alpha & > \beta & > \gamma & > \delta \\ 100\% & 50\% & 26\% & 10\% \end{array}$$

Unlike β-carotene, the quenching of singlet oxygen by α-tocopherol does not involve the energy transfer process. Instead, the tocopherol molecule and the singlet oxygen form a charge-transfer exciplex (Eq. 10-55). The contribution of such a state perturbs the singlet state of oxygen and causes spin-orbit coupling of the singlet charge-transfer state and the triplet charge-transfer state, allowing a formally forbidden "spin-flip" to occur. Over 95% of quenching proceeds via this process ($K_D >> K_p$) (27).

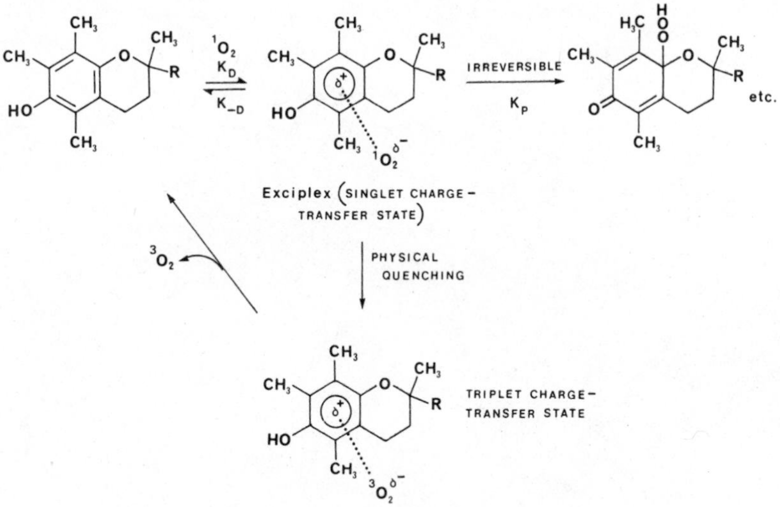

Exciplex (SINGLET CHARGE-TRANSFER STATE)

PHYSICAL QUENCHING

TRIPLET CHARGE-TRANSFER STATE

Eq. 10-55

The products of the irreversible process have been studied. Alpha-tocopherol reacts with singlet oxygen forming tocopherol quinone and quinone epoxide irreversible with the epoxide as the major product (Eq. 10-56). The intermediate, p-hydroperoxydienone, is thermally stable (28).

p—Hydroperoxydienone   α−Tocopherol quinone oxide   α−Tocopherol quinone

Eq. 10-56

SELECTED READINGS
AND REFERENCES

CHAPTER 1. LIPIDS

SELECTED READINGS

Frankel, E. N. 1980. Lipid oxidation. *Prog. Lipid Res.* **19**, 1–22.

Kanner, J., German, J. B., Kinsella, J. E. 1987. Initiation of lipid peroxidation in biological systems. CRC Crit. Rev. Food Sci. & Nutr. **25**, 317–364.

Porter, N. A. 1986. Mechanisms for the autoxidation of polyunsaturated lipids. *Acc. Chem. Res.* **19**, 262–268.

Tilcock, C. P. S. 1986. Lipid polymorphism. *Chem. Phys. Lipids* **40**, 109–125.

Haumann, B. F. 1986. Getting the fat out—researchers seek substitutes for full-fat fat. *JAOCS* **63**, 278–288.

REFERENCES

1. Uri, N. 1961. Physico-chemical aspects of autoxidation. In *Autoxidation and Antioxidants,* vol. 1, ed. W. O. Lundberg, Interscience Publishers, New York and London.

2. Frankel, E. N. 1984. Lipid oxidation: mechanisms, products and biological significance. *JAOCS* **61**, 1908–1917.

3. Gunstone, F. D. 1984. Reaction of oxygen and unsaturated fatty acids. *JAOCS* **61**, 441–444.

4. Ernster, L., and Nordenbrand, K. 1982. Microsomal lipid peroxidation: Mechanism and some biomedical implications. In *Lipid Peroxides in Biology and Medicine,* ed. K. Yagi, Academic Press, New York.

5. Tappel, A. L. 1962. Heme compounds and lipoxidase as biocatalysts. In *Symposium on Foods: Lipids and Their Oxidation,* eds. H. W. Schultz, E. A. Day, and R. O. Sinnhuber, AVI, Westport, Conn.

6. Neff, W. E., Frankel, E. N., and Weisleder, D. 1981. High pressure liquid

chromatography of autoxidized lipids: II. Hydroperoxycyclic peroxides and other secondary products from methyl linolenate. *Lipids* **16**, 439–448.

7. Neff, W. E., Frankel, E. N., and Weisleder, D. 1982. Photosensitized oxidation of methyl linolenate secondary products. *Lipids* **17**, 780–790.

8. Frankel, E. N., Neff, W. E., Selke, E., and Weisleder, D. 1982. Photosensitized oxidation of methyl linoleate: Secondary and volatile thermal decomposition products. *Lipids* **17**, 11–18.

9. Porter, N. A., Lehman, L. S., Weber, B. A., and Smith, K. J. 1981. Unified mechanism for polysaturated fatty acid autoxidation. Competition of peroxy radical hydrogen atom abstraction, β-scission, and cyclization. *J. Am. Chem. Soc.* **103**, 6447–6455.

10. Figge, K. 1971. Dimeric fatty acid $[1-{}^{14}C]$ methyl esters. 1. Mechanism and products of thermal and oxidative-thermal reactions of unsaturated fatty acid esters—literature review. *Chem. Phys. Lipids* **6**, 164–182.

11. Crnjar, E. D., Witchwoot, A., and Nawar, W. W. 1981. Thermal oxidation of a series of saturated triacylglycerols. *J. Agric. Food Chem.* **29**, 39–42.

12. Nawar, W. W. 1985. Thermal and radiolytic decomposition of lipids. In *Chemical Changes in Food During Processing,* ed. T. Richardson and J. W. Finley, AVI, Westport, Conn.

13. Faucitano, A., Locatelli, P., Perotti, A., and Faucitano Martinotti, F. 1972. γ-Radiolysis of crystalline oleic acid. *J.C.S. Perkin II* **1972**, 1786–1791.

14. Sevilla, C. L., Swarts, S., and Sevilla, M. D. 1983. An ESR study of radical intermediates formed by γ-radiolysis of tripalmitin and dipalmitoyl phosphatidylethanolamine. *JAOCS* **60**, 950–957.

15. Larsson, R. 1983. Hydrogenation theory: Some aspects. *JAOCS* **60**, 275–281.

16. Beckmann, H. J. 1983. Hydrogenation practice. *JAOCS* **60**, 282–290.

17. Allen, R. R. 1978. Principles and catalysts for hydrogenation of fats and oils. *JAOCS,* **55**, 792–795.

18. Sreenivasan, B. 1978. Interesterification of fats. *JAOCS* **55**, 796–805.

19. Lutton, E. S. 1972. Lipid structures. *JAOCS* **49**, 1–9.

20. Wiedermann, L. H. 1978. Margarine and margarine oil, formulation and control. *JAOCS* **55**, 823–829.

21. Joyner, N. T. 1953. The plasticizing of edible fats. *JAOCS* **30**, 526–535.

22. Friberg, S. 1976. Emulsion stability. In *Food Emulsions,* ed. S. Friberg, Marcel Dekker, New York.

23. Overbeek, T. Th. G. 1972. *Colloid and Surface Chemistry Part 2: Lyophobic Colloids.* Massachusetts Institute of Technology, Cambridge, Mass.

24. Van Haften, J. L. 1979. Fat-based food emulsifiers. *JAOCS* **56**, 831A–835A.

25. McIntyre, R. T. 1979. Polyglycerol esters. *JAOCS* **56**, 835A–840A.

26. Krog, N. 1977. Functions of emulsifiers in food systems. *JAOCS* **54**, 124–131.

27. Van Nieuwenhuyzen, W. 1976. Lecithin production and properties. *JAOCS* **53**, 425–427.

28. Cherry, J. P., and Gray, M. S. 1981. A review of lecithin chemistry and glandless cottonseed as a potential commercial source. *JAOCS* **58**, 903–913.

29. Krog, N. 1981. Theoretical aspects of surfactants in relation to their use in breadmaking. *Cereal Chem.* **58**, 158–164.

30. Lauridsen, J. B. 1976. Food emulsifiers: Surface activity, edibility, manufacture, composition, and application. *JAOCS* **53**, 400–407.
31. Krog, N. 1975. Interaction between water and surface active lipids in food systems. In *Water Relations of Foods,* ed. R. B. Duckworth, Academic Press, New York.
32. Carlson, T. L.-G., Larsson, K., Dinh-Nguyen, N., and Krog, N. 1979. A study of the amylose-monoglyceride complex by raman spectroscopy. *Stärke* **31**, 222–224.
33. Chung, O. K., Pomeranz, Y., and Finney, K. F. 1978. Wheat flour lipids in breadmaking. *Cereal Chem.* **55**, 598–618.
34. Burton, G. W., and Ingold, K. U. 1984. β-Carotene: An unusual type of lipid antioxidant. *Science* **224**, 569–573.
35. Scott, G. 1985. Antioxidants in vitro and vivo. *Chemistry in Britain* **21**(7), 648–653.
36. Scott, G. 1965. *Atmospheric oxidation and antioxidants.* Elsevier, London and New York.
37. Peers, K. E., Coxon, D. T., and Chan, H. W.-S. 1984. Autoxidation of methyl linolenate: The effect of antioxidants on product distribution. *J. Sci. Food Agric.* **35**, 813–817.
38. Sherwin, E. R. 1976. Antioxidants for vegetable oils. *JAOCS* **53**, 430–436.

CHAPTER 2. PROTEINS

SELECTED READINGS

Feeney, R. E., and Whitaker, J. R. (eds.) 1986. *Protein Tailoring for Food and Medical Uses,* Marcel Dekker, New York.
Feeney, R. E., and Whitaker, J. R. (eds.) 1982. Modification of Proteins: Food, Nutritional, and Pharmacological Aspects. Adv. Chem. Ser. Vol. 198. American Chemical Society, Washington, D.C.
Schmidt, R. H., and Morris, H. A. 1984. Gelation properties of milk proteins, soy proteins, and blended protein systems. *Food Technol.* **38**(5), 85–94, 96.
Asghar, A., and Henrickson, R. L. 1982. Chemical, biochemical, functional, and nutritional characteristics of collagen in food systems. *Adv. Food Res.* **28**, 231–372.
Feeney, R. E., and Whitaker, J. R. 1988. Importance of cross-linking reactions in proteins. *Adv. Cereal Sci. & Technol.* **9**, 21–45.

REFERENCES

1. Richardson, J. S. 1976. Handedness of crossover connections in β sheets. *Proc. Natl. Acad. Sci. USA* **73**, 2619–2623.
2. Birktoft, J. J., and Blow, D. M. 1972. Structure of crystalline α-chymotrypsin. *J. Mol. Biol.* **68**, 187–240.
3. Richardson, J. S. 1977. β-Sheet topology and the relatedness of proteins. *Nature* **268**, 495–500.

4. Levitt, M., and Chothia, C. 1976. Structural patterns in globular proteins. *Nature* **261**, 552–558.
5. Eagland, D. 1975. Protein hydration—its role in stabilizing the helix conformation of the protein. In *Water Relations of Foods,* ed. R. B. Duckworth, Academic Press, New York.
6. Melander, W., and Horváth, C. 1977. Salt effects on hydrophobic interaction in precipitation and chromatography of proteins: An interpretation of the lyotropic series. *Arch. Biochem. Biophys.* **183**, 200–215.
7. Cherry, J. P., and MeWatters, K. H. 1981. Whippability and aeration. In *Protein Functionality in Foods,* ed. J. P. Cherry, ACS Sym. Ser. No. 147, American Chemical Society, Washington, D.C.
8. Phillips, M. C. 1981. Protein conformation at liquid interfaces and its role in stabilizing emulsions and foams. *Food Technol.* **35**(1), 50–57.
9. Schmidt, R. H. 1981. Gelation and coagulation. In *Protetin Functionality in Foods,* ed. Cherry, J. P., ACS Sym. Ser. 147, American Chemical Society, Washington, D.C.
10. Hermansson, A.-M. 1979. Aggregation and denaturation involved in gel formation. In *Functionality and Protein Structure,* ed. A. Pour-El, ACS Sym. Ser. 92, American Chemical Society, Washington, D.C.
11. Whitaker, J. R., and Feeney, R. E. 1983. Chemical and physical modification of proteins by the hydroxide ion. *CRC Crit. Rev. Food Sci. & Nutr.* **19**, 173–212.
12. Liardon, R., and Ledermann, S. 1986. Racemization kinetics of free and protein-bound amino acids under moderate alkaline treatment. *J. Agric. Food Chem.* **34**, 557–565.
13. Hurrell, R. F., and Carpenter, K. J. 1976. Mechanisms of heat damage in proteins. 7. The significance of lysine-containing isopeptides and of lanthionine in heated proteins. *Br. J. Nutr.* **35**, 383–395.
14. Schaich, K. M. 1980. Free radical initiation in proteins and amino acids by ionizing and ultraviolet. *CRC Crit. Rev. Food Sci. & Nutr.* **13**, 89–129.
15. Faraggi, M., and Bettelheim, A. 1977. The reaction of the hydrated electron with amino acids, peptides, and proteins in aqueous solution III. Histidyl peptides. *Radiat. Res.* **71**, 311–324.
16. Braams, R. 1966. Rate constants of hydrated electron reactions with amino acids. *Radiat. Res.* **27**, 319–329.
17. Wilkening, V. G., Lal, M., Arends, M., and Armstrong, D. A. 1968. The cobalt-60 γ-radiolysis of cysteine in deaerated aqueous solutions at pH values between 5 and 6. *J. Phys. Chem.* **72**, 185–190.
18. Rao, P. S., and Hayon, E. 1975. Reaction of hydroxyl radicals with oligopeptides in aqueous solutions. A pulse radiolysis study. *J. Phys. Chem.* **79**, 109–115.
19. Karam, L. R., Dizdaroglu, M., and Simic, M. G. 1984. OH radical-induced products of tyrosine peptides. *Int. J. Radiat. Biol.* **46**, 715–724.
20. Yamamoto, O. 1975. Radiation-induced binding of OH-substituted aromatic amino acids, tyrosine and dopa, mutually and with albumin in aqueous solution. *Radiat. Res.* **61**, 251–260.

21. Yamamoto, O. 1972. Radiation-induced binding of methionine with serum albumin, tryptophan or phenylalanine in aqueous solution. *Int. J. Radiat. Phys. Chem.* **4**, 335–345.

22. Dizdaroglu, M., and Simic, M. G. 1985. Radiation-induced crosslinks between thymine and phenylalanine. *Int. J. Radiat. Biol.* **47**, 63–69.

23. Schaich, K. M. 1980. Free radical initiation in proteins and amino acids by ionizing and ultraviolet radiations and lipid oxidation—Part II: Ultraviolet radiation and photolysis. *CRC Crit. Rev. Food Sci. & Nutr.* **13**, 131–159.

24. Bent, D. V., and Hayon, E. 1975. Excited state chemistry of aromatic amino acids and related peptides. II. Phenylalanine. *J. Am. Chem. Soc.* **97**, 2606–2611.

25. Saito, I., and Matsuura, T. 1985. Chemical aspects of UV-induced cross-linking of proteins to nucleic acid. Photoreaction with lysine and tryptophan. *Acc. Chem. Res.* **18**, 134–141.

26. Gennari, G., Cauzzo, G., and Jori, G. 1974. Further studies on the crystal-crystal-violet-sensitized photooxidation of cysteine to cysteic acid. *Photochem. Photobiol.* **20**, 497–500.

27. Yang, S. F., Ku, H. S., and Pratt, H. K. 1967. Photochemical production of ethylene from methionine and its analogues in the presence of flavin mononucleotide. *J. Biol. Chem.* **242**, 5274–5280.

28. Saito, I, and Matsuura, T. 1977. Peroxidic intermediates in photosensitized oxygenation of tryptophan derivatives. *Acc. Chem. Res.* **10**, 346–352.

29. Schaich, K. M. 1980. Free radical initiation in proteins and amino acids by ionizing and ultraviolet radiation and lipid oxidation—Part III: Free radical transfer from oxidizing lipids. *CRC Crit. Rev. Food Sci. & Nutr.* **13**, 189–244.

30. Yong, S. H., and Karel, M. 1978. Reaction of histidine with methyl linoleate: Characterization of the histidine degradation products. *JAOCS* **55**, 352–357.

31. Yong, S. H., Lau, S., Hsieh, Y., and Karel, M. 1980. Degradation products of L-tryptophan reacted with peroxidizing methyl linoleate. In *Autoxidation in Food and Biological Systems,* ed. M. G. Simic, and M. Karel, Plenum Press, New York.

32. Hibberd. M. G. 1986. Relationships between chemical and mechanical events during muscular contraction. *Ann. Rev. Biophys. Chem.* **15**, 119–161.

33. Goll, D. E., Otsuka, Y., Nagainis, P. A., Shannon, J. D., Sathe, S. K., and Muguruma, M. 1983. Role of muscle proteinases in maintenance of muscle integrity and mass. J. Food Biochemistry **7**, 137–177.

34. Hegarty, G. R., Bratzler, L. J., and Pearson, A. M. 1963. Studies on the emulsifying properties of some intracellular beef muscle proteins. *J. Food Sci.* **28**, 663–668.

35. Ziegler, G. R., and Acton, J. C. 1984. Mechanisms of gel formation by proteins of muscle tissue. *Food Technol.* **38**(5), 77–82.

36. Samejima, K., Ishioroshi, M., and Yasui, T. 1981. Relative roles of the head and tail portions of the molecule in heat-induced gelation of myosin. *J. Food Sci.* **46**, 1412–1418.

37. Schmidt, D. G. 1980. Colloidal aspects of casein. *Neth. Milk Dairy J.* **34**, 42–64.

38. Slattery, C. W., and Evard, R. 1973. A model for the formation and structure of casein micelles from subunits of variable composition. *Biochim. Biophys. Acta* **317**, 529–538.
39. Schmidt, D. G. 1982. Association of caseins and casein micelle structure. In *Developments in Dairy Chemistry* ed. P. F. Fox, Applied Science Publishers, Ltd., London and New York.
40. Dalgleish, D. G. 1979. Proteolysis and aggregation of casein micelles treated with immobilized or soluble chymosin. *J. Dairy Res.* **46**, 653–661.
41. Payens, T. A. J. 1979. Casein micelles: The colloid-chemical approach. *J. Dairy Res.* **46**, 291–306.
42. Papiz, M. Z., Sawyer, L., Eliopoulos, E. E., North, A. C. T., Findley, J. B. C., Siraprasadarao, R., Jone, T. A., Newcomer, M. E., and Kraulis, P. J. 1986. The structure of β-lactoglobulin and its similarity to plasma retinol-binding protein. *Nature* **324**, 383–385.
43. Fox, P. F., and Hoynes, M. C. T. 1975. Heat stability of milk: Influence of colloidal calcium phosphate and β-lactoglobulin. *J. Dairy Res.* **42**, 427–435.
44. Kasarda, D. D., Bernardin, J. E., and Nimmo, C. C. 1976. Wheat proteins. *Adv. Cereal Sci. & Technol.* **1**, 158–236.
45. Tatham, A. S., and Shewry, P. R. 1985. The conformation of wheat gluten proteins. The secondary structures and thermal stabilities of α-, β-, and γ-gliadins. *J. Cereal Sci.* **3**, 103–113.
46. Ewart, J. A. D. 1972. A modified hypothesis for the structure and rheology of glutelins. *J. Sci. Food Agric.* **23**, 687–699.
47. Khan, K., and Bushuk, W. 1979. Studies of glutenin XII. Composition by sodium dodecyl sulfate-polyacrylamide gel electrophoresis of unreduced and reduced glutenin from various isolation and purification procedures. *Cereal Chem.* **56**, 63–68.
48. Graveland, A., Bosveld, P., Lichtendonk, W. J., Marseille, J. P., Moonen, A., J. H. E., and Scheepstra, A. 1985. A model for the molecular structure of the glutenins from wheat flour. *J. Cereal Sci.* **3**, 1–16.
49. Wall, J. S. 1979. The role of wheat proteins in determining baking quality. In *Recent Advances in the Biochemistry of Cereals,* ed. D. L. Laidman and R. G. W. Jones, Academic Press, New York.
50. Pomeranz, Y., and Chung, O. K. 1978. Interaction of lipids with proteins and carbohydrates in breadmaking. *JAOCS* **55**, 285–289.
51. Bell, B. M., Daniels, D. G. H., and Fisher, N. 1977. Physical aspects of the improvement of dough by fat. *Food Chem.* **2**, 57–70.
52. Frazier, P. J., Brimblecombe, F. A., Daniels, N. W. R., and Eggitt, P. W. R. 1977. The effect of lipoxygenase action on the mechanical development of doughs from fat-extracted and reconstituted wheat flour. *J. Sci. Food Agric.* **28**, 247–254.
53. Fraser, D. B., and MacRae, T. P. 1973. *Conformation in Fibrous Proteins and Related Synthetic Polypeptides,* Academic Press, New York, p. 347.
54. Asghar, A., and Henrickson, R. L. 1982. Chemical, biochemical, functional, and nutritional characteristics of collagen in food systems. *Adv. Food Res.* **28**, 231–372.

55. Fujimoto, D., Moriguchi, T., Ishida, T., and Hayashi, H. 1978. The structure of pyridinoline, a collagen crosslink. *Biochem. Biophys. Res. Comm.* **84**, 52–57.
56. Harrington, W. F., and Venkateswara, R. 1970. Collagen structure in solution. I. Kinetics of helix regeneration in single-chain gelatins. *Biochemistry* **9**, 3714–3724.
57. Ottenheym, H. H., and Jenneskens, P. J. 1970. Synthetic amino acids and their use in fortifying foods. *J. Agric. Food Chem.* **18**, 1010–1014.

CHAPTER 3. CARBOHYDRATES

SELECTED READINGS

Danehy, J. P. 1986. Maillard reactions: Nonenzymatic browning in food systems with special reference to the development of flavors. *Adv. Food Res.* **30**, 77–138.
Kester, J. J., and Fennema, O. R. 1986. Edible films and coatings: A review. *Food Technol.* **40**(12), 47–59.
Morris, V. J. 1985. Food gels—roles played by polysaccharides. *Chem. and Ind.,* March 4, 159–164.
Sanderson, G. R. 1981. Polysaccharides in foods. *Food Technol.* **35**(7), 50–57, 83.
von Sonntag, C. 1980. Free-radical reactions of carbohydrates as studied by radiation techniques. *Adv. Carbohydr. Chem. Biochem.* **37**, 7–77.

REFERENCES

1. Capon, B., and Overend, W. G. 1960. Constitution and physicochemical properties of carbohydrates. *Adv. Carbohydr. Chem.* **15**, 11–51.
2. de Wit, G., Kieboom, A. P. G., and van Bekkum, H. 1979. Enolization and isomerization of monosaccharides in aqueous alkaline solution. *Carbohydr. Res.* **74**, 157–175.
3. Feather, M. S., and Harris, J. F. 1973. Dehydration reactions of carbohydrates. *Adv. Carbohydr. Chem. Biochem.* **28**, 161–224.
4. Mizuno, T., and Weiss, A. H. 1974. Synthesis and utilization of formose sugars. *Adv. Carbohydr. Chem. Biochem.* **29**, 173–227.
5. Anet, E. F. L. J. 1964. 3-Deoxyglycosuloses (3-deoxyglycosones) and the degradation of carbohydrates. *Adv. Carbohydr. Chem.* **19**, 181–218.
6. Feather, M. S., and Harris, J. F. 1970. On the mechanism of conversion of hexoses into 5-(hydroxymethyl)-2-furaldehyde and metasaccharinic acid. *Carbohydr. Res.* **15**, 304–309.
7. Čern, M., and Stanek, J. Jr. 1977. 1,6-Anhydro derivatives of aldohexoses. *Adv. Carbohydr. Chem. Biochem.* **34**, 24–177.
8. Hodge, J. E. 1967. Origin of flavor in foods—non-enzymatic browning reactions. In *The Chemistry and Physiology of Flavors,* ed. H. W. Schultz, E. A. Day, and L. M. Libbey, AVI, Westport, Conn.
9. Kort, M. J. 1970. Reactions of free sugars with aqueous ammonia. *Adv. Carbohydr. Chem. Biochem.* **25**, 311–349.

10. Hodge, J. E. 1955. The Amadori rearrangement. *Adv. Carbohydr. Chem.* **10**, 169–205.

11. Isbell, H. S., and Frush, H. I. 1958. Mutarotation, hydrolysis and rearrangement reactions of glycosylamine. *J. Org. Chem.* **23**, 1309–1319.

12. Burton, H. S., and McWeeny, D. J. 1964. Non-enzymatic browning: Routes to the production of melanoidins from aldoses and amino-compounds. *Chem. and Ind.*, March 14, 462–463.

13. Namiki, M., and Hayashi, T. 1983. A new mechanism of the Maillard reaction involving sugar fragmentation and free radical formation. In *The Maillard Reaction in Foods and Nutrition*, ed. G. R. Waller, and M. S. Feather, ACS Sym. Ser. 215, American Chemical Society, Washington, D.C.

14. Kato, H., and Tsuchida, H. 1981. Estimation of melanoidin structure by pyrolysis and oxidation. *Prog. Food. Nutr. Sci.* **5**, 147–156.

15. Angal, S. J. 1980. Sugar-cation complexes—structure and applications. *Chem. Soc. Rev.* **9**(4) 415–428.

16. Rendleman, J. A., Jr. 1973. Ionization of carbohydrates in the presence of metal hydroxides and oxides. In *Carbohydrates in Solution*, ed. H. S. Isbell, Adv. Chem. Ser. 117, American Chemical Society, Washington, D.C.

17. French, A. D., and Murphy, V. G. 1977. Computer modeling in the study of starch. *Cereal Foods World* **22**(2), 61–70.

18. Osaka, Z. N. 1978. Studies on starch granules. *Starch/Stärke* **30**, 105–111.

19. Olkku, J. 1978. Gelatinization of starch and wheat flour starch—a review. *Food Chem.* **3**, 293–317.

20. Luallen, T. E. 1985. Starch as a functional ingredient. *Food Technol.* **39**(1), 59–63.

21. Littlecott, G. W. 1982. Food gels—the role of alginates. *Food Technol. Aust.* **34**, 412–418.

22. Grant, G. T., Morris, E. R., Rees, D. A., Smith, P. J. C., and Thorn, D. 1973. Biological interactions between polysaccharides and divalent cations: The egg-box model. *FEBS Lett.* **32**, 195–198.

23. Oakenfull, D., and Scott, A. 1984. Hydrophobic interaction in the gelation of high methyoxyl pectins. *J. Food Sci.* **49**, 1093–1098.

24. Walkinshaw, M. D., and Arnott, S. 1981. Conformations and interactions of pectins II. Models for junction zones in pectinic acid and calcium pectate gels. *J.Mol. Biol.* **153**, 1075–1085.

25. Anon. 1984. *Marine Colloids, Introductory Bulletin A-1.* FMC Corporation, Philadelphia, Pennsylvania.

26. Glicksman, M. 1979. Gelling hydrocolloids in food product applications. In *Polysaccharides in Food*, ed. J. M. V. Blanshard and J. R. Mitchell, Butterworths, London.

27. Ledward, D. A. 1979. Protein-polysaccharide interactions. In *Polysaccharides in Food*, ed. J. M. V. Blanshard and J. R. Mitchell, Butterworths, London.

28. Dea, I. C. M. 1979. Interactions of ordered polysaccharide structure-synergism and freeze-thaw phenomona. In *Polysaccharides in Food*, ed. J. M. V. Blanshard and J. R. Mitchell, Butterworths, London.

29. Rees, D. A. 1972. Polysaccharide gels. *Chem. and Ind.* **16**, 630–636.

30. Klug, E. D. 1970. Hydroxypropylcellulose. *Food Technol.* **24**(1), 51–54.
31. Anon. 1984. *Cellulose gum: Sodium carboxymethyl cellulose, chemical and physical properties.* Hercules Incorporated, Wilmington, Delaware.
32. Symes, K. C. 1980. The relationship between the covalent structure of the *Xanthomonas* polysaccharide (xanthan) and its function as a thickening, suspending and gelling agent. *Food Chem.* **6**, 63–76.
33. Betz, D. A. 1979. Xanthan gum, a biosynthetic polysaccharide for the food industry. *Food Technol. Aust.* **31**, 11–16.
34. Whistler, R. L., and Richards, E. L. 1970. Hemicelluloses. In *The Carbohydrates Chemistry and Biochemistry,* Second Edition, ed. W. Pigman and D. Horton, Academic Press, New York.
35. Neukom, H. 1976. Chemistry and properties of the non-starchy polysaccharides (NSP) of wheat flour. *Lebensm.-Wiss. u.-Technol.* **9**, 143–148.
36. Hoseney, R. S. 1984. Functional properties of pentosans in baked foods. *Food Technol.* **38**(1), 114–116.

CHAPTER 4. COLORS

SELECTED READINGS

Livingston, D. J., and Brown, W. D. 1981. The chemistry of myoglobin and its reactions. *Food Technol.* **35**(5), 244–252.

Gordon, H. T., and Bauernfeind, J. C. 1982. Carotenoids as food colorants. *CRC Crit. Rev. Food Sci. & Nutr.* **18**, 59–91.

Francis, F. J. 1987. Lesser-known food colorants. *Food Technol.* **41**(4), 62–68.

Ilker, R. 1987. In-vitro pigment production: An alternative to color synthesis. *Food Technol.* **41**(4), 70–72.

REFERENCES

1. Sinsheimer, R. L. 1955. Ultraviolet absorption spectra. In *Radiation Biology* Vol. II: Ultraviolet and Related Radiations, ed. A. Hollaender, McGraw-Hill, New York.
2. Bauernfeind, J. C. 1972. Carotenoid, vitamin A precursors and analogs in foods and feeds. *J. Agric. Food Chem.* **20**, 456–473.
3. Zechmeister, L. 1962. *Cis-Trans Isomeric Carotenoids, Vitamins A and Arylpolyenes,* Academic Press, New York.
4. Land, D. G. 1962. Stability of plant pigments. In *Recent Advances in Food Science, Proceeding,* Vol. II, ed. J. Hawthorn and J. M. Leitch, Butterworths, London.
5. Schweiter, M., Englert, G., Rigassi, N., and Vetter, W. 1969. Physical organic methods in carotenoid research. *Pure & Appl. Chem.* **20**, 365–420.
6. Foote, C. S., Chang, Y. C., and Denny, R. W. 1970. Chemistry of singlet oxygen XI. Cis-trans isomerization of carotenoids by singlet oxygen and a probable quenching mechanism. *J. Am. Chem. Soc.* **92**, 5218–5219.

7. Isler, O. 1979. History and industrial application of carotenoids and vitamin A (1). *Pure & Appl. Chem.* **51**, 447–462.
8. Preston, H. D., and Rickard, M. D. 1980. Extraction and chemistry of annatto. *Food Chem.* **5**, 47–56.
9. Watanabe, S., Sakamura, S., and Obata, Y. 1966. The structures of acylated anthocyanins in eggplant and Perilla and the position of acylation. *Agri. Biol. Chem.* **30**, 420–422.
10. Timberlake, C. F. 1980. Anthocyanins—occurrence, extraction and chemistry. *Food Chem.* **5**, 69–80.
11. Timberlake, C. F., and Bridle, P. 1968. Flavylium salts resistant to sulfur dioxide. *Chem. and Ind.* **1968**, 1489.
12. Hoshino, T., Matsumoto, U., and Goto, T. 1981. Self-association of some anthocyanins in neutral aqueous solution. *Phytochemistry* **20**, 1971–1976.
13. Somers, T. C. 1971. The polymeric nature of wine pigments. *Phytochemistry* **10**, 2175–2186.
14. Timberlake, C. F., and Bridle, P. 1976. Interactions between anthocyanins, phenolic compounds, and acetaldehyde and their significance in red wines. *Am. J. Enol. Vitic.* **27**, 97–105.
15. Hrazdina, G., and Franzese, A. J. 1974. Oxidation products of acylated anthocyanins under acidic and neutral conditions. *Phytochemistry* **13**, 231–234.
16. Schwartz, S. J., and von Elbe, J. H. 1983. Identification of betanin degradation products. *Z. Lebensm Unters Forsch* **176**, 448–453.
17. North, R. S. 1973. Caramel—the versatile colouring. *The Flavour Industry* **4**, 337–338.
18. Newsome, R. L. 1986. Food Colors, Food Technol. **40**(7), 49–56.
19. Cotton, F. A., and Wilkinson, G. 1980. *Inorganic Chemistry,* Wiley, New York.
20. Gouterman, M. 1961. Spectra of porphyrins. *J. Mol. Spectroscopy* **6**, 138–163.
21. Adar F. 1978. Electronic absorption spectra of hemes and hemoproteins. In *The Porphyrins,* Vol. III. Physical Chemistry, Part A., ed. D. Dolphin, Academic Press, New York.
22. James, B. R. 1978. Interaction of dioxygen with metalloporphyrins. In *The Porphyrins,* Vol. V. Physical Chemistry, Part C., ed. D. Dolphin, Academic Press, New York.
23. Nicholls, P. 1961. The formation and properties of sulphmyoglobin and sulphcatalase *Biochem. J.* **81**, 374–383.
24. Fox, J. B., Jr., and Thomson, J. S. 1963. Formation of bovine nitrosylmyoglobin I. pH 4.5–6.5. *Biochemistry* **2**, 465–470.
25. Dickerson, R. E. 1964. X-ray analysis and protein structure. In *The Proteins,* Vol. 2, Second Edition, ed H. Neurath, Academic Press, New York.
26. Giddings, G. G. 1977. The basis of color in muscle foods. *CRC Crit. Rev. Food Sci. & Technol.* **9**(1), 81–114.
27. Perutz, M. F. 1978. Hemoglobin structure and respiratory transport. *Sci. Am.* **239**, 92–125.
28. Castro, C. E., Wade, R. S., and Belser, N. O. 1978. Conversion of oxyhemoglobin to methemoglobin by organic and inorganic reductants. *Biochemistry* **17**, 225–231.

29. Nicol, D. J., Shaw, M. K., and Ledward, D. A. 1970. Hydrogen sulfide production by bacteria and sulfmyoglobin formation in prepacked chilled beef. *Appl. Microbiology* 19, 937–939.
30. Tarladgis, B. G. 1962. Interpretation of the spectra of meat pigment, II*-cured meats. The mechanism of colour fading. *J. Sci. Food Agric.* 13, 485–491.
31. Houssier, C., and Sauer, K. 1970. Circular dichroism and magnetic circular dichroism of the chlorophyll and photochlorophyll pigments. *J. Am. Chem. Soc.* 92, 779–791.
32. Katz, J. J., Shipman, L. L., Cotton, T. M., and Janson, T. R. 1978. Chlorophyll aggregation: Coordination interaction in chlorophyll monomers, dimers, and oligomers. In *The Porphorins,* Vol V, Physical Chemistry, Part C, ed. D. Dolphin, Academic Press, New York.
33. Katz, J. J., Norris, J. R., and Shipman, L. I. 1977. Models for reaction center and antenna chlorophyll. *Brookhaven Sym. Biol.* 28, 16–55.
34. Humphrey, A. M. 1980. Chlorophyll. *Food Chem.* 5, 57–67.

CHAPTER 5. ENZYMES

SELECTED READINGS

Whitaker, J. R. 1985. Mechanisms of oxidoreductases important in food component modification. In *Chemical Changes in Food during Processing,* ed. T. Richardson and J. W. Finley, AVI, Westport, Conn.
Polgár, L., and Halász, P. 1982. Current problems in mechanistic studies of serine and cysteine proteinases. *Biochem. J.* 207, 1–10.
Solomon, E. I. 1983. Electronic and geometric structure-function correlations of the coupled binuclear copper active site. *Pure & Appl. Chem.* 55, 1069–1088.
Malmström, B. G. 1982. Enzymology of oxygen. *Ann. Rev. Biochem.* 51, 21–59.
Klibanov, A. M. 1986. Enzymes that work in organic solvents. *Chemtech* June 1986, 354–359.

REFERENCES

1. Pariza, M. W., and Foster, E. M. 1983. Determining the safety of enzymes used in food processing. *J. Food Protection* 46(5), 453–468.
2. Drenth, J., Jansonius, J. N., Koekoek, R., and Wolthers, B. G. 1971. Papain, X-ray structure. In *The Enzymes,* Vol. III, ed. P. D. Boyer, Academic Press, New York.
3. Lewis, S. D., Johnson, F. A., and Shafer, J. A. 1981. Effect of cysteine-25 on the ionization of histidine-159 in papain as determined by proton nuclear magnetic resonance spectroscopy. Evidence for a His-159-Cys-25 ion pair and its possible role in catalysis. *Biochemistry* 20, 48–51.
4. Polgár, L. 1973. On the mode of activation of the catalytically essential sulfhydryl group of papain. *Eur. J. Biochem.* 33, 104–109.
5. Storer, A. C., and Carey, P. R. 1985. Comparison of the kinetics and mecha-

nism of the papain-catalyzed hydrolysis of esters and thiono esters. *Biochemistry* **24**, 6808–6818.

6. Heller, M. J., Walder, J. A., and Klotz, I. M. 1977. Intramolecular catalysis of acylation and deacylation in peptides containing cysteine and histidine. *J. Am. Chem. Soc.* **99**, 2780–2785.

7. Kang, C. K., and Rice, E. E. 1970. Degradation of various meat fractions by tenderizing enzymes. *J. Food Sci.* **35**, 563–565.

8. Galliard, T., and Chan, H. W.-S. 1980. Lipoxygenase. In *The Biochemistry of Plants, A Comprehensive Treatise,* Vol. 4, ed. P. K. Stumpf and E. E. Conn, Academic Press, New York.

9. Yoon, S., and Klein, B. P. 1979. Some properties of pea lipoxygenase isoenzymes. *J. Agric. Food Chem.* **27**, 955–962.

10. Egmond, M. R., Vliegenthart, J. F. G., and Boldingh, J. 1972. Stereospecificity of the hydrogen abstraction at carbon atom n-8 in the oxygenation of linoleic acid by lipoxygenases from corn germs and soya beans. *Biochem. Biophys. Res. Comm.* **48**, 1055–1060.

11. van Os, C. P. A., Rijke-Schilder, G. P. M., van Halbeek, H., Verhagen, J., and Vliegenthart, J. F. G. 1981. Double dioxygenation of arachidonic acid by soybean lipoxygenase-1. *Biochim. Biophys. Acta* **663**, 177–193.

12. Slappendel, S., Veldink, G. A., Vliegenthart, J. F. G., Aasa, R., and Malmström B. G. 1981. EPR spectroscopy of soybean lipoxygenase-1. Description and quantification of the high-spin Fe (III) signals. *Biochim. Biophys. Acta* **667**, 77–86.

13. Cheesbrough, T. M., and Axelrod, B. 1983. Determination of the spin state of iron in native and activated soybean lipoxygenase 1 by paramagnetic susceptibility. *Biochemistry* **22**, 3837–3840.

14. deGroot, J. J. M. C., Veldink, G. A., Vliegenthart, J. F. G., Boldingh, J., Wever, R., and van Gelder, B. F. 1975. Demonstration by EPR spectroscopy of the functional role of iron in soybean lipoxygenase-1. *Biochim. Biophys. Acta* **377**, 71–79.

15. Garssen, G. J., Vliegenthart, J. F. G., and Boldingh, J. 1971. An anaerobic reaction between lipoxygenase, linoleic acid and its hydroperoxides. *Biochem. J.* **22**, 327–332.

16. Garssen, G. J., Vliegenthart, J. F. G., and Boldingh, J. 1972. The origin and structures of dimeric fatty acids from the anaerobic reaction between soya-bean lipoxygenase, linoleic acid and its hydroperoxide. *Biochem. J.* **130**, 435–442.

17. Gardner, H. W. 1979. Stereospecificity of linoleic acid hydroperoxide isomerase from corn germ. *Lipids* **14**, 208–211.

18. Vick, B. A., and Zimmerman, D. C. 1981. Lipoxygenase, hydroperoxide isomerase, and hydroperoxide cyclase in young cotton seedlings. *Plant Physiol.* **67**, 92–97.

19. Galliard, T., Phillips, D. R., and Reynolds, J. 1976. The formation of *cis*-3-nonenal, *trans*-2-nonenal and hexanal from linoleic acid hydroperoxide isomers by a hydroperoxide cleavage enzyme system in cucumber (*Cucumis sativus*) fruits. *Biochim. Biophys. Acta* **441**, 181–192.

20. Schieberle, P., Grosch, W., Kexel, H., and Schmidt, H.-L. 1981. A study of

oxygen isotope scrambling in the enzymic and non-enzymic oxidation of linoleic acid. *Biochim. Biophys. Acta* **666**, 322–326.

21. Strothkamp, K. G., Jolley, R. L., and Mason, H. S. 1976. Quaternary structure of mushroom tyrosinase. *Biochem. Biophys. Res. Comm.* **70**, 519–524.

22. Lerch, K. 1982. Primary structure of tyrosinase from *Neurospora crassa*. *J. Biol. Chem.* **257**, 6414–6419.

23. Himmelwright, R. S., Eickman, N. C., Lubien, C. D., Lerch, K., and Solomon, E. I. 1980. Chemical and spectroscopic studies of the binuclear copper active site of *Neurospora* tyrosinase: Comparison to hemocyanins. *J. Am. Chem. Soc.* **102**, 7339–7344.

24. Winkler, M. E., Lerch, K., and Solomon, E. I. 1981. Competitive inhibitor binding to the binuclear copper active site in tyrosinase. *J. Am. Chem. Soc.* **103**, 7001–7003.

25. Wilcox, D. E., Porras, A. G., Hwang, Y. T., Lerch, K., Winkler, M. E., and Solomon, E. I. 1985. Substrate analogue binding to the coupled binuclear copper active site in tyrosinase. *J. Am. Chem. Soc.* **107**, 4015–4027.

26. Palmer, J. K. 1963. Banana polyphenoloxidase. Preparation and properties. *Plant Physiol.* **38**, 508–513.

27. James, T. L., Edmondson, D. E., and Husain, M. 1981. Glucose oxidase contains a disubstituted phosphorus residue. Phosphorus-31 nuclear magnetic resonance studies of the flavin and nonflavin phosphate residues. *Biochemistry* **20**, 617–621.

28. Gibson, Q. H., Swoboda, B. E. P., and Massey, V. 1964. Kinetics and mechanism of action of glucose oxidase. *J. Biol. Chem.* **239**, 3927–3934.

29. Bright, H. J., and Appleby, M. 1969. The pH dependence of the individual steps in the glucose oxidase reaction. *J. Biol. Chem.* **244**, 3625–3634.

30. Weibel, M. K., and Bright, H. J. 1971. The glucose oxidase mechanism. *J. Biol. Chem.* **246**, 2734–2744.

31. Chan, T. W., and Bruice, T. C. 1977. One and two electron transfer reactions of glucose oxidase. *J. Am. Chem. Soc.* **99**, 2387–2389.

32. Stankovich, M. T., Schopfer, L. M., and Massey, V. 1978. Determination of glucose oxidase oxidation-reduction potentials and the oxygen reactivity of fully reduced and semiquinoid forms. *J. Biol. Chem.* **253**, 4971–4979.

33. Dawson, H. G., and Allen, W. G. 1984. *The Use of Enzymes in Food Technology*, Miles Laboratories, Inc., Biotech Products Division, Elkhart, In.

34. Thoma, J. A., Spradlin, J. E., and Dygert, S. 1971. Plant and animal amylases. In *The Enzymes*, Vol. V, Third Edition, ed. P. D. Boyer, Academic Press, New York.

35. Takeda, Y., and Hizukuri, S. 1981. Re-examination of the action of sweet-potato beta-amylase on phosphorylated $(1 \rightarrow 4)$-α-D-glucan. *Carbohydr. Res.* **89**, 174–178.

36. Chung, H., and Friedberg, F. 1980. Sequence of the N-terminal half of *Bacillus amyloliquefaciens* α-amylase. *Biochem. J.* **185**, 387–395.

37. Payan, F., Haser, R., Pierrot, M., Frey, M., and Astier, J. P. 1980. The three-dimensional structure of α-amylase from porcine pancreas at 5Å resolution—the active site location. *Acta Cryst.* **B36**, 416–421.

38. Thoma, J. A. 1968. A possible mechanism for amylase catalysis. *J. Theoret. Biol.* **19**, 297–310.
39. Lai, H.-L., Butler, L. G., Axelrod, B. 1974. Evidence for a covalent intermediate between α-glucosidase and glucose. *Biochem. Biophys. Res. Comm.* **60**, 635–640.
40. Robyt, J. F. 1984. Enzymes in the hydrolysis and synthesis of starch. In *Starch: Chemistry and Technology,* Second Edition, ed. R. L. Whistler, J. N. Bemiller, and E. F. Paschall, Academic Press, New York.
41. Thoma, J. A., and Koshland, D. E., Jr. 1960. Competitive inhibition by substrate during enzyme action. Evidence for the induced-fit theory. *J. Am. Chem. Soc.* **82**, 3329–3333.
42. Robyt, J. F., and French, D. 1970. Multiple attack and polarity of action of porcine pancreatic α-amylase. *Arch. Biochem. Biophys.* **138**, 662–670.
43. Allen, J. D., and Thoma, J. A. 1979. Multimolecular substrate reactions catalyzed by carbohydrases. *Aspergillus oryzae* α-amylase degradation of maltooligosaccharides. *Biochemistry* **17**, 2338–2344.
44. Robyt, J. F., and French, D. 1970. The action pattern of procine pancreatic α-amylase in relationship to the substrate binding site of the enzyme. *J. Biol. Chem.* **10**, 3917–3927.
45. Allen, J. D., and Thoma, J. A. 1976. Subsite mapping of enzymes. *Biochem. J.* **159**, 121–132.
46. Rexová-Benková, L., and Markovič, O. 1976. Pectic enzymes. *Adv. Carbohydr. Chem. Biochem.* **33**, 323–385.
47. Miller, L., and MacMillan, J. D. 1971. Purification and pattern of action of pectinesterase from *Fusarium oxysporum* f. sp. *vasinfectum. Biochemistry* **10**, 570–576.
48. Chan, H. T., and Tam, S. N. Y. 1982. Partial separation and characterization of papaya endo- and exo-polygalacturonase. *J. Food Sci.* **47**, 1478–1483.
49. Pressey, R., and Avants, J. K. 1973. Separation and characterization of endopolygalacturonase and exopolygalacturonase from peaches. *Plant Physiol.* **52**, 252–256.
50. Liu, Y. K., and Luh, B. S. 1978. Purification and characterization of endopolygalacturonase from *Rhizopus arrhizus. J. Food Sci.* **43**, 721–726.
51. Lubomira Rexová-Benková, L. 1973. The size of the substrate-binding site of an *Aspergillus niger* extracellular endopolygalacturonase. *Eur. J. Biochem.* **39**, 109–115.
52. Nitta, Y., Mizushima, M., Hiromi, K., and Ono, S. 1971. Influence of molecular structures of substrates and analogues on Taka-amylase A catalyzed hydrolyses. *J. Biochem.* **69**, 567–576.
53. Pressey, R., and Avants, J. K. 1975. Modes of action of carrot and peach exopolygalacturonases. *Phytochemistry* **14**, 957–961.
54. Press, J., and Ashwell, G. 1963. Polygalacturonic acid metabolism in bacteria. *J. Biol. Chem.* **238**, 1571–1576.
55. Jensen, R. G., Gerrior, S. A., Hagerty, M. M., and McMahon, K. E. 1978. Preparation of acylglycerols and phospholipids with the aid of lipolytic enzymes. *JAOCS* **55**, 422–427.

56. Wells, M. A., and DiRenzo, N. A. 1983. Glyceride digestion. In *The Enzymes,* Vol. XVI, Third Edition, ed. P. D. Boyer, Academic Press, New York.
57. Larsson, A., and Erlanson-Albertsson, C. 1981. The identity and properties of two forms of activated colipase from porcine pancreas. *Biochim. Biophys. Acta* **664,** 538–548.
58. Chapus, C., Sari, H., Semeriva, M., and Desnuelle, P. 1975. Role of colipase in the interfacial adsorption of pancreatic lipase at hydrophilic interfaces. *FEB Letters* **58,** 155–158.
59. Brockerhoff, H. 1973. A model of pancreatic lipase and the orientation of enzymes at interfaces. *Chem. Phys. Lipids* **10,** 215–222.
60. Sémériva, M., and Desnuelle, P. 1979. Pancreatic lipase and colipase. An example of heterogeneous biocatalysis. *Adv. Enzymol.* **48,** 320–371.
61. Quinn, D. M. 1985. Solvent isotope effects for lipoprotein lipase catalyzed hydrolysis of water-soluble *p*-nitrophenyl esters. *Biochemistry* **24,** 3144–3149.
62. Jensen, R. G., DeJong, F. A., and Clark, R. M. 1983. Determination of lipase specificity. *Lipids* **18,** 239–253.
63. Tsujisaka, Y., Okumura, S., and Iwai, M. 1977. Glyceride synthesis by four kinds of microbial lipase. *Biochim. Biophys. Acta* **489,** 415–422.
64. Linfield, W. M., Barauskas, R. A., Sivieri, L., Serota, S., and Stevenson, R. W., Jr. 1984. Enzymatic fat hydrolysis and synthesis. *JAOCS* **61,** 191–195.
65. Macrae, A. R. 1983. Lipase-catalyzed interesterification of oils and fats. *JAOCS* **60,** 291–294.

CHAPTER 6. FLAVORS

SELECTED READINGS

Lancet, D., and Pace, U. 1987. The molecular basis of odor recognition. *TIBS* **12**(2), 63–66.

Belitz, H.-D., and Wieser, H. 1985. Bitter compounds: Occurrence and structure-activity relationships. *Food Rev. Int.* **1**(2), 271–354.

Shahidi, F., and Rubin, L. J. 1986. Meat flavor volatiles: A review of the composition, techniques of analysis, and sensory evaluation. *CRC Crit. Rev. Food Sci. & Nutr.* **24,**141–243.

Hebard, C. E., Flick, G. J., and Martin, R. E. 1982. Occurrence and significance of trimethylamine oxide and its derivatives in fish and shellfish. In *Chemistry & Biochemistry of Marine Food Products,* ed. R. E. Martin, G. T. Flick, C. E. Hehard and D. R. Ward, AVI, Westport, Conn.

Whitaker, R. J., and Evans, D. A. 1987. Plant biotechnology and the production of flavor compounds. *Food Technol.* **41**(9), 86–101.

REFERENCES

1. Belitz, H.-D., Chen, W., Jugel, H., Stempfl, H., Treleano, R., and Wieser, H. 1983. Quantitative structure activity relationships of bitter tasting compounds. *Chem. and Ind.,* January 3, 23.

2. Amoore, J. E. 1970. *Molecular Basis of Odor.* Charles C. Thomas, Springfield, Ill.

3. Beets, M. G. J. 1971. Relationship of chemical structure to odor and taste. In *Proceedings, Third International Congress on Food Science and Technology,* Institute of Food Technologists, Chicago, Ill.

4. Rizzi, G. P. 1972. A mechanistic study of alkylpyrazine formation in model systems. *J. Agric. Food Chem.* **20**, 1081–1085.

5. Rizzi, G. P. 1969. The formation of tetramethylpyrazine and 2-isopropyl-4, 5-dimethyl-3-oxazoline in the strecker degradation of DL-valine with 2,3-butanedione. *J. Org. Chem.* **34**, 2002–2004.

6. Tressl, R., Rewicki, D., Helak, B., and Kamperschröer, H. 1985. Formation of pyrrolidines and piperidines on heating L-proline with reducing sugars. *J. Agric. Food Chem.* **33**, 924–928.

7. Mills, F. D., Baker, B. G., and Hodge, J. E. 1969. Amadori compounds as nonvolatile flavor precursors in processed foods. *J. Agric. Food Chem.* **17**, 723–727.

8. Takken, H. J., van der Linde, L. M., deValois, P. J., van Dort, H. M., and Boelens, M. 1976. Reaction products of α-dicarbonyl compounds, aldehydes, hydrogen sulfide, and ammonia. In *Phenolic, Sulfur, and Nitrogen Compounds in Food Flavors,* ACS Sym. Ser. 26, American Chemical Society, Washington D.C.

9. Sanderson, G. W., and Graham, H. N. 1973. On the formation of black tea aroma. *J. Agric. Food Chem.* **21**, 576–585.

10. Sanderson, G. W., Co, H., and Gonzalez, J. G. 1971. Biochemistry of tea fermentation: The role of carotenes in black tea aroma formation. *J. Food Sci.* **36**, 231–236.

11. Gunst, F., and Verzele, M. 1978. On the sunstruck flavor of beer. *J. Inst. Brew.* **84**, 291–292.

12. Tressl, R., Friese, L., Fendesack, F., and Koppler, H. 1978. Gas chromatographic–mass spectrometric investigation of hop aroma constituents in beer. *J. Agric. Food Chem.* **26**, 1422–1426.

13. Tressl, R., and Silwar, R. 1981. Investigation of sulfur-containing components in roasted coffee. *J. Agric. Food Chem.* **29**, 1078–1082.

14. Block, E. 1985. The chemistry of garlic and onions. *Scientific American* **252**(3), 114–119.

15. Boelens, M., deValois, P. J., Wobben, H. J., and van der Gen, A. 1971. Volatile flavor compounds from onion. *J. Agric. Food Chem.* **19**, 984–991.

16. Govindarajan, V. S. 1977. Pepper—chemistry, technology, and quality evaluation. *CRC Crit. Rev. Food Sci. & Nutr.* **9**, 115–225.

17. Todd, P. H., Jr., Bensinger, M. G., and Biftu, T. 1977. Determination of pungency due to capsicum by gas-liquid chromatography. *J. Food Sci.* **42**, 660–680.

18. Connell, D. W. 1970. The chemistry of the essential oil and oleoresin of ginger (*Zingiber officinale* Roscoe). *The Flavour Industry* **1**(10), 677–693.

19. Johnson, J. D., and Vora, J. D. 1983. Natural citrus essences. *Food Technol.* **37**(12), 92–93, 97.

20. Demole, E., Enggist, P., and Ohloff, G. 1982. 1-*p*-Menthene-8-thiol: A powerful flavor impact constituent of grapefruit juice (*Citrus paradisi* Macfayden). *Helv. Chim. Acta* **65**, 1785–1794.
21. Wilson, C. W. III, and Shaw, P. E. 1981. Importance of thymol, methyl *N*-methylanthranilate, and monoterpene hydrocarbons to the aroma and flavor of mandarin cold-pressed oils. *J. Agric. Food Chem.* **29**, 494–496.
22. Hasegawa, S., and Maier, V. P. 1983. Solutions to the limonin bitterness problem of citrus juices. *Food Technol.* **37**(6), 73–77.
23. Horowitz, R. M., and Gentili, B. 1979. Taste and structure relations of flavonoid compounds. In *Proceedings of The Fifth International Congress of Food Science and Technology,* ed. H. Chiba, Elsevier Scientific Publishing Co., New York.
24. Nursten, H. E. 1978. Flavor chemistry of fruits and vegetables. In *Agricultural and Food Chemistry: Past, Present, Future,* ed. R. Teranishi, AVI, Westport, Conn.
25. Takken, H. J., van der Linde, L. M., Boelens, M., and van Dort, J. M. 1975. Olfactive properties of a number of polysubstituted pyrazines, *J. Agric. Food Chem.* **23**, 638–642.
26. Pareles, S. R., and Chang, S. S. 1974. Identification of compounds responsible for baked potato flavor. *J. Agric. Food Chem.* **22**, 339–340.
27. Maga, J. A. 1981. Mushroom flavor. *J. Agric. Food Chem.* **29**, 1–4.
28. Tressl, R., Bahri, D., Holzer, M., and Kossa, T. 1977. Formation of flavor components in asparagus. 2. Formation of flavor components in cooked asparagus. *J. Agric. Food Chem.* **25**, 459–463.
29. Gold, H. J., and Wilson, C. W. III. 1963. The volatile flavor substances of celery. *J. Food Sci.* **28**, 484–488.
30. MacLeod, G., and Seyyedain-Ardebili, M. 1981. Natural and simulated meat flavors (with particular reference to beef). *CRC Crit. Food Sci. & Nutr.* **14**, 308–437.
31. van den Ouwedand, G. A. M., and Peer, H. G. 1975. Components contributing to beef flavor. Volatile components produced by the reaction of 4-hydroxy-5-methyl-3(2*H*)-furanone and its thio analog with hydrogen sulfide. *J. Agric. Food Chem.* **23**, 501–505.
32. Wilson, R. A., and Katz, I. 1974. Synthetic meat flavors. *The Flavour Industry* **5**, 30–35, 38.

CHAPTER 7. SWEETENERS

SELECTED READINGS

Dziezak, J. D. 1986. Sweeteners 3. Alternatives to cane and beet sugar. *Food Technol.* **40**(1), 116–128.
Newsome, R. L. 1986. Sweeteners: Nutritive and non-nutritive. *Food Technol.* **40**(8), 196–206.

Kinghorn, A. D. 1986. Sweetening agents of plant origin. *CRC Crit. Rev. Plant Sci.* **4**(2), 79–120.

REFERENCES

1. Shallenberger, R. S. 1983. The chiral principles contained in structure-sweetness relations. *Food Chem.* **12**, 89–107.
2. Shallenberger, R. S. 1979. Taste and chemical structure. In *Proceedings of The Fifth International Congress of Food Science and Technology,* ed. H. Chiba, Kodansha, Ltd., Tokyo, and Elsevier Scientific Publishing Co., New York.
3. Boch, K. and Lemieux, R. U. 1982. The conformational properties of sucrose in aqueous solution: Intramolecular hydrogen-bonding. *Carbohydr. Res.* **100**, 63–74.
4. Homler, B. E. 1984. Properties and stability of aspartame. *Food Technol.* **38**(7), 50–55.
5. Hatada, M., Jancarik, J., Graves, B., and Kim, S.-H. 1985. Crystal structure of aspartame, a peptide sweetener. *J. Am. Chem. Soc.* **107**, 4279–4282.
6. Lelj, F., Tancredi, T., Temussi, P. A., and Toniolo, C. 1976. Interaction of α-L-aspartyl-L-phenylalanine methyl ester with the receptor site of the sweet taste bud. *J. Am. Chem. Soc.* **98**, 6669–6675.
7. Pautet, F., and Nofre, C. 1978. Correlation of chemical structure and taste in the cyclamate series and the steric nature of the chemoreceptor site. *Z. Lebensm. Unters.-Forsch.* **166**, 167–170.
8. Horowitz, R. M., and Gentili, B. 1974. Dihydrochalcone sweeteners. In *Sweeteners,* ed. G. E. Inglett, AVI, Westport, Conn.
9. Inglett, G. E. 1969. Dihydrochalcone sweeteners—sensory and stability evaluation. *J. Food Sci.* **34**, 101–103.
10. Anon. 1984. *Magnasweet,* Macandrews & Forbes Co., Camden, N.J.
11. Inglett, G. E. 1974. Sweeteners in perspective. *Cereal Sci. Today* **19**(7), 259–261, 292–295.
12. Linko, P., Saijonmaa, T., Heikonen, M., and Kreula, M. 1980. Lactitol. In *Carbohydrate Sweeteners in Foods and Nutrition,* eds. P. Koivistonen and L. Hyvönen, Academic Press, New York.
13. Horn, H. E. 1981. Corn sweeteners: Functional properties. *Cereal Foods World* **26**(5), 219–223.
14. deVos, A. M., Hatada, M., van der Wel, H., Krabbendam, H., Peerdeman, A. F., and Kim, S.-H. 1985. Three-dimensional structure of thaumatin I, an intensely sweet protein. *Proc. Natl. Acad. Sci. USA* **82**, 1406–1409.
15. Iyengar, R. B., Smits, P., van der Ouderaa, F., van der Wel, H., van Brouwershaven, J., Ravestein, P., Richters, G., and van Wassenaar, P. D. 1979. The complete amino-acid sequence of the sweet protein Thaumatin I. *Eur. J. Biochem.* **96**, 193–194.
16. Kuninaka, A. 1967. Flavor potentiator. In *The Chemistry and Physiology of Flavors,* ed. Schultz, H. W., AVI, Westport, Conn.
17. Kuninaka, A. 1981. Taste and flavor enhancers. In *Flavor Research, Recent*

Advances, eds. R. Teranishi, R. A. Flath, and H. Sugisawa, Marcel Dekker, New York.

CHAPTER 8. NATURAL TOXICANTS

SELECTED READINGS

Ames, B. N. 1983. Dietary carcinogens and anticarcinogens. *Science* **221**, 1256–1264.

Singer, B., and Kúsmierek, J. T. 1982. Chemical mutagenesis. *Ann. Rev. Biochem.* **52**, 655–693.

Hecht, S. S., Melikian, A. A., and Amin, S. 1986. Methylchrysenes as probes for the mechanism of metabolic activation of carcinogenic methylated polynuclear aromatic hydrocarbons. *Acc. Chem. Res.* **19**, 174–180.

Guengerich, P., and Liebler, D. C. 1985. Enzymatic activation of chemicals to toxic metabolites. *CRC Crit. Rev. Toxicol.* **14**(3), 259–307.

REFERENCES

1. Conn, E. E. 1969. Cyanogenic glycosides. *J. Agric. Food Chem.* **17**, 519–526.
2. Conn, E. E. 1981. Unwanted biological substances in foods: Cyanogenic glycosides. In *Impact of Toxicology on Food Processing,* eds. J. C. Ayres and J. C. Kirschman, AVI, Westport, Conn.
3. Sinden, S. L., and Webb, R. E. 1972. Effect of variety and location on the glycoalkaloid content of potatoes. *Am. Potato J.* **49**, 334–338.
4. Osman, S. F. 1983. Glycoalkaloids in potatoes. *Food Chem.* **11**, 235–247.
5. Roddick, J. G. 1979. Complex formation between solanaceous steroidal glycoalkaloids and free sterols in vitro. *Phytochemistry* **18**, 1467–1470.
6. Fenwick, G. R., Heaney, R. K., and Mullin, W. J. 1983. Glucosinolates and their breakdown products in food and food plants. *CRC Crit. Rev. Food Technol.* **18**, 123–200.
7. van Etten, C. H., Daxenbichler, M. E., and Wolff, I. A. 1969. Natural glucosinolates (thioglucosides) in foods and feeds. *J. Agric. Food Chem.* **17**, 483–491.
8. Benn, M. 1977. Glucosinolates. *Pure & Appl. Chem.* **49**, 197–210.
9. Roberts, H. R., and Barone, J. T. 1983. Biological effects of caffeine, history and use. *Food Technol.* **37**(9), 32–39.
10. von Borstel, R. W. 1983. Biological effects of caffeine. *Food Technol.* **37**(9), 40–43, 46.
11. Tarka, S. M., Jr. 1982. The toxicology of cocoa and methylxanthines: A review of the literature. *CRC Crit. Rev. Toxicol.* **9**, 275–310.
12. Anon. 1983. Caffeine. *Food Technol.* **37**(4), 87–91.
13. Snyder, S. H., Katims, J. J., Annau, Z., Bruns, R. F., and Daly, J. W. 1981. Adenosine receptors and behavioral action of methylxanthines. *Proc. Natl. Acad. Sci. USA* **78**, 3260–3264.

14. Bell, E. A. 1980–1981. The structure and biosynthesis of lathyrogens and related compounds. *Food Chem.* **6**, 213–222.
15. Wieland, T. 1968. Poisonous principles of mushrooms of the genus *Amanita*. *Science* **159**, 946–952.
16. Gigliotti, H. J., and Levenberg, B. 1964. Studies in the γ-glutamyltransferase of *Agaricus bisporus*. *J. Biol. Chem.* **239**, 2274–2284.
17. Ross, A. E., Nagel, D. L., and Toth, B. 1982. Evidence for the occurrence and formation of diazonium ions in the *Agaricus bisporus* mushroom and its extracts. *J. Agric. Food Chem.* **30**, 521–525.
18. Simpson, L. L. 1981. The origin, structure, and pharmacological activity of botulinum toxin. *Pharmacol. Rev.* **33**, 155–188.
19. Simpson, L. L. 1986. Molecular pharmacology of botulinum toxin and tetanus toxin. *Ann. Rev. Pharmacol. Toxicol.* **26**, 427–453.
20. Lovenberg, W. 1974. Psycho- and vasoactive compounds in food substances. *J. Agric. Food Chem.* **22**, 23–26.
21. Smith, T. A. 1980–1981. Amines in food. *Food Chem.* **6**, 169–200.
22. Taylor, S. L. 1986. Histamine food poisoning: Toxicology and clinical aspects. *CRC Crit. Rev. Toxicol.* **17**, 91–128.
23. Swenson, D. H., Miller, J. A., and Miller, E. C. 1975. The reactivity and carcinogenicity of aflatoxin B_1-2,3-dichloride, a model for the putative 2,3-oxide metabolite of aflatoxin B_1. *Cancer Res.* **35**, 3811–3823.
24. Croy, R. G., and Wogan, G. N. 1981. Temporal patterns of covalent DNA adducts in rat liver after single and multiple doses of aflatoxin B_1. *Cancer Res.* **41**, 197–203.
25. Patterson, D. S. P., and Roberts, B. A. 1970. The formation of aflatoxins B_{2a} and G_{2a} and their degradation products during the in vitro detoxification of aflatoxin by livers of certain avian and mammalian species. *Food Cosmet. Toxicol.* **8**, 527–538.
26. Norred, W. P. 1982. Ammonia treatment to destroy aflatoxins in corn. *J. Food Protection* **45**, 972–976.
27. Fretheim, K. 1983. Polycyclic aromatic hydrocarbons in grilled meat products—a review. *Food Chem.* **10**, 129–139.
28. Huberman, E., Sachs, L., Yang, S. K., and Gelboin, H. V. 1976. Identification of mutagenic metabolites of benzo[a]pyrene in mammalian cells. *Proc. Natl. Acad. Sci. USA* **73**, 607–611.
29. King, H. W., Osborne, M. R., Beland, F. A., Harvey, R. G., and Brookes, P. 1976. (±)-7 α, 8 β-dihydroxy-9β, 10 β-epoxy-7,8,9,10-tetrahydrobenzo[a]pyrene in an intermediate in the metabolism and binding to DNA of benzo[a]pyrene. *Proc. Natl. Acad. Sci. USA* **73**, 2679–2681.
30. Hashimoto, Y., Shudo, K., and Okamoto, T. 1984. Mutagenic chemistry of heteroaromatic amines and mitomycin C. *Acc. Chem. Res.* **17**, 403–408.
31. Hashimoto, Y., Shudo, K., and Okamoto, T. 1980. Activation of a mutagen, 3-amino-methyl-5H-pyrido[4,3-b]indole. Identification of 3-hydroxyamino-1-methyl-5H-pyrido[4,3-b]indole and its reaction with DNA. *Biochem. Biophys. Res. Comm.* **96**, 355–362.

32. Chow, Y. L. 1973. Nitrosamine photochemistry: Reactions of aminium radicals. *Acc. Chem. Res.* **6**, 354–360.

33. Fan, T.-Y, and Tannenbaum, S. R. 1973. Fractors influencing the rate of formation of nitrosomorpholine from morpholine and nitrite: Acceleration by thiocyanate and other anions. *J. Agric. Food Chem.* **21**, 237–240.

34. Davies, R., Massey, R. C., and McWeeny, D. J. 1980–1981. The catalysis of the *N*-nitrosation of secondary amines by nitrosophenols. *Food Chem.* **6**, 115–122.

35. Coleman, M. H. 1978. A model system for the formation of *N*-nitrosopyrrolidine in grilled or fried bacon. *J. Food Technol.* **13**, 55–69.

36. Bharucha, K. R., Cross, C. K., and Rubin, L. J. 1979. Mechanism of *N*-nitrosopyrrolidine formation in bacon. *J. Agric. Food Chem.* **27**, 63–69.

37. Crosby, N. T., and Sawyer, R. 1976. N-nitrosamines: a review of chemical and biological properties and their estimation in foodstuffs. *Adv. Food Res.* **22**, 1–56.

38. Wakabayashi, K., Ochiai, M., Saito, H., Tsuda, M., Suwa, Y, Nagao, M., and Sugimura, T. 1983. Presence of 1-methyl-1,2,3,4-tetrahydro-β-carboline-3-carboxylic acid, a precursor of a mutagenic nitroso compound, in soy sauce. *Proc. Natl. Acad. Sci. USA* **80**, 2912–2916.

39. Bosin, T. R., Krogh, S., and Mais, D. 1986. Identification and quantitation of 1,2,3,4-tetrahydro-β-carboline-3-carboxylic acid and 1-methyl-1,2,3,4-tetrahydro-β-carboline-3-carboxylic acid in beer and wine. *J. Agric. Food Chem.* **34**, 843–847.

40. Yang, D., Tannenbaum, S. R., Buchi, G., and Lee, G. C. M. 1984. 4-Chloro-6-methoxyindol is the precursor of a potent mutagen (4-chloro-6-methoxy-2-hydroxy-1-nitroso-indolin-3-one oxime) that forms during nitrosation of the fava bean (*Vicia faba*). *Carcinogenesis* **5**, 1219–1224.

41. Archer, M. C. 1982. Reactive intermediates from nitrosamines In *Biological Reactive Intermediates—II, Advances in Experimental Medicine and Biology,* Vol. 136B, ed. R. Snyder, Plenum Press, New York.

CHAPTER 9. ADDITIVES

SELECTED READINGS

McWeeny, D. J. 1982. Reactions of some food additives during storage. *Food Chem.* **9**, 89–101.

Walker, R. 1984. Biological and toxicological consequences of reactions between sulfur (IV) oxoanions and food components. *Food Chem.* **15**, 127–138.

Hotchkiss, J. H. 1987. Nitrate, nitrite, and nitroso compounds in foods. *Food Technol.* **41**(4), 127–134.

Anselme, J.-P. 1979. The organic chemistry of *N*-nitrosamines: A brief review. In *N-Nitrosamines,* ed. J.-P. Anselme, ACS Sym. Ser. 101, American Chemical Society, Washington, D.C.

REFERENCES

1. Ellinger, R. H. 1972. *Phosphates as Food Ingredients,* CRC Press, Cleveland, Ohio.
2. Anon. 1984. *Food Phosphates,* Monsanto Nutritional Chemicals Division, St. Louis, Mo.
3. Irani, R. R., and Morgenthaler, W. W. 1963. Iron sequestration by phosphates. *JAOCS* **40**, 283–285.
4. Hamm, R. 1971. Interaction between phosphates and meat proteins. In *Phosphates in Food Processing,* eds. J. M. Deman and P. Melnychyn, AVI, Westport, Conn.
5. Offer, G., and Trinick, J. 1983. On the mechanism of water holding in meat: the swelling and shrinking of myofibrils. *Meat Science* **8**, 245–281.
6. Conn, J. F. 1981. Chemical leavening systems in flour products. *Cereal Food World* **26**(3), 119–123.
7. Clusker, J. P. 1980. Citrate conformation and chelation: Enzymatic implications. *Acc. Chem. Res.* **13**, 345–352.
8. Schmidt, T. R. 1983. *The Use of Citric Acid in The Canned Fruit and Vegetable Industry,* Miles Laboratories, Inc., Biotech Products Division, Elkhart, In.
9. Irwin, W. E. 1983. *The Use of Citric Acid in The Beverage Industry,* Miles Laboratories, Inc., Biotech Products Division, Elkhart, In.
10. Strouse, J., Layten, S. W., and Strouse, C. E. 1977. Structural studies of transition metal complexes of triionized and tetraionized citrate. Models for the coordination of the citrate ion to transition metal ions in solution and at the active site of aconitase. *J. Am. Chem. Soc.* **99**, 562–572.
11. Freese, E., Sheu, C. W., and Galliers, E. 1973. Function of lipophilic acids as antimicrobial food additives. *Nature* **241**, 321–325.
12. Shen, C. W., Konings, W. N., and Freese, E. 1972. Effects of acetate and other short-chain fatty acids on sugar and amino acid uptake of *Bacillus substilis. J. Bacteriol.* **111**, 525–530.
13. Hunter, D. R., and Segel, I. H. 1973. Effect of weak acids on amino acid transport by *Penicillium chrysogenum:* Evidence for a proton or charge gradient as the driving force. *J. Bacteriol.* **113**, 1184–1192.
14. Warth, A. D. 1977. Mechanism of resistance of *Saccharomyces Bacilli* to benzoic, sorbic and other weak acids used as food preservatives. *J. Appl. Bacteriol.* **43**, 215–230.
15. Guthrie, J. P. 1979. Tautomeric equilibria and pK_a values for "sulfurous acid" in aqueous solution: A thermodynamic analysis. *Can. J. Chem.* **57**, 454–457.
16. Wedzicha, B. L. 1985. *Chemistry of Sulfur Dioxide in Foods,* Elsevier Applied Science Publ., London and New York.
17. Ingles, D. L. 1962. The formation of sulphonic acids from the reaction of reducing sugars with sulphite. *Aust. J. Chem.* **15**, 342–349.
18. Wedzicha, B. L., and McWeeney, D. J. 1974. Non-enzymatic browning reactions of ascorbic acid and their inhibition. The production of 3-deoxy-4-sulphopentosulose in mixtures of ascorbic acid, glycine and bisulphite ion. *J. Sci. Food Agric.* **25**, 577–587.

19. Schimz, K.-L. 1980. The effect of sulfite on the yeast *Saccharomyces cerevisiae. Arch. Microbiol.* **125**, 89–95.
20. Hayatsu, H., Wataya, Y., and Kal, K. 1970. The addition of sodium bisulfite and uracil and to cytosine. *J. Am. Chem. Soc.* **92**, 724–726.
21. Shapiro, R., Welcher, M., Nelson, V., and DiFate, V. 1976. Reaction of uracil and thymine derivatives with sodium bisulfite. Studies on the mechanism and reduction of the adduct. *Biochem. Biophys. Acta* **425**, 115–124.
22. Shapiro, R., DiFate, V., and Welcher, M. 1974. Deamination of cytosine derivatives by bisulfite. Mechanism of the reaction. *J. Am. Chem. Soc.* **96**, 906–912.
23. Shapiro, R., and Gazit, A. 1976. Crosslinking of nucleic acids and proteins by bisulfite. In *Protein Crosslinking, Biochemical and Molecular Aspects,* ed. M. Friedman, Plenum Press, New York.
24. Hayon, E., Treinin, A., and Wilf, J. 1972. Electronic spectra, photochemistry, and autoxidation mechanism of the sulfite-bisulfite-pyrosulfite system. The SO_2^-, SO_3^-, SO_4^- and SO_5^- radicals. *J. Am. Chem. Soc.* **94**, 47–57.
25. Yang, S. F. 1970. Sulfoxide formation from methionine or its sulfide analogs during aerobic oxidation of sulfite. *Biochemistry* **9**, 5008–5014.
26. Wedzicha, B. L., and Lamikanra, L. 1983. Sulfite mediated destruction of β-carotene: The partial characterization of reaction products. *Food Chem.* **10**, 275–283.
27. Lizada, M. C. C., and Yang, S. F. 1981. Sulfite-induced lipid peroxidation. *Lipids* **16**, 189–194.
28. Hayatsu, H. 1969. The oxygen-catalyzed reaction between 4-thiouridine and sodium sulfite. *J. Am. Chem. Soc.* **91**, 5693–5694.
29. Gunnison, A. F. 1981. Sulphite toxicity: A critical review of in vitro and in vivo data. *Food Cosmet. Toxicol.* **19**, 667–682.
30. Gunnison, A. F., Dulak, L., Chiang, G., Zaccardi, J., and Farruggella, T. J. 1981. A sulphite oxidase-deficient rat model: Subchronic toxicology. *Food Cosmet. Toxicol.* **19**, 221–232.
31. Til, H. P., Feron, V. J., and deGroot, A. P. 1972. The toxicity of sulphite. I. Long term feeding and multigeneration studies in rats. *Food Cosmet. Toxicol.* **10**, 291–310.
32. Taylor, S. L., and Bush, R. K. 1986. Sulfites as food ingredients. Food Technol. **40**(6), 47–52.
33. Cohen, H. J., Drew, R. T., Johnson, J. L., and Rajagopalan, K. V. 1973. Molecular basis of the biological function of molybdenum. The relationship between sulfite oxidase and the acute toxicity of bisulfite and SO_2. *Proc. Natl. Acad. Sci. USA* **70**, 3655–3659.

CHAPTER 10. VITAMINS

SELECTED READINGS

Walsh, C. 1980. Flavin coenzymes: At the crossroads of biological redox chemistry. *Acc. Chem. Res.* **13**, 148–155.

McCay, P. B. 1985. Vitamin E: Interactions with free radicals and ascorbate. *Ann. Rev. Nutr.* **5**, 323–340.

Burton, G. W., and Ingold, K. U. 1986. Vitamin E: Application of the principles of physical organic chemistry to the exploration of its structure and function. *Acc. Chem. Res.* **19**, 194–201.

Liao, M.-L., and Seib, P. A. 1987. Selected reactions of L-ascorbic acid related to foods. *Food Technol.* **41**(11), 104–107, 111.

REFERENCES

1. Abrahamson, E. W. 1975. Dynamic processes in vertebrate rod visual pigments and their membranes. *Acc. Chem. Res.* **8**, 101–106.

2. Drujan, B. D. 1971. Determination of vitamin A. *Methods in Enzymology* **XVII**, Part C.

3. Krinsky, N. T. 1979. Carotenoid protection against oxidation. *Pure & Appl. Chem.* **51**, 649–660.

4. Dwivedi, B. K., and Arnold, R. G. 1973. Chemistry of thiamine degradation in food products and model systems: A review. *J. Agric. Food Chem.* **21**, 54–60.

5. Doerge, D. R., and Ingraham, L. L. 1980. Kinetics of thiamine cleavage by bisulfite ion. *J. Am. Chem. Soc.* **102**, 4828–4830.

6. Zoltewiez, J. A., Uray, G., and Kauffman, G. M. 1980. Evidence for an intermediate in nucleophilic substitution of a thiamin analogue. Change from first- to second-order kinetics in sulfite ion. *J. Am. Chem. Soc.* **102**, 3653–3654.

7. van Dort, H. M., van der Linde, L. M., and de Rijke, D. 1984. Identification and synthesis of new odor compounds from photolysis of thiamin. *J. Agric. Food Chem.* **32**, 454–457.

8. Hevesi, L., and Bruice, T. C. 1973. Reactions of sulfite with isoalloxazines. *Biochemistry* **12**, 290–297.

9. Heelis, P. F. 1982. The photophysical and photochemical properties of flavins (isoalloxazines). *Chem. Soc. Rev.* **11**(1), 15–39.

10. Martell, A. E. 1982. Reaction pathways and mechanisms of pyridoxal catalysis. *Adv. Enzymol.* **53**, 163–199.

11. Halpern, J. 1985. Mechanisms of coenzyme B_{12}-dependent rearrangements. *Science* **227**, 869–875.

12. Johnson, A. W. 1980. Vitamin B_{12}. Retrospect and prospects. *Chem. Soc. Rev.* **9**(2), 125–141.

13. Kluger, R., and Adawadkar, P. D. 1976. A reaction proceeding through intramolecular phosphorylation of a urea. A chemical mechanism for enzymic carboxylation of biotin involving cleavage of adenosine 5'-triphosphate. *J. Am. Chem. Soc.* **98**, 3741–3742.

14. O'Keefe, S. J., and Knowles, J. R. 1986. Enzymatic biotin-mediated carboxylation is not a concerted process. *J. Am. Chem. Soc.* **108**, 328–329.

15. Laroff, G. P., Fessenden, R. W., and Schuler, R. H. 1972. The electron spin resonance spectra of radical intermediates in the oxidation of ascorbic acid and related substances. *J. Am. Chem. Soc.* **94**, 9062–9073.

16. Packer, J. E., Slater, T. F., and Willson, R. L. 1979. Direct observation of a free radical interaction between vitamin E and vitamin C. *Nature* **278**, 737–738.
17. Kurata, T., and Sakurai, Y. 1967. Degradation of L-ascorbic acid and mechanism of nonenzymatic browning reaction. Part II. Non-oxidative degradation of L-ascorbic acid including the formation of 3-deoxy-L-pentosone. *Agric. Biol. Chem.* **31**, 170–176.
18. Kurata, T., and Fujimaki, M. 1976. Formation of 3-keto-4-deoxypentosone and 3-hydroxy-2-pyrone by the degradation of dehydro-L-ascorbic acid. *Agric. Biol. Chem.* **40**, 1287–1291.
19. Kurata, T., Fujimaki, M., and Sakurai, Y. 1973. Red pigment produced by the reaction of dehydro-L-ascorbic acid with α-amino acid. *Agric. Biol. Chem.* **37**, 1471–1477.
20. Martell, A. E. 1982. Chelates of ascorbic acids. Formation and catalytic properties. In *Ascorbic Acid: Chemistry, Metabolism, and Uses,* eds. P. A. Seib and B. M. Tolbert, Adv. Chem. Ser. 200, American Chemical Society, Washington D.C.
21. Mair, G., and Grosch, W. 1979. Changes in glutathione content (reduced and oxidized form) and the effect of ascorbic acid and potassium bromate on glutathione oxidation during dough mixing. *J. Sci. Food Agric.* **30**, 914–920.
22. DeLuca, H. F., and Schnoes, H. K. 1983. Vitamin D: Recent advances. *Ann. Rev. Biochem.* **52**, 411–439.
23. Wing, R. M., Okamura, W. H., Rego, A., Pirio, M. R., and Norman, A. W. 1975. Studies on vitamin D and its analogs. VII. Solution conformations of vitamin D_3 and $1\alpha,25$-dihydroxyvitamin D_3 by high-resolution proton magnetic resonance spectroscopy. *J. Am. Chem. Soc.* **97**, 4980–4985.
24. Jones, H., and Rasmusson, G. H. 1980. Recent advances in the biology and chemistry of vitamin D. In *Progress in The Chemistry of Organic Natural Products,* eds. W. Herz, H. Grisebach, and G. W. Kirby, Springer-Verlag Wien, New York.
25. Gruger, E. H., Jr., and Tappel, A. L. 1970. Reactions of biological antioxidants: 1. Fe(III)-catalyzed reactions of lipid hydroperoxides with α-tocopherol. *Lipids* **5**, 326–331.
26. Burton, G. W., and Ingold, K. U. 1981. Autoxidation of biological molecules. 1. The antioxidant activity of vitamin E and related chain-breaking phenolic antioxidants in vitro. *J. Am. Chem. Soc.* **103**, 6472–6477.
27. Gorman, A. A., Gould, I. R., Hamblett, I., and Standen, M. C. 1984. Reversible exciplex formation between singlet oxygen, $^1\Delta g$, and vitamin E. Solvent and temperature effect. *J. Am. Chem. Soc.* **106**, 6956–6959.
28. Clough, R. L., Yee, B. G., and Foote, C. S. 1979. Chemistry of singlet oxygen. 30. The unstable primary product of tocopherol photooxidation. *J. Am. Chem. Soc.* **101**, 683–686.

Appendix 1

GENERAL KINETICS OF OLEFIN AUTOXIDATION

Initiation: $\quad$ LH $\xrightarrow{k_i}$ L· + H· (Rate of initiation = R_i) $\qquad$ (1)

Propagation: $\quad$ L· + O$_2$ $\xrightarrow{ko}$ LOO· $\qquad$ (2)

$\qquad\qquad$ LOO· + LH $\xrightarrow{k_p}$ LOOH + L· $\qquad$ (3)

Termination: $\quad$ LOO· + LOO· $\xrightarrow{k_t}$ $\qquad$ (4)

$\qquad\qquad$ LOO· + L· $\xrightarrow{k_t{'}}$ $\left.\vphantom{\begin{array}{c}1\\1\\1\end{array}}\right\}$ Nonradical products $\qquad$ (5)

$\qquad\qquad$ L· + L· $\xrightarrow{k_t{''}}$ $\qquad$ (6)

Applying steady-state conditions,

$$\partial[\text{L·}]/\partial t = R_i - k_o[\text{L·}][\text{O}_2] + k_p[\text{LOO·}][\text{LH}] - k_t{'}[\text{LOO·}][\text{L·}]$$
$$- k_t{''}[\text{L·}]^2 = 0 \qquad (7)$$

$$\partial[\text{LOO·}]/\partial t = k_o[\text{L·}][\text{O}_2] - k_p[\text{LOO·}][\text{LH}] - k_t[\text{LOO·}]^2$$
$$- k_t{'}[\text{LOO·}][\text{L·}] = 0 \qquad (8)$$

401

Addition of (7) and (8),

$$R_i - 2k_{t'}[\text{LOO·}][\text{L·}] - k_t[\text{LOO·}]^2 - k_{t''}[\text{L·}]^2 = 0 \tag{9}$$

Assuming $k_t = k_{t'} = k_{t''}$,

$$R_i/k_t = [\text{LOO·}]^2 + 2[\text{LOO·}][\text{L·}] + [\text{L·}]^2$$
$$= ([\text{LOO·}] + [\text{L·}])^2$$
$$(R_i/k_t)^{1/2} = [\text{LOO·}] + [\text{L·}] \tag{10}$$

But $k_o[\text{L·}][\text{O}_2] = k_p[\text{LOO·}][\text{LH}]$,

$$[\text{L·}] = \frac{k_p[\text{LOO·}][\text{LH}]}{k_o[\text{O}_2]} \tag{11}$$

Substituting (11) into (10),

$$\left(\frac{R_i}{k_t}\right)^{1/2} = [\text{LOO·}] + \frac{k_p[\text{LOO·}][\text{LH}]}{k_o[\text{O}_2]}$$
$$= [\text{LOO·}]\ \frac{k_o[\text{O}_2] + k_p[\text{LH}]}{k_o[\text{O}_2]}$$
$$[\text{LOO·}] = \left(\frac{R_i}{k_t}\right)^{1/2} \frac{k_o[\text{O}_2]}{k_p[\text{LH}] + k_o[\text{O}_2]} \tag{12}$$

Since $\partial[\text{LOOH}]/\partial t = k_p[\text{LOO·}][\text{LH}]$, by substituting (12)

$$-\frac{\partial[\text{O}_2]}{\partial t} = \frac{\partial[\text{LOOH}]}{\partial t} = \left(\frac{R_i}{k_t}\right)^{1/2} \frac{k_p[\text{LH}]k_o[\text{O}_2]}{k_p[\text{LH}] + k_o[\text{O}_2]} \tag{13}$$

The termination reactions (5) and (6) become negligible at high pressures of oxygen (above 100 mm Hg for most olefins), since the peroxy radical (LOO·) is the dominant species. Applying steady-state conditions,

$$\partial[\text{L·}]/\partial t = R_i - k_o[\text{L·}][\text{O}_2] + k_p[\text{LOO·}][\text{LH}] = 0 \tag{14}$$

$$\partial[\text{LOO·}]/\partial t = k_o[\text{L·}][\text{O}_2] - k_p[\text{LOO·}][\text{LH}] - k_t[\text{LOO·}]^2 = 0 \tag{15}$$

Addition of (14) and (15),

$$R_i - k_t[\text{LOO·}]^2 = 0$$

$$[LOO\cdot] = \left(\frac{R_i}{k_t}\right)^{1/2}$$

(16)

Since $\partial[LOOH]/\partial t = k_p[LOO\cdot][LH]$, by substituting (16),

$$-\frac{\partial[O_2]}{\partial t} = \frac{\partial[LOOH]}{\partial t} = \left(\frac{R_i}{k_t}\right)^{1/2} k_p[LH]$$

(17)

Appendix 2
SINGLET OXYGEN

Oxygen atom has an electronic configuration $1s^2 2s^2 2p_x^2 2p_y^1 2p_z^1$. The diatomic molecule has eight 2p electrons. Six of these occupy the $\sigma 2p$, $\pi_y 2p$, and $\pi_z 2p$ orbitals (two paired electrons each, Pauli exclusion principle). The $\pi_y^* 2p$ and $\pi_z^* 2p$ antiorbitals are degenerate (with equivalent energy levels), each having one electron with parallel spin (Hund's rule). The molecular orbital scheme for O_2 is shown in Fig. S2.1.

The molecule contains two unpaired electrons with parallel spin in the highest molecular orbital (HOMO), each having a magnetic moment. The two magnetic fields interact in three ways:

1. Reinforce each other to augment an extended field.
2. Counteract to decrease an external field.
3. Cancel each other.

The state of possessing two unpaired electrons in the HOMO is called a triplet state, commonly referred to as having a multiplicity of 3. Multiplicity is given by the Eq. S2-1.

$$S = 2s + 1$$

Eq. S2-1

where S = multiplicity and s = total spin (an electron has spin of $1/2$). For the oxygen molecule,

$$S = 2(+1/2 + 1/2) + 1 = 3$$

Eq. S2-2

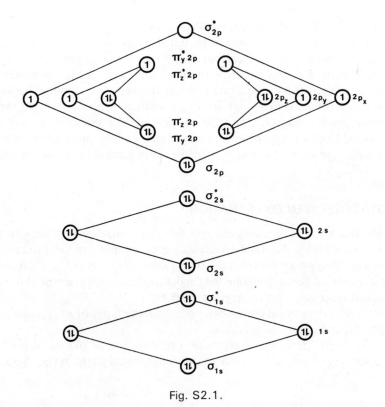

Fig. S2.1.

The ground state of the oxygen molecule (O_2) is thus a triplet, which is unusual, since most molecules have singlet ground states. The two excited states of the oxygen molecule are singlets which are 24 and 37 kcal/mole above the ground state. The electronic configurations of the HOMO of the two singlets are represented in Fig. S2.2.

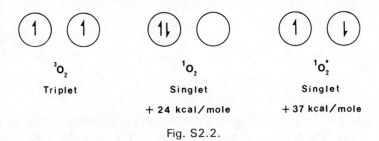

Fig. S2.2.

The $^1O_2{}^*$ state is extremely unstable and decays rapidly to the $^1O_2$ state. The singlet oxygen $^1O_2$ has one unoccupied π^*2p orbital and therefore is very electrophilic, seeking electrons to fill the empty orbital. The singlet species, therefore, reacts readily with unsaturated fatty acids, for example.

Singlet oxygen can be generated by (1) the chemical reaction between H_2O_2 and strong oxidizing species (e.g., sodium hypochlorite, triphenyl phosphate ozonide) and (2) the action of light in the presence of a sensitizer such as chlorophyll, flavin, heme compounds, porphyrins, and some synthetic colors in food systems. It is in the latter reaction that we are most interested.

PHOTOSENSITIZED OXYGENATION

A sensitizer absorbs light and changes from the singlet ground state to the singlet excited state. Intersystem crossing occurs rapidly to yield the triplet state. The energy transfer from the triplet state of the sensitizer is captured by the triplet oxygen to generate the singlet oxygen. (This is a type II photosensitized reaction; refer to Appendix 3.)

The singlet oxygen generated can undergo three types of oxygenation reaction, depending on the reactant:

1. The "ene" reaction—oxygenation of unsaturated olefins to give allylic hydroperoxides, with a shift of the double bond position (Eq. S2-3.1).

Eq. S2-3

2. The 4 + 2 cycloaddition—oxygenation of cyclic dienes, polycyclic aromatics, and heterocyclic compounds to form cyclic peroxides analogous to the Diels-Alder reaction (Eq. S2-3.2).

3. The "dioxetane" reaction—certain olefins react with oxygen by 1,2-cycloaddition to form dioxetane intermediate, which cleaves to carbonyl compounds. This reaction requires the olefins activated by amino or alkoxy groups, and the absence of very active allylic hydrogen in the molecule (Eq. S2-3.3).

PHYSICAL AND CHEMICAL QUENCHING

In the equation in the previous section, the singlet oxygen is converted to its triplet state by transferring the energy to the formation of hydroperoxides and cycloaddition products. This type of energy quenching, in which the quencher is removed with the formation of new products, is referred to as chemical quenching (or destructive quenching). A general equation for this type of quenching is given in Eq. S2-4.

$$Q \;+\; {}^1O_2 \;\longrightarrow\; QO_2 \;+\; \text{other oxidation products} \qquad \text{Eq. S2-4}$$

Another type of quenching process is called physical quenching (or degenerative quenching). The energy is transferred from $^1O_2$ to the quencher, resulting in the latter molecule being converted to the excited triplet state, which subsequently decays to the ground state (Eq. S2-5). The structure of the quencher molecule is not chemically changed. Both α-tocopherol and β-carotene are effective physical quenchers of the $^1O_2$ molecule, although α-tocopherol has been shown to form a charge-transfer exciplex with singlet oxygen. The latter mechanism is discussed in more detail in Chapter 10.

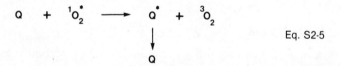

$$Q \;+\; {}^1O_2^* \;\longrightarrow\; Q^* \;+\; {}^3O_2$$
$$\downarrow$$
$$Q$$

Eq. S2-5

Appendix 3

WHERE DO THE RADICALS COME FROM?

There are three main processes by which free radicals may be produced:

1. Photolysis
2. Radiolysis
3. Molecular homolysis

PHOTOLYSIS

A molecule can absorb a quantum of energy accompanied by a transition to the higher energy state. Depending on the wavelength, there can be various kinds of transitions, (1) rotation, vibration, and excitation by infrared light; and (2) electronic excitation by ultraviolet and visible light. Higher-energy radiation may cause ionization of the molecule and will be discussed under radiolysis (Fig. S3.1).

Photodissociation

Most aldehydes and ketones under a photodissociation reaction (Eq. S3-1). If γ-hydrogen is present, intramolecular hydrogen abstraction usually occurs, resulting in β-scission (Eq. S3-2).

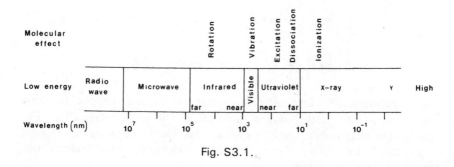

Fig. S3.1.

$$R-\overset{\overset{\displaystyle O}{\|}}{C}-R' \xrightarrow[n-\pi^*]{h\nu} \left[R-\overset{\overset{\displaystyle O}{\|}}{C}-R'\right]^* \longrightarrow R-\overset{\overset{\displaystyle O}{\|}}{C}\cdot \ + \ R'\cdot \qquad \text{Eq. S3-1}$$

$$R-\overset{\overset{\displaystyle O}{\|}}{C}-\underset{\alpha}{CH_2}\underset{\beta}{CH_2}\underset{\gamma}{CH_2}R' \longrightarrow \left[\begin{array}{c} H \quad R' \\ \backslash \quad / \\ O\cdot \quad CH \\ \diagdown \quad / \\ R-C \quad CH_2 \\ \diagdown \quad / \\ CH_2 \end{array}\right]^* \rightleftarrows \left[\begin{array}{c} R' \\ | \\ \cdot CH \\ OH \quad / \\ \diagdown \quad / \\ R-C \quad CH_2 \\ \diagdown \quad / \\ CH_2 \end{array}\right]^*$$

Eq. S3-2

$$R-\overset{\overset{\displaystyle O}{\|}}{C}-CH_3 \longleftarrow R-\overset{\overset{\displaystyle OH}{|}}{C}=CH_2 \ + \ CH_2=CHR'$$

Photosensitized Reactions

Light irradiation causes electronic excitation of sensitizers, such as chlorophyll, flavin, heme compounds, and porphyrins, from the ground state to the excited states ($^1Sen^*$ and $^3Sen^*$). The excited singlet decays rapidly, while the triplet can initiate two types of reactions.

Type I Reaction.   The sensitizer triplet reacts directly with another molecule in the form of hydrogen abstraction (Eq. S3-3), or electron transfer (Eq. S3-4). The radical products can abstract H· from another molecule and initiate other free-radical chain reactions.

$$^3\text{Sen}^* + \text{RH} \longrightarrow \cdot\text{Sen-H} + \text{R}\cdot \qquad \text{Eq. S3-3}$$

$$^3\text{Sen}^* + \text{R} \longrightarrow \text{Sen}^{\cdot -} + \text{R}^{\cdot +} \qquad \text{Eq. S3-4}$$

Type II Reaction.   The ^{3}Sen* can transfer the energy to oxygen, resulting in the excitation of the oxygen molecule to the singlet state ($^1\text{O}_2$) (Eq. S3-5). The electrophilic singlet oxygen is reactive to unsaturated systems, for example, and forms oxygenated addition or substitution products with many organic compounds.

$$^3\text{Sen}^* + {}^3\text{O}_2 \longrightarrow {}^0\text{Sen} + {}^1\text{O}_2 \qquad \text{Eq. S3-5}$$

RADIOLYSIS

The ionization potentials of most molecules of biological interest are within the range of 10 eV. In practice, radiation chemical studies are carried out with radiation energy in the range of 0.1–20 MeV (Fig. S3.2).

In biological and food systems, where water is the major constituent, most damage is caused by the indirect effect produced by radiolysis of water. The reactive species produced by the radiolysis of water are (1) hydroxy radical ($\cdot$OH), (2) hydrogen atom (H$\cdot$), and (3) hydrated electron (e_{aq}^-). The G values (number of molecules or radicals formed per 100 eV of energy absorbed) are 2.8, 0.7, and 2.7, respectively. With the typical irradiation

	MeV	Charge
ELECTROMAGNETIC		
x – ray	0.3 – 0.5	0
γ – ray	0.2 – 2.0	0
PARTICLES		
electron	1 – 30	–1
β – particle	0.01 – 2	–1
α – particle	1 – 45	+2
neutron	2 – 30	0
VISIBLE LIGHT	0.4 – 3 (eV)	
ULTRAVIOLET	3 – 6 (eV)	

Fig. S3.2.

energy levels in the MeV range, the energy of radiation is in excess of the bond energies or ionization or excitation potentials of the absorbing molecule.

The chemical changes of water can be expressed by Eq. S3-6.

IONIZATION	$H_2O \longrightarrow H_2O^{+\cdot} + e^-$	10^{-18} sec
EXCITATION	$H_2O \longrightarrow H_2O^*$	10^{-15}
ION–MOLECULAR RXN	$H_2O^{+\cdot} + H_2O \longrightarrow H_3O^+ + \cdot OH$	10^{-14} Eq. S3-6
DISSOCIATION	$H_2O^* \longrightarrow H\cdot + \cdot OH$	10^{-13}
HYDRATION	$e^- + nH_2O \longrightarrow e^-_{aq}$	10^{-13}

Both $H\cdot$ and e_{aq}^- are strong reducing agents. The hydrated electron reacts by the following two types of reactions:

1. *Electron capture* (Eq. S3-7)
2. *Dissociation* (Eq. S3-8)

$$e^-_{aq} + R \longrightarrow R^{+\cdot} \qquad \text{Eq. S3-7}$$

$$e^-_{aq} + RH \longrightarrow H\cdot + R^- \qquad \text{Eq. S3-8}$$

The hydroxy radical, a strong oxidizing agent, undergoes the following types of reactions.

1. *Electron transfer* (Eq. S3-9)
2. *Hydrogen abstraction* (Eq. S3-10)
3. *Addition* (Eq. S3-11)

$$HO\cdot + R^- \longrightarrow OH^- + R\cdot \qquad \text{Eq. S3-9}$$

$$HO\cdot + RH \longrightarrow H_2O + R\cdot \qquad \text{Eq. S3-10}$$

$$HO\cdot + R \longrightarrow \cdot ROH \qquad \text{Eq. S3-11}$$

MOLECULAR HOMOLYTIC DECOMPOSITION

The most studied system in this category is the reaction in the decomposition of lipid hydroperoxides. Peroxide bonds $-O-O-$ are relatively weak. For example, t-butyl peroxide has a bond energy of 37 kcal/mole compared with the ~ 90 kcal/mole for the normal $C-C$ bond. Homolytic scission of the $-O-O-$ occurs when sufficient heat energy is applied.

In the biological and food systems, decomposition is largely induced by other radicals in the system, and the resulting chain reactions are fast even at room temperature (Eq. S3-12).

$$R\cdot + ROOH \longrightarrow ROO\cdot + RH \qquad \text{Eq. S3-12}$$

Decomposition is also accelerated by other molecules that can form hydrogen bonding and enhance bond dissociation (Eq. S3-13).

$$\left(\begin{array}{c} R-O-OH \\ \vdots \\ H-O-O-R' \end{array} \right) \longrightarrow RO\cdot + R'OO\cdot + H_2O \qquad \text{Eq. S3-13}$$

INDEX

Absorption spectrum
  chlorophyll, 191
  conjugation, 149
  myoglobin, 185
  substituent effect, 152
  transition energy, 148
Acesulfame K, 264, 271
Acetylation, 38
Acetylcholine, 286
Acid
  monosaccharides and, 111–113
  vitamin B_1 and, 342
  see also pH
Actin, 80
Action pattern
  amylases, 218–221
  exo-polygalacturonases, 223
Acyl-enzyme intermediate
  lipase, 228
  papain, 199
Acyl transfer reaction, 229–230
Acylium ion, 15
Additives, 314–334
  antimicrobial short-chain acid derivatives,
    324–325
  citric acid, 321–324
  generally, 314
  phosphates, 314–321
  sulfite, 326–334
  vitamin D, 365
Adipic acid anhydride, 127

Adsorption
  emulsifier functions, 41–42
  proteins, 61
Aflatoxins, 299–302
Agaritine, 295
Aggregation
  β-lactoglobulin, 91
  milk micelles, 89–90
  protein gel, 62–63
  carrageenan gel, 136
Aglycoside sweeteners, 276
Alanine, 54
Alcohol(s)
  ester derivatives of, 34–36
  wheat protein classification, 92
Alcohol fractionation, 37
Alcoholate, 124
Aldose-ketose rearrangement, 108–109
Alginate, 129–132
  applications of, 131–132
  chemical structure of, 129–130
  gelling, 130–131
Alkali
  monosaccharides and, 108–111
  vitamin B_1 and, 342
  see also Acid; pH
Alkali degradation, 63–66
  beta-elimination, 64
  hydrolysis, 64
  racemization, 66
Alkaline
  sugar/metal ion complexes, 124

Alkaline (*cont.*)
  vitamin B$_2$ and, 348
Alkyldiazonium ion, 313
Allicin, 247
Alliin, 247–248
Alliinase, 248
Alloxazine, 349
(+)-S-Allyl-L-cysteine sulfoxide, 247–248
Allyl-2-propenethiosulfinate, 247–248
Allylic radical, 10
Allysine, 99–100
Alpha-amylase. *See* Amylase
Alumina hydrate [Al(OH)$_3$], 171
Allura red AC, 170
Alpha-caseins, 87
Amadori rearrangement, 113, 115, 117–119
Amatoxins, 294–295
Amines, 297–299
Amino acids
  collagen, 97–99
  heterocyclic amines, 304
  sweeteners, 267–270
  toxicants, 292–294
  water and, 53–54
  see also Protein(s); Protein structure
α-Aminoadipic acid-δ-semialdehyde, 100
1-Amino-1-deoxy-2-α-D-fructopyranose
    (Amadori product), 114, 115
2-Amino-2-deoxy-α-D-glucopyranose, 118,
    120
1-Amino-1-deoxy-2-ketose, 117
1-Amino-3-oxalylamino propionic acid, 294
Aminosulfonates, 270–271
Amygdalin, 284
Amylases, 215–221
  action pattern, 218–220
  characteristics of, 215–216
  corn sweeteners, 277
  induced-fit model, 218
  industrial uses, 221
  multimolecular process, 220–221
  reaction mechanism, 216–217
Amylopectin, 125–126
Amylose, 124–125
Anhydro sugars, 112–113
Annatto, 159
Anthocyanins, 160–166
  condensation, 164
  generally, 160
  oxidation, 165–166

pH effect on color, 160–162
  self-association/copigmentation, 164
  sulfur dioxide decoloration, 162–163
Anthracene, 151
Antimicrobial action, 328
Antimicrobial short-chain acid derivatives,
    324–325
Antioxidants, 44–47
β-Apo-8'-carotenal, 154, 157, 337
Aqueous lamella, 62
Arabinogalactans, 143–144
Arabinoxylan, 143–144
Ascorbic acid (vitamin C), 359–365
  browning, 326
  polyphenol oxidase, 210
Ascorbigen, 287, 289
Asparagusic acid, 259
ε-N-(β-Aspartyl)-lysine, 66
Aspartame, 264, 267–270
ATP (Adenosine triphosphate)
  energy conversion in muscle, 81–82
  rigor mortis, 83
Autoxidation
  described, 2–4
  general kinetics, 401–403
  olefin, 401–403
  oxymyoglobin, 184–185
  stereochemistry of product hydroperox-
    ides in, 11–12
Azetidine-2-carboxyline acid, 293

Bayer-Villiger oxidation, 165
Beer flavor, 244–246
Benzene, 57
Benzillic acid type rearrangement, 109–110
Benzenoid, 260
Benzo[a]anthracene, 303
Benzoate, 325
Benzo[b]fluoranthene, 303
Benzoin-type condensation, 341
Benzo[a]pyrene, 303
Benzoyloxyphenylacetate, 164, 167
Beta-amylase. *See* Amylases
Beta-carotene
  isomerization, 155
  singlet oxygen quench, 157
  structure, 154
  sulfite additives, 331
  vitamin A, 335–336, 339, 340
Beta-casein, 87

Beta-elimination, 64
Beta-lactoglobulin, 91
Beta-meander, 50
Beta-pleated sheet, 49
Betacyanin, 166
Betamic acid, 168
Betaxanthins, 166
Betanain, 166–168
Beta-oxidation, 237–238
Beverage caffeine, 290
Beverage flavors, 242–247
  beer, 244–246
  coffee, 246–247
  tea, 242–244
Biotin, 356–358
Bisulfite
  additive, 326
  anthocyanin, 162–163
  flavin, 348
  Maillard reaction, 121
  thiamin and, 343
Bitter taste, 232
Bixin, 154
Bleaching, 206
Botulinum neurotoxin, 296–297
Bread
  flour into dough transformation, 95–97
  lipoxygenase bleaching, 206
  staling of, 206
Brilliant blue FCF, 170
Bromelain, 199, 200
1,3-Butadiene, 149–150
Butylated hydroxyanisole (BHA), 45–47
Butylated hydroxytoluene (BHT), 45–47

Caffeic acid, 164, 247
Caffeine, 289–292
3-Caffeolyquinic acid, 247
Calcium, 130–131
Canthaxanthin, 154, 157
Capsaicin, 251
Capsaicinoids, 250–251
Caramel, 169
Carbinol base, 161–162
Carbohydrates, 105–146
  acid action on monosaccharides, 111–113
  alginate, 129–132
  alkali action on monosaccharides, 108–111
  carrageenan, 134–138

cellulose, 138–141
generally, 105
glycosidic linkage, 105–108
hemicellulose, 143–146
nonenzymatic browning (Maillard reaction), 113–122
pectin, 132–134
starch, 124–129
sugar/metal ion complexes, 123–124
xanthum gum, 141–142
Carbonium ion, 105–106
Carbonyl compounds, 75
Carboxymethyl cellulose, 139–140
Carcinogenic polycyclic aromatic hydrocarbons, 302–304
Carotene
  photosensitized oxidation, 4
  quenching, 339
  structure, 336
  with sulfite, 330–331
Carotenoids, 153–158
  color additive role, 157–158
  epoxides, 156
  generally, 153–155
  isomerization, 155–156
  photochemical reactions, 157
  tea oxidation, 244–245
  thermal degradation, 156–157
Carrageenan, 134–138
  chemical structure of, 134–135
  gelling, 135–137
  locust bean gum synergism, 137–138
  protein interaction, 137
Caseins
  α-Casein, β-Casein, κ-Casein, 87
  interaction with carrageenan, 137
  interaction with β-lactoglobulin, 91
  micelle, 88–89
  proteolysis, 89–90
Catechin
  condensation, 164, 165
  tea flavor, 243
Catechol oxidase. See Polyphenol oxidase
Catecholase. See Polyphenol oxidase
Cellulose, 138–141
  chemical structure of, 138
  gelling, 139–141
Certified color additives, 169–170
α-Chaconine, 285
Chalcone
  anthocyanidin, 162–163

Chalcone (*cont.*)
  naringin, 256
  sweetener, 271–273
Character-impact compounds, 234
Chelated iron
  ADP, 5
  perferryl ion, 6
Chemical oxidation, 74–75
Chili flavor, 250
Chlorin, 189
Chlorogenic acid, 247
Chlorophyll, 187–193
  derivatives of, 191–193
  described, 187–188
  dimers and oligomers, 189–191
  magnesium-ligand coordination, 188–189
  oxidation and reduction, 193
Chlorophyllide production, 192
Cholecalciferol (vitamin D), 365–369
Choline acetyltransferase, 297
Chorismic acid, 235, 236
Chroman ring, 369
Chrysene, 303
Chymosin, 89–90
trans-Cinnamic acid, 235–236
Cinnamon flavor, 252–253
Cinnamaldehyde, 262
Cis-trans isomerization, 19–21
Citral, 255
Citric acid, 321–324
Citrous red No. 2, 170
Coagulation (milk), 89–90
Coalescence, 30
Cocoa powder, 38
Coffee flavor, 246–247
Colipase, 226–227
Collagen, 97–104
  amino acid composition of, 97–99
  cross-linking, 99–101
  described, 97
  gelatin and, 102–104
  triple helix structure of, 97
Collagenase, 200
Colloidal calcium phosphate, 88–89
Colorants, 147–193
  annatto, 159
  anthocyanins, 160–166
  betanain, 166–168
  caramel, 169
  carotenoids, 153–158

chlorophyll, 187–193
  conjugation, 149–151
  coordination chemistry, 171–174
  dyes and lakes, 169–171
  generally, 147
  light absorption, 147–149
  metallopophyrin, 174–178
  myoglobin, 178–187
  substituent effects, 151–153
  U.S. Food and Drug Administration, 169
Complexation
  lipid-starch, 43
  sugar with metal, 123
Condensation
  amylases, 220
  anthocyanins, 164
Conformation of proteins, 54–60
Conjugated diene, 3–4, 14
Conjugated tricarbonyl system, 360
Conjugation, 149–151
Coordination chemistry, 171–174
Cooxidation, 206
Copigmentation, 164
Copper, 206
Copper-pheophytin, 192
Corn sweeteners, 277–278
Corn syrup, 169
Counterion atmosphere, 31–32
Coulomb's law, 51, 57
Creaming, 29
Cross-linking
  collagen, 99–101
  pentosans, 146
  photolysis, 72–73
  radiolysis, 69–71
  protein-lipid, 75–76
  starch, 127–128
Crossover connection, 49, 50
Crystal habit of fat, 24–26
Crystal structure, 23–24
Cured meat
  nitrosamines in, 309–310
  nitrosylmyoglobin, 186–187
  see also Meat; Meat protein
Cyanidin, 160
Cyanocobalamin, 353
Cyanogenic glycosides, 283–285
Cyanohydrin, 284
Cyclamate, 264, 270
Cycloaddition, 406–407

Cyclic adenosine 3', 3'-monophosphate
    phosphodiesterase, 291-292
Cyclic dimer, 13
Cyclic esters-hopflavor, 245-246
Cyclodopa-5-O-glycoside, 168
Cysteine, 73

γ-Decalactone, 257
Dehydroalanine, 64-66
Dehydroascorbic acid, 359
7-Dehydrocholesterol, 367
Dehydro-L-scorbamic acid, 363
Dehydrolysinonorleucine, 100
Delphinidin, 160
Delphinidin-3-[4-(p-coumaroyl)-L-
    rhamnosyl(1,6)glucosideo]-5-
    glucoside, 160-161
Denaturation, 83
5'-Deoxyadenosylcobalamin, 354
3-Deoxyglycosulose
    alkali degradation, 110
    ascorbic degradation, 362
    flavor formation, 239
    Maillard reaction, 114-115
    reaction with sulfite, 326-327
Deoxymyoglobin, 179. See also Myoglobin
Desmin, 80-81
Deuteroflavin, 349
Dialdehyde phenolate ion, 300
N,N'-Dialkyldihydropyrazine, 120
Dialkylpyrazinium compound, 119
α,γ-Diaminobutyric acid, 293
Diazonium ion, 295
3,4-Dideoxy-glycosulos-3-ene
    Maillard reaction, 114-115
    reaction with sulfite, 327
3,4-Dideoxy-4-sulfo-D-glycosulose, 121
Dienoic dimer, 13
Dielectric constants, 57
Dietary fats and oils. See Lipids
Dihydrocapsaicin, 251
Dihydrochalcone, 271-274
7,8-Dihydro-7,8-epoxybenzo[a]pyrene, 303
8,9-Dihydro-8-(N[7]-guanyl)-9-hydroxy-
    aflatoxin B_l, (AFB$_l$-N[7]-GUA), 300
8-Ds,15-Ls-dihydroperoxy-5-cis,9-trans, 11-
    cis,13-trans-epicosatetraenoic acid,
    202
Dihydrothiachrome, 342
5,6-Dihydrouracil-6-sulfonate, 328-329

Diglycerinate, 22
Dimers, 12-14, 189-191
Dimethylanthranilate, 254-255
7,7'-Dimethyl-6,8-dioxabicyclo[3.2.1]
    octane, 246
2,5-Dimethyl-4-hydroxy-3(3H)-furanone,
    257
2,6-Dimethylnaphthalene, 156
Dinitroferrohemochrome, 187
Dioxetane reaction, 340, 407
Dioxindole-3-alanine, 68, 77
Dipeptides, 267-270
Diperoxide, 8-9
Diphenol, 209-210
Disproportionation, 29-30
Disulfide bonds
    dough, 95
    protein gel, 63
    protein structure, 53
Dithiacyclopentene, 259
Dithiazine, 242
DNA
    peptide toxicants, 295
    protein and, 69-70
Dopamine, 297-298
Dopamine hydroxylase, 361
Double displacement reaction, 216-217
Dough
    ascorbic acid, 365
    dough reaction rate, 320
    flour transformed into, 95-97
    pentosans, 146
Dyes, 169-171

Egg box model, 131
Electric double layer, 40-41
Electronic structure, 174-178
Electrostatic interaction
    dipole-dipole, 51-52, 55
    ion-dipole, 51-52, 55
    in emulsion, 31-32
α-Eleostearic, 18
Emulsification (proteins), 60-62
Emulsifier functions
    adsorption at interface, 41-42
    complexation with starch, 43
    electric double layer, 40-41
    fat cyrstallization control, 44
    liquid-crystalline interface formation, 42-
        43
    protein interactions, 43-44, 60-61

Emulsifiers, 33–38
    complexation with starch, 43
    ester derivatives of alcohols, 34–36
    fat crystallization, 44
    generally, 33
    interaction with protein, 43
    lecithin, 36–38
    monoglyceride derivatives, 33–34
    monoglycerides, 33
Emulsifier-water system, 38–40
Emulsions, 27–32
    breakdown of, 29–30
    electrostatic repulsion, 31–32
    formation of, 29
    generally, 27–28
    meat, 85
    phosphate, 319
    surface tension/area, 28–29
    Van der Waals attraction, 31
Endo-polygalacturonases, 222–223
1,2-Enaminol, 114–115
Endocytosis, 296–297
"ene" Reaction, 4, 340, 406
Enediol, 108–109, 111–112, 362
Energy conversion, 81–82
Enolization
    aldose-ketose rearrangement, 108
    acid, 111
    Maillard reaction, 114–119
    vitamin C, 362
Enthalpy, 54–55
Enzymatic browning
    inhibition, 327–328
    polyphenol oxidase, 210
Enzymes, 194–230
    amylases, 215–221
    flavors and, 238
    generally, 194
    glucose oxidase, 211–214
    lipolytic enzymes, 225–230
    lipoxygenase, 199–206
    listing of, 196–197
    papain, 194–195, 198–199
    pectic enzymes, 221–225
    polyphenol oxidases, 206–210
    vitamin B_2 and, 344–348
    vitamin B_6 and, 351–353
    vitamin B_{12} and, 354–355
Epicatechin, 243
Epicatechin gallate, 243

Epichlorohydrin, 127
Epigallocatechin gallate, 243
Epoxides, 156
Epoxide hydrase, 303
Epoxy-cation intermediate, 205
5,6-Epoxy-β-ionene, 244
Ergocalciferol (vitamin D), 365–369
Erythrosine, 170
Ester derivative of alcohols, 34–36
Ethanol, 57
Ethyl-2-methylbutyrate, 257
Ethyl-*trans*-2-*cis*-4-decadienoate, 257
Ethoxylated monoglyceride, 34
Eugenol, 252
Exciplex, 373
Exo-polygalacturonases, 223–224

Fat crystallization control, 44
Fast green FCF, 170
Fats. *See* Lipids
FDA. *See* U.S. Food and Drug Administration (FDA)
Ferrihemochrome, 186
Ferrohemochrome, 186
Ferulic acid, 247
FeO_2^{2+} complex, 181–182
Ficin, 199, 200
Flavanol, 242, 244–245
Flavylium cation, 160–163
Flavor(s), 231–263
    beverages, 242–247
    character-impact compounds, 234
    citric acid additives, 321–322
    fruits, 253–257
    generally, 231
    meat flavor, 259–262
    microencapsulation of, 262–263
    odor, 232–234
    origin of, 234–242
    spice, 247–253
    taste sensation, 231–232
    vegetables, 257–259
Flavor potentiators (sweeteners), 280–282
Flocculation, 30
Flour, 95–97. *See also* Wheat proteins
Foaming, 60–62
Food additives. *See* Additives
Formylkynurenine, 77
13-Formyl trideca-9,11-dienoate, 8

Free radicals
    general, 408–412
    lipid, 2–3
    sulfite, 330
    vitamin E, 371–373
α-D-Fructopyranosylamine, 118, 120
Fruit(s)
    dihydrochalcone, 271
    pectic enzymes, 225
Fruit flavors, 253–257
Furan derivatives, 111–112, 260
Furanocoumarin, 299
5,8-Furanoxide, 156
Furfural, 112
2-Furylmethanethiol, 247

Galactan, 143
Galactomannan, 142
Galactose-4-sulfate-3,6-anhydro-galactose, 135
Galactose-2-sulfate-galactose-2,6-disulfate, 135
Galactose-4-sulfate-3,6-anhydro-galactose-2-sulfate, 135
Gallocatechin, 243
Garlic flavor, 247–249
Gelatin, 102–104
Gelatinization (starch), 126–127
Gelation
    meat protein, 85–86
    polysaccharides, 105
Gel formation, 62–63
Gelling
    alginate, 130–131
    carrageenan, 135–137
    cellulose, 139–141
    pectin, 133–134
    pentosans, 143, 146
    protein, 62–63
    xanthan gum, 142
Ginger flavor, 250–252
Gingerol, 250, 252
Gliadins, 92–93
Globin, 183–184
Glucoamylase, 277
Glucobrassicin, 258
Glucomannan, 143
Glucose oxidase, 211–214
    characteristics of, 211
    flavin, 346

industrial uses of, 214
    reaction mechanism, 212
    two-electron transfer mechanism, 212–214
β-Glucosidase, 284
Glucosinolates, 258, 286–289
Glu-P-1, 305
γ-L-Glutamyl-S-allyl-L-cysteine sulfoxide, 249
γ-Glutamylaminopropionitrile, 293–294
ε-N-(γ-Glutamyl)-lysine, 66
γ-Glutamyltransferase, 295
Gluten, 92. See also Wheat proteins
Glutenins, 93–95
Glycerol monostearate, 42
Glycoaldehyde alkylimine, 119
Glycoalkaloids, 285–286
Glycophore, 264–266
Glycosidic linkage, 105–108
Glycosulos-3-ene, 121, 239
Glycosylamine formation, 113
Glycyrrhizin, 274–275
Greek-key pattern, 51
Grignard reaction, 157–158
Guansine 5′-monophosphate, 280–281
Guluronic acid, 130

Hairpin connection, 49, 50
Heat
    betanain, 168
    caramel, 169
    gelation, 102–104
    gel formation, 63
    isopeptide formation, 66
    meat emulsion formation, 85, 86
    milk protein, 90–91
    protein structure, 54–55
    starch, 127
    wheat proteins, 96
    see also entries under Thermal
Heme-hydroperoxide, 7
Heme iron, 179–183
Hemicellulose, 143–146
Hemochromes, 185–186
2-Heptenal, 9
Hexanal, 9
Hexaquoiron counterion, 323
Heynes rearrangement, 118
Heterocyclic amines, 304–306
Histamine, 297, 298

Histamine-N-methyl-transferase, 299
Histidine
  back-bonding and, 183
  photolysis, 71
Homoarginine, 293
Homocapsaicin, 251
Humulene epoxide, 246
Humulol, 246
Humulone (α-acid), 244–245
Hydrated electron
  description, 411
  reaction with amino acids, 67
Hydrogen bonding
  alcoholate, 124
  description, 52–53
  pectin, 133
  protein and water, 54–55
  tripartite model, 265–267
Hydrogenation, 18–21
  cis-trans isomerization, 19–21
  generally, 18
  mechanism, 18–19
  selectivity, 21
Hydrogen peroxide
  anthocyanins, 164–166
  protein oxidation, 75
Hydrolysis
  glycoside, 105–107
  lecithin, 37–38
  monosaccharide, 108–113
  protein, 64
Hydroperoxide
  autoxidation, 2–4
  heme, 7
  lipoxygenase, 204–205
  mono-, 7
  photosensitized oxidation, 4–5
  protein reactions with, 75–78
  secondary products, 7–10
  stereochemistry of, 11–12
  with metal ions, 5–7
  with sulfite, 331
Hydroperoxide isomerase, 204–205
Hydroperoxide lyase, 205
Hydroperoxy cyclic peroxides, 8–9
13-Hydroperoxy-9-cis,11-trans-octadeca
  dienoic, 11
13-Ls-Hydroperoxy-9-cis,11-trans-
  octadecadienoic acid, 201
Hydrophobic interaction
  description, 53

thermodynamic, 54
  casein submicelles, 88
  pectin, 133
Hydropropylmethyl cellulose, 139
Hydroxy radical
  description, 411
  lipid oxidation, 5
  with amino acids, 68–69
Hydroxyallysine, 99
Hydroxycyclohexadienyl radical, 69–70
11-Hydroxy-12:13-epoxy-9-cis-octadecenoic
  acid, 202
Hydroxylation, 207–208
Hydroxylysine, 97
Hydroxymethyl furfural, 115
4-Hydroxymethyl-phenylhydrazine, 295
Hydroxynitrile lyase, 284
α-Hydroxy-N-nitroso compound, 311
Hydroxyproline, 99
Hydroxypropyl phosphate, 128
5-Hydroxytryptamine, 297–298
Hypoglycine, 294

Imidazolium-ion pair, 198
Indicaxanthin, 166, 168
Indoglucosinolate, 287, 289
Induced-fit model, 218
Inosine 5'-monophosphate, 280–281
Interesterification, 21–22
Ionic strength, 63
Ionization, 198
Ionone, 244
Iron
  chelated, 5, 6
  lipoxygenase, 202
  myoglobin, 179–180
Iron-dioxygen complex, 184
Isoalloxazine, 349
Isobetanin, 168
3-Isobutylidene phthalide, 259
3-Isobutyl-2-methoxypyrazine, 259
Isochromane, 233–234
Isomerization
  carotenoids, 155–156
  hydrogenation, 19–21
Isopeptides, 66
Isoprene pathway, 235–237
Isothiocyanate, 257, 287–288
2-Isovalidene phthalide, 259

Kappa-carrageenan, 136
Kappa-casein
  described, 87
  kappa-carrageenan and, 136
Ketide unit, 236
Ketimine, 253
α-Keto fatty acid, 205
Kynurenine, 77

Lactitol, 277
Lactone, 260
Lamella
  between oil droplets, 29-30
  mesophase, 39
  protein film, 61-62
Lanthionine, 65-66
Lakes, 169-171
Leavening, 319-320
Lecithin, 36-38. See also Phospholipids
Lemienx effect, 267
Leucine, 54
Leucodeuteroflavin, 349
Levulinic acid, 111, 112,
Ligand
  exogenic, 208
  ionic, 183
  neutral, 180
  orbital, 172
  polyatomic, 172
  transition, 173-174
Light absorption, 147-149
δ-Limonene, 254
Limonin, 256
Limonoate A-ring lactone, 256
Linoleate
  lipid oxidation, 3-7, 18, 27
  lipoxygenase, 201
Linolenate, 3, 18
Lipase, 37, 238
Lipid oxidation, 2-12
  autoxidation, 2-4
  metal ion role in, 5-7
  photosensitized oxidation, 4-5
  secondary products in, 7-10
  stereochemistry of autoxidation, 11-12
Lipid oxidation products, 75-78
Lipids, 1-47
  antioxidants, 44-47
  emulsifiers, 33-38

emulsifiers in stabilization functions, 40-44
emulsions, 27-32
generally, 1
hydrogenation, 18-21
interesterification, 21-22
liquid-crystalling mesophase in emulsifier-water system, 38-40
oxidation, 2-12. See also Lipid oxidation
oxidative thermal reactions, 15
plasticity of fat, 26-27
radiolysis of, 15-18
thermal reactions, 12-15
triglyceride polymorphism, 23-26
Lipolytic enzymes, 225-230
  acyl transfer reaction, 229-230
  colipase role, 226-227
  generally, 225
  mechanism of catalysis, 227-228
  pancreatic lipase, 225-226
  specificity, 228-229
Lipoxygenase, 199-206
  aerobic reaction mechanism, 202-203
  anaerobic reaction mechanism, 203-204
  cooxidation, 206
  flavors and, 238
  generally, 199-200
  hydroperoxides, 204-205
  iron in, 202
  regiospecificity/stereospecificity, 201-202
  soybean lipoxygenase I, 200
Liquid-crystalline interface formation, 42-43
Liquid-crystalline mesophase, 38-40, 42
Locust bean gum
  carrageenan synergism, 137-138
  xanthan and, 142
Lossen rearrangement, 257, 287
Lumisterol, 368
Lupulone (β-acid), 245
Lycopene, 154
Lysinoalanine, 64-66

Macropeptide, 89-90
Magnasweet. See Glycyrrhizin
Magnesium, 187, 188-189
Maillard reaction
  carbohydrates, 105
  flavors, 238-239, 240

Maillard reaction (*cont.*)
  glucose oxidase prevents, 214
  sulfite additives, 326–327
  see also Nonenzymatic browning
Malonaldehyde, 9–10
Malvidin, 160, 163
Malvidin-3,5-diglucoside, 164, 166–167
Mannuronic, 129–131
Margarine, 38
Meat
  ascorbic acid, 365
  cured meats, 186–187, 309–310
  enzyme treatment, 200
  papain, 199
Meat flavor, 259–262
  chemistry of, 259–260
  simulated, 260–262
Meat proteins, 78–86
  emulsion formation, 85
  energy conversion, 81–82
  gelation, 85–86
  muscle macroscopic structure, 78, 79
  muscle proteins, 78, 80–81
  pH, 83
  postmortem tenderness, 84
  rigor mortis, 82–83
  see also Protein(s)
MeIQ, 304, 306
MeIQ$_x$, 305, 306
Melanin, 210
Melanoidin formation, 121–122
1-para-Menthene-8-thiol, 254
Menthofuran, 252
Menthol, 252
Menthone, 252
3-Mercapto-3-methylpentan-2-one, 262
Mesophase. *See* Liquid-crystalline mesophase
Metabolic pathway, 234–237
Metal ions
  lipid oxidation, 5–7
  sugars and, 123–124
  vitamin C, 363–364
Metalloporphyrin, 174–178
Metaphosphates, 315–316
Methacrylic acid, 249
Methanol, 57
Methionine, 73–74
2-Methyl-4-amino-5-hydroxymethyl pyrimi-
  dine, 342

2-Methyl-2-butene-1-thiol, 245
2-Methylbutyl acetate, 257
Methylcellulose, 139
Methyl 1,2-dithiolane-4-carboxylate, 259
Bis-(2-Methyl-3-furyl)disulfide, 262
4-Methyl-5(2-hydroxyethyl)thiazole, 262
γ-Methylene glutamic acid, 293
2-Methyl-3-mercapto-4,5-dihydrofuran,
  345
Methyl octanoate, 9
Methyl 10-oxo-8-decenoate, 9
Methyl 9-oxononanoate, 9
1-Methyl-1,2,3,4-tetrahydro-β-carboline-3-
  carboxylic acid, 310
Methylxanthines, 289–292
Mevolonic acid, 235–236
Milk and milk proteins, 86–91
  alpha-caseins, 87
  beta-caseins, 87
  casein micelle, 88–89
  coagulation, 89–90
  heat stability, 90–91
  kappa-casein, 87
  vitamin D fortification, 365
Miraculin, 278
M-line, 84
Monellin, 278–279
15,15'-mono-*cis*-β-carotene, 155
Monoenoic dimer, 13
Monoglyceride derivatives, 33–34
Monoglycerides, 33
Monohydroperoxide, 7
Monophenol, 207–208
Monosaccharides
  acid action on, 111–113
  alkali action on, 108–111
  carbohydrates, 105
Monosodium glutamate (MSG),
  280–282
Monostearin-amylose helical complex,
  43
Monoterpenoids, 252
Muscle
  energy conversion in, 81–82
  macroscopic structure of, 78, 79
  postmortem tenderness, 84
  proteins of, 78, 80–81
  rigor mortis, 82–83, 84
  see also Meat proteins
Multiplicity, 404

Mycotoxins, 299–302
Myoglobin, 178–187
  absorption spectrum, 185
  globin's role in, 183–184
  heme iron, 179–183
  hemochromes, 185–186
  molecular structure, 178–179
  nitrosylmyoglobin, 186–187
  oxymyoglobin autoxidation, 184–185
Myomesin, 78
Myosin
  description, 78–79
  muscle contraction, 81–82
  gelation, 85–86

Naringin, 256, 271–272
Natural toxicants, 283–313
  amines, 297–299
  amino acids, 292–294
  cyanogenic glycosides, 283–285
  generally, 283
  glucosinolates, 286–289
  glycoalkaloids, 285–286
  heterocyclic amines, 304–306
  methylxanthines, 289–292
  mycotoxins, 299–302
  nitrosamines, 306–313
  peptides, 294–295
  polycyclic aromatic hydrocarbons, 302–304
  proteins, 296–297
Nebulin, 80
Neohesperidin, 271–272
Neohesperidoside, 256
Neurotoxins, *botulinum,* 296–297
Neutralizing value, 320
Neutral solution, 123–124
Niacin, 358–359
Nicotinic acid, 358
S-Nitrocysteine, 308–309
para-Nitrophenol, 308–309
Nitrosamines, 306–313
Nitrosammonium ion, 308
N-Nitrosodiethylamine, 309
N-Nitrosodimethylamine (NDMA), 309
Nitrosonium ion, 307
N-Nitrosopiperidine, 309
N-Nitrosopyrrolidine (NPYR), 309–310
Nitrosylmyoglobin, 186–187
cis-3-Nonenal, 205

Nonenzymatic browning
  carbohydrates, 113–122
  chemistry of reactions, 115–120
  described, 113–115
  secondary reactions, 120–122
  sulfite additives, 326–327
  see also Maillard reaction
Norbixin, 159
Nonprotein amino acid, 292–294
Nordihydrocapsaicin, 251
Nucleotides, 280–282
NutraSweet. *See* Aspartame

Octahedral coordination, 171–172, 177
3-Octene-2-one, 9
Odor, 232–234. *See also* Flavor(s)
Oil (fruit flavors), 253–257
Oils, *see* Lipids
Oleate, 3–5, 18, 27
Olefin autoxidation, 401–403
Olfactory receptor sites, 233
Oligomers, 189–191
Oligosaccharides, 105
Onion flavor, 247–249
Organic solvents, 57
Ornithinoalanine, 65–66
Orthophosphates, 316, 321
Oxalyldiaminopropionic acids, 294
2-Oxalylamino-3-aminopropionic acid, 294
Oxazoles, 239, 260
Oxazolidine-2-thione, 289
Oxazoline, 239
Oxazolinide ion, 239
Oxidation
  anthocyanins, 164–166
  chlorophylls, 193
  diphenol, 209–210
  monophenol, 207–208
  protein, 73–75
  see also Lipid oxidation
Oxidative thermal reactions, 15
Oxodienoic acid, 203
10-Oxo-13-hydroxy-11-*trans*-octadecenoic acid, 205
9-Oxononanoic acid, 205
Oxycarbonium ion mechanism, 216
Oxy-heme radical, 7
Oxymyoglobin, 180, 184–185. *See also* Myoglobin

Palmitic acid, 25–26
Pancreatic lipase, 225–226
Papain, 194–195, 198–199, 220
Para-kappa-casein, 89–90
Paraben, 324
Pectate lyases, 224–225
Pectic enzymes, 221–225
  generally, 221–222
  industrial uses, 225
  pectate lyases, 224–225
  pectinesterase, 222
  polygalacturonases, 222–224
Pectin, 132–134
  chemical structure of, 132
  classification of, 133
  gelling, 133–134
Pectinesterase, 222
Pelargonidin, 160
1,4-Pentadiene system, 4, 11, 200–201
Pentane, 8
Pentosans, 143–146
Peonidin, 160
Pepper flavor, 250
Peppermint, 252
Peptides, 294–295
Perferryl ion, 6
Perhydroxyl radical ($HO_2\cdot$), 6
Peroxydienone, 44
Peroxy epoxide mechanism, 340
Peroxy radical, 2, 7, 15, 44
pH
  amylases, 219
  anthocyanins, 160–162
  gel formation, 63
  meat protein, 83
  phosphates, 318
  protein structure, 56–57
  soybean lioxygenase I, 200
  vitamin B_1 and, 342
Phallotoxins, 294–295
Phe-P-1, 305
Phenolase. See Polyphenol oxidases
Phenoxyl radical, 46
Phenylalanine
  amino acids, 54
  photolysis, 71
4-"Phenyl" anthocyanin, 164–165
2-Phenyl-benzopyrylium, 160
Pheophorbide, 192
Pheophytin, 192

Phosphates, 314–321
Phosphatidylcholine, 36–38
Phosphatidylethanolamine, 36–38
Phosphatidylinositol, 36–38
O-Phosphobiotin, 357
Phospholipids, 36–37
Phosphorus oxychloride, 127
Photochemical reactions. See Radiolysis;
  Photolysis; Photosensitized reaction
Photolysis
  description, 408–410
  nitrosamine, 312
  protein, 71–73
  riboflavin, 349–350
  thiamin, 344
  vitamin B_{12}, 355
Photosensitized oxidation
  beta-carotene, 157, 339
  carotenoids, 157
  description, 409–410
  "ene" reaction, 4–5
  protein, 73–74
  riboflavin, 350
Phthalides, 258–259
β-Pinene, 254–255
Pipecolic acid, 293
Piperine, 250
pK_a amino acids, 56
Plasticity (fats), 26–27
Poisoning. See Natural toxicants
Polycyclic aromatic hydrocarbons,
  302–304
Polygalacturonases, 222–224
  endo-polygalacturonases, 222–223
  exo-polygalacturonases, 223–224
Polyglycerol, 35–36
Polyglycerol monostearate, 40
Polyphenolase. See Polyphenol oxidases
Polyhydric alcohol sweeteners, 276
Polyketide pathway, 235
Polymorphism of triglycerides. See Triglyc-
  eride polymorphism
Polypeptide chain, 99
Polyphenol oxidase
  characteristics, 207
  reaction mechanism, 207–210
  secondary reaction products, 210
Polysaccharides
  carbohydrates, 105
  carboxymethylcellulose and, 140

Polysorbate, 35, 40, 42
Polysulfide, 249
Polysulfide heterocyclic compounds, 242
Porphyrin-Fe^{2+}-O$_2$Fe^{2+}-porphyrin, 183
Porphin (free base), 180
Porphyrin
  free base, 175
  metal, 176
Postmortem tenderness, 84
Progoitrin, 258
syn-Propanethiol S-oxide, 248
2-Propenesulfenic acid, 248
trans-(+)-S-(1-Propenyl)-L-cysteine sulfoxide, 248
Propionate, 325
Propylene glycol monostearate, 35
Propyl gallate, 45-47
Protein(s), 48-104
  carrageenan interaction, 137
  chemical reaction, 63-78
  emulsification and foaming, 60-62
  emulsifier interaction with, 43-44
  gel formation, 62-63
  generally, 48
  organized systems, 78-104
  structure of, 49-60
  sweeteners, 278-280
  toxicants, 296-297
Protein chemical reaction, 63-78
  alkali degradation, 63-66
  carbonyl compounds, 75
  chemical oxidation, 74-75
  heat-induced isopeptide formation, 66
  lipid oxidation products and, 75-78
  photolysis, 71-73
  photosensitized oxidation, 73-74
  radiolysis, 67-71
Protein structure, 49-60
  conformation change, 54-60
  generally, 49-53
  water role in, 53-54, 55
Protein systems, 78-104
  collagen, 97-104. See also Collagen
  meat proteins, 78-86. See also Meat proteins
  milk proteins, 86-91. See also Milk proteins
  wheat proteins, 92-97. See also Wheat proteins
Proteolysis, 89, 90

Protocatechuic acid, 163
Protocollagen proline hydroxylase, 361
Protophorphyrin, 180
Pseudobase
  anthocyanin, 161
  thiamin, 342
Pseudoplastic flow, 140, 142
Pyrazine, 239, 259-260
β-Pyrazol-1-ylalane, 293
Pyridine, 247, 260
Pyridoxol (vitamin B$_6$), 351-353
Pyrimidines, 328
Pyrochlorin, 188-189
Pyrones, 240
Pyrophosphates, 315-316
Pyroporphin, 189
Pyrrole, 121, 193, 239, 247, 260
Pyrrolidines, 240
Pyrrolines, 240

Quenching
  beta-carotene, 157, 339
  tocopherol, 407
Quinoidal base, 161, 162
Quinone methine, 371

Racemization, 66
Radiolysis
  description, 410-411
  lipids, 15-18
  proteins, 67-71
  vitamin D, 368
Reduction, 115, 118, 193
Reductones, 240
11-cis-Retinal, 337
Retinol (vitamin A), 335-340
Retrogradation, 127
Retunidin, 160
Reverse turns, 49-50
Rhodanase, 285
Rhodopsin, 337
Riboflavin (vitamin B$_2$), 344-351
Rigor mortis, 82-83, 84
Ripening (fruits), 225
Rootkatone, 254
Rutinose, 271-272

Saccharin, 264, 271
Saccharinic acid, 109–110
Salt effects
    protein–"salting-in" and "salting-out",
        58–60
Salt-soluble proteins, 85
Salty taste, 232
Scission products, 7–8
Seaweed, 129
Sedimentation, 29
Selectivity, 21
Self-association, 164
Semiquinone radical, 346
Serotonin, 297–298
Sesquiterpene, 250
Sesquiterpenoid, 246
Shagaol, 252
Shikimic pathway, 234–235
Shortening effect, 96
Sinalbin, 258
Singlet oxygen, 404–407
Sinigrin, 258
Site-fitting theory, 232
Sodium stearoyl-2-lactylate, 40, 42
Solanidine, 285
α-Solanine, 285
Sorbate, 324–325
Sorbitan monostearate, 40
Sorbitan monoglyceride, 34
Sour taste, 231–232
Soybean lioxygenase, 200
Spice flavor, 247–253
    chili, 250
    cinnamon, 252–253
    garlic/onion, 247–249
    ginger, 250–252
    pepper, 250
    peppermint, 252
Staling of bread, 43
Starch, 124–129
    chemical modification, 127–129
    chemical structure of, 124–126
    complexation with, 43
    gelatinization, 126–127
    retrogradation, 127
Stearate, 18, 20, 25
Stearoyl-2-lactylate, 35, 43
Stereochemistry, 11–12
Steroids, 366–367
Steviol, 275

Stevioside, 264, 275–276
Strecker degradation
    ascorbic acid, 363
    Maillard reaction, 122
    tea oxidation, 244
Substituent effect, 151–153
Succinylated monoglyceride, 34
Sucrose, 266, 267
Sugar(s), 123–124
Sugar alcohol sweeteners, 276–277
Sulfhydryl-disulfide exchange
    β-lactoglobulin, 91
    in dough, 95
    protein function, 53
    gel, 63
Sulfite
    food additives, 326–334
    Maillard reaction, 121
    vitamin B_2 and, 348
Sulfite oxidase, 334
Sulfur dioxide, 162–163
Superoxide anion, 5–6
Surface tension/area, 28–29
Sweeteners, 264–282
    amino acids/dipeptides, 267–270
    aminosulfonates, 270–271
    corn sweeteners, 277–278
    dihydrochalcone, 271–274
    flavor potentiators, 280–282
    generally, 264
    glycyrrhizin, 274–275
    molecular theory of, 264–267
    proteins, 279–280
    stevioside, 275–276
    sugar alcohol, 276–277
Sweet taste, 232

Tartrazine, 170
Tachysterol, 368
Taste sensation, 231–232
Tea flavor, 242–244
Termolecular-shift binding, 221
γ-Terpinene, 254–255
Tertiary butylhydroquinone, 45–47
2,2,7,7-Tetramethyl-1,6-
    dioxaspiro[4.4]none-3,8-diene, 246
2,3,5,6-Tetramethyl-4-methoxyphenol,
    372
Thaumatin, 278
Theaflavin, 243, 245

Thearubigins, 243, 245
Theobromine, 290
Theophylline, 290
Thermal degradation, 156–157
Thermal reactions
  lipids, 12–15
  oxidative, 15
  vitamin D, 369
  see also Heat
Thiamine (vitamin B₁), 340, 344
Thiapane, 260–261
Thiazole, 241
Thiazoline, 241, 260
Thiochrome, 343
Thiocyanate, 257, 287–288
β-Thioglucose, 286–287
Thioglucoside glucohydrolase, 286
Thio-disulfide intermediate. See Sulfhydryl-disulfide
Thiophene, 249, 260
Thiosulfonate, 249
Thixotropic flow, 130, 140
Thymol, 254–255
Titin, 80
α-Tocopherol quinone, 370–371
Tocopherols, 158, 369–374
Tropocollagen, 97
Toxicants. See Natural toxicants
Transglycosylation, 221
Trideca-9,11-dienoate, 8
Triglyceride polymorphism (lipids), 23–26
  crystal habit of fat, 24–26
  crystal structure, 23–24
Trigonal bypyramidal intermediate, 207–208
Trimethylamine-N-oxide, 298
1,3,7-Trimethyuric acid, 291
Tripartite model, 265
Triple helix, 97
Trisaccharides, 142
Trithiane, 242
1,2,3-Trithiane-5-carboxylic acid, 259
Trithiolane, 242
Tropomyosin
  energy conversion, 82
  muscle protein, 80
Troponin
  energy conversion, 82
  muscle protein, 80
  postmortem tenderness, 84

Trp-P-1, 305
Trypsin, 200
Tyramine, 297–298
Tyrosinase. See Polyphenol oxidase
Two-electron transfer, 212–214
Tyrosine, 72

Uncertified color additives, 169
Ureido, 356
Uridine-4-sulfonate, 332
U.S. Food and Drug Administration (FDA), 169

Valencene, 253–254
Vanillyl alkylamine, 250
Valine, 54
Van der Waals forces
  emulsions, 31, 40
  liquid-crystalline mesophase, 39
Vegetable flavor, 257–259
Vinyl-β-ionol, 158
5-Vinyloxazolidine-2-thione, 289
Vitamins, 335–374
  A (retinol), 335–340
  B₁ (thiamin), 340–344
  B₂ (riboflavin), 344–351
  B₆ (pyridoxol), 351–353
  B₁₂, 353–356
  biotin, 356–358
  C (ascorbic acid), 359–365
  D (ergocalciferol, cholecalciferol), 365–369
  E, 369–374
  generally, 335
  niacin, 358–359

Water
  dielectric constant, 57
  liquid-crystalline mesophase, 38–40
  phosphate, 318–319
  protein structure and, 53–54, 55
  water-holding capacity, 83
  wheat protein classification, 92
Water holding capacity, 316, 318–319
Wheat proteins, 92–97
  flour into dough transformation, 95–97

Wheat proteins, *(cont.)*
  gliadine, 92–93
  glutenins, 93–95
Wittig reaction, 157–158

Xanthan gum, 141–142
Xanthophylls, 154
Xathiazinone oxide, 270
Xylan, 143
Xylitol, 277

Yeast, 221
Ylid, 341

Zeaxanthin, 154
Zingerone, 252
Zingiberene, 250
Z-disk, 84